Managing and Modelling Complex Projects

NATO ASI Series

Advanced Science Institutes Series

A Series presenting the results of activities sponsored by the NATO Science Committee, which aims at the dissemination of advanced scientific and technological knowledge, with a view to strengthening links between scientific communities.

The Series is published by an international board of publishers in conjunction with the NATO Scientific Affairs Division

A Life Sciences **B Physics**	Plenum Publishing Corporation London and New York
C Mathematical and Physical Sciences **D Behavioural and Social Sciences** **E Applied Sciences**	Kluwer Academic Publishers Dordrecht, Boston and London
F Computer and Systems Sciences **G Ecological Sciences** **H Cell Biology** **I Global Environmental Change**	Springer-Verlag Berlin, Heidelberg, New York, London, Paris and Tokyo

PARTNERSHIP SUB-SERIES

1. Disarmament Technologies	Kluwer Academic Publishers
2. Environment	Springer-Verlag / Kluwer Academic Publishers
3. High Technology	Kluwer Academic Publishers
4. Science and Technology Policy	Kluwer Academic Publishers
5. Computer Networking	Kluwer Academic Publishers

The Partnership Sub-Series incorporates activities undertaken in collaboration with NATO's Cooperation Partners, the countries of the CIS and Central and Eastern Europe, in Priority Areas of concern to those countries.

NATO-PCO-DATA BASE

The electronic index to the NATO ASI Series provides full bibliographical references (with keywords and/or abstracts) to more than 50000 contributions from international scientists published in all sections of the NATO ASI Series.
Access to the NATO-PCO-DATA BASE is possible in two ways:

– via online FILE 128 (NATO-PCO-DATA BASE) hosted by ESRIN, Via Galileo Galilei, I-00044 Frascati, Italy.

– via CD-ROM "NATO-PCO-DATA BASE" with user-friendly retrieval software in English, French and German (© WTV GmbH and DATAWARE Technologies Inc. 1989).

The CD-ROM can be ordered through any member of the Board of Publishers or through NATO-PCO, Overijse, Belgium.

Series 4: Science and Technology Policy – Vol. 17

Managing and Modelling Complex Projects

edited by

Terry M. Williams
Strathclyde University,
Glasgow, U.K.

Kluwer Academic Publishers

Dordrecht / Boston / London

Published in cooperation with NATO Scientific Affairs Division

Proceedings of the NATO Advanced Research Workshop on
Managing and Modelling Complex Projects
Kiev, Ukraine
November 13–15, 1996

A C.I.P. Catalogue record for this book is available from the Library of Congress.

ISBN 0-7923-4844-3

Published by Kluwer Academic Publishers,
P.O. Box 17, 3300 AA Dordrecht, The Netherlands.

Sold and distributed in the U.S.A. and Canada
by Kluwer Academic Publishers,
101 Philip Drive, Norwell, MA 02061, U.S.A.

In all other countries, sold and distributed
by Kluwer Academic Publishers,
P.O. Box 322, 3300 AH Dordrecht, The Netherlands.

Printed on acid-free paper

Printed in the Netherlands

Table of Contents

PREFACE

This volume contains the proceedings of an Advanced Research Workshop funded by the NATO Science Programme, entitled “Managing and modelling complex projects”. The Workshop took place on 13th-15th November 1996 in Kiev, Ukraine. The financial assistance and encouragement of the NATO Science Committee is warmly acknowledged.

The Workshop was attended by 54 invited participants from 15 countries, representing a heterogenous mix of interests. 13 Key Papers were presented, and around half of the duration of the Workshop consisted of a series of Open Discussion sessions.

The Workshop was co-directed by Dr. Terry Williams, of Strathclyde University, Glasgow, UK, the editor of this volume, and by Prof. Sergey Bushuyev, of Kiev Technical University, Ukraine. Very considerable acknowledgements must be made to Prof. Bushuyev and his team in the Ukraine, who made all of the local arrangements. Alexandre Rodrigues acted as co-organiser.

Question and answer sessions, and Open Discussion sessions, have been summarised by the editor from tape-recordings made at the Workshop. However, these summaries have not been viewed or in any way authorised by the participants, and apologies are made for any errors or mis-representation by the editor.

SECTION 1
INTRODUCTION

Technology advances and development occurs more and more by means of major projects. Developing new technological products, building new capital assets, undertaking new and unique large-scale enterprises, or implementing solutions to approach the large national and international problems we face, all require major projects to be undertaken.

In recent years, these technological products and systems have become more complex. This means that projects developing and building the products, and projects implementing large-scale solutions, have also become correspondingly more complex. At the same time, projects have become more time-constrained: product development times and market opportunity slots are shrinking and major clients expect capital assets built and solutions implemented quicker. Furthermore, there are frequently heavy penalties for late delivery, either explicitly in a contract or implicit in the lost opportunity. This shortening of project-times exacerbates the increasing project complexity. These trends are apparent in all major projects, but possibly particularly in Design and Manufacture and in Software projects, such as carried out by the defence and redeployed defence industries [1].

For such projects, traditional project management (PM) methods (based on decomposing projects into constituent entities) and project management structures have proved inadequate. Projects have resulted in over-runs, over-spends and project failures, and that this continues to be true rather than improving. This is expensive for the implementers of the projects, and damaging to the institutions sponsoring the projects. Furthermore, the trends in modern projects discussed above are exacerbating these problems. For industry to maintain competitiveness, there is clearly a need for more effective means of managing these projects, supported by more effective means of understanding and modelling these projects, in order to provide effective estimating, planning and control. Furthermore, this need is imperative, and becoming more urgent as projects continuously increase in complexity.

A number of approaches have been taken to such work, but it has been difficult to take a view as to the most promising approach, nor has any synthesis been

T. M. Williams (ed.), Managing and Modelling Complex Projects, 1–3.

attempted (either between the approaches or with lower-level operational methods). It is thus clearly the right time to take an overall view. This workshop brought together the leading experts in various main thrusts of research, both applied academics and practitioners, in order that such an overall view could be taken. All of the key speakers are internationally known in their particular field.

Industries in the Co-operation Partner (CP) countries (the "Eastern bloc") are particularly required to undertake complex major projects, under considerable time-constraints, at the present time. Recent history has meant that Western countries have had more motivation to develop such techniques. This Workshop, as well as furthering global research, was intended to help in the two-way transfer of knowledge between Western and CP participants.

54 Carefully selected invited participants from 15 countries representing academics, consultants, contractors and clients attended. The format of the Workshop was not that of a conference: an important feature was a series of open discussion sessions to develop the ideas in the 13 Key Papers. Key Speaker presentations occupied around half of the time, and these open discussion sessions the remainder. The discussions were far-ranging, and are reported here roughly as they occurred, although this sometimes strained the conference structure.

After a "welcome" to the Ukraine, the first technical paper in the Workshop considered what constitutes a "complex" project. Then the Workshop looked at a variety of approaches, both techniques to manage the projects and techniques for modelling projects to inform such management.

- Modelling techniques were considered first. PERT-type methods are inadequate because they do not incorporate management actions, adequately consider uncertainty or consider effects that are systemic or "soft". The first two papers extended PERT methods, the first using controlled networks, the second diffusion networks to consider uncertainty. The third paper covered risk analysis techniques, considering how time risk impacts on cost and what drives time risk. Finally, the System Dynamics technique was presented, which provides a simulation model of projects, capturing systemic effects and effects that are "soft".
- Three papers looked at PM in particular domains: the first new technology projects, the second software development projects, which because of their special attributes need special management techniques; the third paper covered infrastructure projects, but dealt specifically with the human factors involved in PM.

- Two papers considered corporate structures, one from the West looking at a joint venturing scheme, and the second from a CP country considering extensively the differences between PM in a market and a transition economy.
- Two papers looked at management techniques, again one from a Western and one from a CP country. The first dealt with the use of "alternative" teams in strategic management. The second described a hierarchical approach to PM.
- Finally in the conference, techniques for managing portfolios or programmes of projects were discussed, and the distinctive differences of this problem from one-project PM.
- The final section in this volume contains some brief conclusions.

Reference

1. WILLIAMS, T.M., (1993) Taking the Risk Manager to 2020. Proc. 1st British Project Management Colloquium, Henley-on-Thames, UK December 1993.

SECTION 2
THE NEED FOR THE WORKSHOP

This Section sets the scene for the Workshop. Following a brief "welcome" by the Head of the National Agency for the Reconstruction and Development of Ukraine, Roman Schpek, the Section contains just one paper, by the Director of the Workshop, Dr. Terry Williams of Strathclyde University, U.K. It explores what "complexity" implies when applied to projects, and attempts to justify the Workshop itself - that is, it claims that projects are becoming more complex, and that traditional project Management tools are becoming less adequate to deal with such projects.

Although this introductory session includes no Discussion Session, these premises did not go entirely unchallenged during the later Discussion Sessions. However, the definition of Complexity provided a useful foundation for the Workshop, and the premises gave a useful impetus to discussion.

WELCOME TO UKRAINE

Roman V. Shpeck
Head of National Agency for Reconstruction & Development of Ukraine

Let me open the International Workshop "Managing and Modelling Complex Projects".

Ukraine is going down the route of reforms to the market economy. One of the most important conditions for the successful development of Ukraine is integration into the political and economic world system.

Regarding the formation of the European Community, Ukraine has a flexible position concerning the development of international co-operation, with already existent political, military, economical and other professional organisations. It is also concerned with organisations such as NATO.

Some words about the objectives and basic direction of activity of the recently-founded Ukrainian National Agency for Economic Development and Reconstruction. The agency was created to organise and co-ordinate the programmes and projects of the economic development of Ukraine. One of the most important functions of the agency, as a compound part of the Ukrainian Government, is the co-ordination of technical assistance and provision of effective interaction of Ukrainian scientists, businessmen and representatives of state and private enterprises with foreign colleagues and partners.

The experience of many countries shows that using contemporary methods and tools for project management, and the presence of a critical mass of project managers, is one of the most important conditions of economic development of our society. Because of that, such a workshop is well-timed and of immediate interest. It will promote exchange of experience, new techniques and methods and models of project management between scientists and specialists from NATO and Eastern European countries.

T. M. Williams (ed.), Managing and Modelling Complex Projects, 7–8.

In the workshop, the following problems will be discussed:

- modelling tools to support analysis
- corporate structures
- domain specific examples
- human factors
- management techniques.

I hope that the workshop will be fruitful, and that it will promote development of methods and tools for project management as well as put new scientific and practical connections between project management specialists.

Concerning this Workshop, I would like on behalf of the Ukrainian Government to thank the NATO Scientific & Environmental Affairs Division, the International Project Management Association, the Ukrainian Project Management Association and the University of Strathclyde, Scotland, for their support of this event, and to wish a fruitful outcome.

THE NEED FOR NEW PARADIGMS FOR COMPLEX PROJECTS

T.M. WILLIAMS
University of Strathclyde, Glasgow, UK

This paper forms the opening session of a workshop on managing and modelling complex projects. In order to introduce the theme, it discusses what constitutes project complexity. It aims to be inclusive rather than exclusive. Highlighted are structural complexity, the number and interdependence of elements (following a paper by Baccarini) and uncertainty in goals and means (following a paper by Turner and Cochran). The paper considers whether these aspects can be operationalised. It gives some ideas about why project complexity might be considered to be increasing, in particular the increasing complexity of products being developed, and moves towards shorter timescales. Finally, it notes some areas where new methods are needed, following the structure of the Workshop programme.

1. Introduction

This paper forms the opening session of the NATO Advanced Research Workshop *Managing and Modelling Complex Projects*. The Workshop is based on a series of premises: projects are becoming increasingly complex; traditional project management methods are proving inadequate; new methods of analysis and management are needed. The aim of the Workshop is to look at new methods and new experiences, both in the West and the new economic environment in Co-operation Partner countries. In order to consider these premises, and to look at new methods, it is essential at the start to look at whether we can define complexity, to provide a framework for the succeeding discussion.

While many project managers use the term a *complex* project, there is no clear definition about what is meant - beyond the general acceptance that it is something more than simply a "big" project. This paper does not aim to give a definitive definition of complexity. Instead, in order to facilitate rather than restrict the open discussions that it is hoped will be a feature of this Workshop, the paper explores a number of various aspects of the idea of project complexity.

T. M. Williams (ed.), Managing and Modelling Complex Projects, 9–18.

It aims to be inclusive rather than exclusive, to encourage discussion of all of the dimensions of complexity. The paper considers whether these aspects can be operationalised, and gives some ideas about why project complexity might be considered to be increasing. Finally, it notes some areas where new methods are needed, following the structure of the Workshop programme.

By a fortunate coincidence, an excellent review of project complexity was published in the *International Journal of Project Management* only 3 months before the Workshop by Baccarini [1], and this can be used to provide a useful starting-point for this discussion. He stresses the importance of the concept of complexity to the Project manager, and its role in the strategic management of projects. He also states as a given, referencing Morris and Hough [2], that "complex projects demand an exceptional level of management, and that the application of conventional systems developed for ordinary projects have been found to be inappropriate for complex projects".

2. What is complexity?

2.1 STRUCTURAL COMPLEXITY

Baccarini [1] proposes a definition of project complexity as "consisting of many varied interrelated parts", which he operationalises in terms of *differentiation* - the number of varied elements - and *interdependency* - the degree of interrelatedness between these elements (or connectivity). These measures are to be applied in respect to various project dimensions, and he discusses two of them:

(i) In terms of organisational complexity, "differentiation" would mean the number of hierarchical levels, number of formal organisational units, division of tasks, number of specialisations etc; "interdependency" would be the degree of operational interdependencies between organisational elements.

(ii) In terms of technological complexity, "differentiation" would mean the number and diversity of inputs, outputs, tasks or specialities; "interdependency" would be the interdependencies between tasks, teams, technologies or inputs.

This clearly defines an important element of project complexity, perhaps the element which we think of most often when we consider a "complex" project: complexity is concerned with the underlying structure of the project. For this paper, we can call this "Structural Uncertainty".

For projects such as design-and-manufacture, or design-and build, a major source of *project* (structural) complexity is *product* (structural) complexity. A project to develop a more complex product must normally be a more complex (in this sense) project, but it is useful to distinguish this cause and effect of complexity. Product (structural) complexity, in the Baccarini sense, is the number of subsystems of a product and their interrelationships. When modelling or analysing such a project, these interrelationships are quantified and then measures of complexity can be propounded. As an example, take the analysis of a large design and manufacture engineering project for a Channel Tunnel sub-contractor, undertaken as part of a Delay and Disruption litigation by a team led by Eden [3]. Here, the system was divided into around 50 sub-systems, and a "cross-impact matrix" P developed, giving the probability p_{ij} that a change in system i will affect system j. From this could be defined, for example,

- sequential complexity: the likely length of a sequence of interactions;
- feedback complexity (the probability that a change in system i eventually affects system i).

Having measured the product complexity, such measures can be used to investigate aspects of project complexity: for example, in evaluating the effect of client changes (how many other changes need to be made, or how many hitherto frozen systems to "un-freeze").

Clearly, merely counting interdependencies is not sufficient, the nature of those interdependencies is also important (as Baccarini points out). In a well-known paper, Thompson [4] looked at interdependencies and identified three types: pooled (in which each element gives a discrete contribution to the project, each element proceeding irrespective of the other elements), sequential (one element's output becomes another's input) and reciprocal. In the last of these types, each element's output becomes inputs for other elements, so the actions of each must be modified to the actions of others. It is this last type of interdependency which particularly produces complexity. In particular, reciprocal interdependencies will produce the type of feedback effects found in the Channel Tunnel work, and described in [5], where client comments caused work to return to the designer, then causing changes to other elements which themselves return to be re-worked; such effects, of course, run counter to the assumptions of most first-generation tools (e.g. PERT), which assume a steady progress through a project. These reciprocal effects can clearly and obviously be seen in the case of a set of designers designing a product subject to a cross-impact matrix as described above. But less easily modelled reciprocal interdependencies occur, for example when there are functional aspects affected by, and affecting, many activities (e.g. the weight of the product, electro-magnetic interactions, safety issues - in any of these, a significant occurrence in one activity can have major effects across all of a project), or when events occur which affect many project elements (e.g. loss of a prototype, change in regulations - the possible list is endless). Clearly, the more complex the type of interdependency, the greater the added complexity. While the above is a general managerial definition, one driver in the project-management domain causing an increase in reciprocal interdependencies is the rise in concurrent engineering, which is be discussed further below.

While these are the main aspects of structural complexity, concerned with the multiplicity of project elements, two other aspects should be noted:

- Virtually all projects are by definition multi-objective, with conflicting goals (either constraints or optimisation). This adds complexity, as the effects of activities on all goals has to be assessed, and trade-offs have to consider the balancing effects of other activities.
- Virtually all projects have a multiplicity of stake-holders, not only the obvious - client, project manager(s) and project team - but also owner, champion, the public, sometimes public bodies, and so on. This will add complexity in a similar manner to the multiplicity of goals.

2.2 UNCERTAINTY

Jones [6], in discussing the social psychology of conflict within project management, defines 'technical complexity' as a three-fold concept: the variety of tasks, the degree of interdependencies within these tasks, and "the instability of the assumptions upon which the tasks are based". The first two of these, of course, are the same as Baccarini's two aspects that we have termed Structural Complexity above. The third, however, is one example of *Uncertainty*, which brings a further element to the idea of complexity - although we can perhaps widen the scope beyond his example.

The idea of Uncertainty is discussed in the paper by Turner and Cochran [7]. They classify projects by two parameters: how well defined the goals are, and how well-defined are the methods of achieving those goals - a classification arrived at by a number of authors. They then identify four distinct types of project, depending on whether the goals are well- or ill-defined, and whether the methods are well- or ill-defined, and suggest different management and particularly different project start-up methods for the four types. It is suggested in this present paper that these two types of uncertainty bring two dimensions of added complexity to projects.

Dealing with the second dimension first, uncertainty in the methods to carry out the projects will add complexity. Turner and Cochran point out that, if methods are uncertain, the fundamental building-blocks of project management will not be known: the WBS, the tasks required to complete the job and their sequence, the Organisational Breakdown Structure, etc; and even when they are planned, the plan will be subject to change. Clearly, then, some of the characteristics of Structural Complexity will occur here: as the team structures the work and refines the methods, there will be considerable interdependencies between project sub-teams; as methods are tried and re-planned, feed-back loops naturally occur, and so on.

Uncertainty in methods is well known to be related to complexity. For example, Shenhar [8] undertook a field study to distinguish among good management styles and practices for different types of projects (for engineering projects). He classified projects by two parameters: system scope (assemblies, systems arrays - equivalent to our idea of Structural Complexity) and "technological uncertainty" (the newness of the technology to be used - i.e. the uncertainty in the methods). We should note here that "Uncertainty" is being used in the broad sense, including both those elements which are stochastic but also those resulting from a lack of knowledge (i.e. epistemic uncertainties); thus, a project where a body of knowledge exists (e.g. building an air-liner) is less complex than a State-of-the-Art project where there is no experience (e.g. building a HOTOL craft to go to the atmosphere edge).

But while uncertainty in the methods brings about one element of added complexity, even when the methods are known, another dimension of added complexity comes about when there is uncertainty in the goals. Turner and Cochran use Software Development projects as a typical of projects whose methods are known but whose goals are uncertain, since users' requirements are difficult to specify, and are often changed when initial prototypes are seen. The essential difficulty with such projects is that the requirements are not frozen, and uncertainty or changes in some requirements will mean that interfacing elements also need to change, and again we have cross-impacts, re-work, feedback-loops - an increase in the features of structural complexity. Indeed, the "freezing" and "un-freezing" of sub-systems formed a central part of the Channel Tunnel model described in [5]. Many of these results in practice are called "Delay and Disruption". (Wozniek's [9] "clarity of scope definition" is presumably similar to Goals Uncertainty.)

A key element of the added complexity brought about by the changes or modifications that results from uncertainty in goals, is that these changes often cause two separate increases in complexity:

- not only does the action of making the changes increase the project complexity,
- but also the individual changes themselves often combine to increase the product complexity and thus the project complexity. Continuously adding elements will eventually mean that it is extremely difficult to put in any more cable-ways, or fit all the elements into a constrained space, or work out the pipe-work.

In evaluating such a project, then, not only does the product complexity have to be taken into account, but also the increase in product complexity throughout the life of the project [3], remembering that the effect on the project of many changes is more than the sum of the effects of each change individually [5].

Both of these uncertainty measures are probably difficult to operationalise into a quantifiable parameter. The vagueness of the goals might be measurable by how long it would take to establish whether the goals were satisfied; changes in goals could perhaps be measured in terms of contract changes. It is not obvious to this author how uncertainty in methods could be operationalised, beyond vague terms such as "newness of technology".

3. Increasing complexity

It appears to be an accepted fact that the complexity of projects is increasing (even if complexity has not often been defined), and this is part of the *raison d'etre* of this Workshop. Baccarini's review begins with the words "Construction projects are invariably complex and since World War II have become progressively more so". Helbrough [10] states as a given that "increased complexity of projects and the project environment have meant that despite improved methods, many projects still fail to meet expectations."

Williams [11] points to two compounding causes for projects increasing in complexity. The first derives from the relationship between product complexity and project complexity discussed above. As new products are developed which extend, or improve upon, previous generations of product, those products become more complex, because of extra functionality, or reduction in physical size, or closer intra-connectivity, or similar reasons. This means that the projects developing those products are appearing to increase in Structural complexity, with a larger number of project elements and, in particular, a greater degree of inter-element connectivity. It should perhaps be said that this view is supported by anecdotal evidence only: without a clear definition of complexity, the author is aware of no quantitative evidence collected to investigate this; furthermore, the author's view is based on design-and-manufacture and software projects, and it isn't clear whether this applies also to, say, the civil engineering domain.

The second cause compounds this increasing structural complexity. Projects have tended to become more time-constrained [12], and the ability to deliver a project quickly is becoming an increasingly important element in winning a bid. Furthermore, there is an increasing emphasis on tight contracts, using prime contractorship to pass time-risk onto the contractor, frequently with heavy liquidated damages for lateness. As projects become shorter in duration, this enforces parallelism and concurrency, which by definition increases project complexity further. The increasing desire to reduce "time to market" times, and the subsequent development of the field of Concurrent Engineering (which aims to support the integrated, concurrent design of products and their related processes, including manufacture and support) is well-described in textbooks such as Syan and Menon [13].

It is not obvious that there has been any significant increase in Goal Uncertainty - indeed, the tendency towards fixed-cost, prime contracts, with a tightly-written (perhaps performance-based) specification. suggests that, at least in the early 1990's, there was perhaps a general reduction in Goal Uncertainty. But many writers point to the increasing rate of change in technology (classically Toffler [14]), suggesting that uncertainty in Methods may well be increasing. But this is speculation that requires proper study.

Covering all of the above, Laufer [15] characterised the last 4 decades of project management by an evolution of models appropriate to changing dominant project characteristics: he characterises the 1960s by scheduling (control), for simple, certain projects; the 1970s by teamwork (integration) and the 1980s for reducing uncertainty (flexibility), both for complex, uncertain projects, and the 1990s by simultaneity (dynamism) for complex, uncertain and quick projects - in other words, the very elements that we have defined as "Complexity".

4. Methods

Having made a tentative definition of complexity, it is clear that classical project management techniques are unsuitable for dealing with such projects. For structural uncertainty, decomposition models do not account for the compounding effects when individual perturbations accumulate in a project [5]; they cannot deal with feed-back loops [3]; they do not account for the systemic, holistic effects that are present in structurally complex projects [11]; they are not able to deal with goal- or method-uncertainty [7].

What are needed, then, are new ways of looking at modern, complex projects, new models and techniques for analysing them, new methods for managing them - in fact, new paradigms to underlie our approach to them. A number of such approaches have been taken, but it is difficult to take a view as to the most promising approach, nor has much synthesis been attempted (either between the approaches or with lower-level operational methods). This paper notes some of these areas in which work is on-going, but it is not the purpose of this paper to investigate such approaches - indeed, it is the aim of this Workshop to do this, by bringing together 50 internationally known personnel, from different backgrounds and with different agendas, hence the emphasis on round-table open discussions.

In order to support the management function, in particular planning, forecasting, monitoring and control, analysts must be able to model complex projects.

- Models can improve on classical methods, retaining the bottom-up decomposition into project elements. Network models can be improved to include stochastic effects, or the effects of management decisions. Models of time and cost risk can be developed by modelling the combination of many risk elements. Simulation models can be built that simulate the behaviour of many project elements of different types in combination.
- Alternatively, top-down holistic models can be built - perhaps the most successful technique being System Dynamics. While such models usually fail to capture the detail desired by operational management, they allow a strategic overview, and allow modelling of systemic effects that bottom-up methods miss out.
- Traditional methods capture only "hard" quantitative data. It has become clear that "softer" ideas must also be included in project models, if they are to be a useful representation of the real project. Soft systems methods, and "soft" Operational Research methods have proved useful. Some of this data can be used in some holistic modelling techniques, particularly System Dynamics.

Management techniques must similarly adapt to the new environment. Jones [6] explains how an increase in project complexity leads to an increase in internal conflicts within the project, so management methods and style must adapt to deal with such conflict. Changes need to be made to the internal management structures within projects; in particular, the use of multi-disciplinary teams is becoming more widespread. Laufer's study [15], discussed above, concludes that for the new type of project, that we have defined by a wide definition of complexity, you should have a project management style based on elements such as integration, systemic management, simultaneous management, the use of teams, and managing functional plans simultaneously and interdependently *inter al.* Looking wider than one project, new views have to be taken of the multi-project environment, programme management. Complexity finally needs to be considered in the establishment of joint ventures and other inter-corporate arrangements.

5. Conclusion

Dalchar [16] claims that "Contemporary project management practise is characterised by: late delivery, exceed budgets, reduced functionality and questioned quality. As the complexity and scale of attempted projects increases, the ability to bring these projects to a successful completion dramatically decreases." This Workshop is to take up this challenge, and looks to see how, differently, we should tackle complex projects. As the first small step, this paper has tried to ask what constitutes complexity. In particular, it has highlighted structural complexity, the number and interdependence of elements (following Baccarini) and uncertainty in goals and means (following Turner and Cochran). Complexity is increasing as all of these elements increase, exacerbated by simultaneity resulting from tightening project deadlines.

Acknowledgements The help of Prof. Eden, Dr. Ackermann and Dr. Monk of the Department of Management Science, Strathclyde University, in providing their thoughts on the above topic is gratefully acknowledged.

References

1. Baccarini, D. (1996) The concept of project complexity - a review, *International Journal of Project Management* **14**, 201-204.
2. Morris, P.W.G. and Hough, G.H. (1987) *The anatomy of major projects: a study of the reality of project management*, Wiley, Chichester.
3. Ackermann, F., Eden, C. and Williams, T. (1996) A persuasive approach to Delay and Disruption - using "mixed methods". *Interfaces* Nov-Dec 1996.
4. Thompson, J.D. (1967) *Organizations in Action*. McGraw-Hill, New York.
5. Williams, T.M., Eden, C.L., Ackermann, F.R., and Tait, A. (1995). The effects of design changes and delays on project costs. *Journal of the Operational Research Society 46,* 7, 809-818
6. Jones, R.E. and Deckro, R.F. (1993) The social psychology of project management conflict. *European Journal of Operational Research* **64**, 216-228.
7. Turner, J.R. and Cochrane, R.A. (1993) Goals-and-methods matrix: coping with projects with ill defined goals and/or methods of achieving them. *International Journal of Project Management* **11**, 93-102.
8. Shenhar, A.J. and Dvir, D. (1995) Managing technology projects: a contingent exploratory approach. *Proceedings of the 28th Annual Hawaii International Conference on Systems Sciences.* pub. IEEE, 494-503.
9. Wozniek, T.M. (1993) Significance vs capability: 'Fit for use' project controls. *American Association of Cost Engineers International (Trans)* (Conference proceedings) Dearborn, Michigan, A.2.1-8
10. Helbrough, B. (1995) Computer assisted collaboration - the fourth dimension of project management? *International Journal of Project Management* **13**, 329-333.
11. Williams, T.M. (1995) Holistic methods in project management. *Proceedings of INTERNET Symposium*, St.Petersburg, Russia, 14-16 September 1995. ISBN 5-87115-011-x pages 332-336.
12. Clarke, L. (1994) *The essence of change* Prentice-Hall, Herts, UK.
13. Syan, C.S. and Menon, U. (eds) (1994) *Concurrent Engineering: concepts, implementation and practice.* Chapman and Hall, London.
14. Toffler, A. (1971). *Future Shock.* Bodley Head, London
15. Laufer, A., Denker, G.R. and Shenhar, A.J. (1996) Simultaneous management: the key to excellence in capital projects. *International Journal of Project Management* **14**, 189-199.
16. Dalcher, D. (1993) The new project management mindset for the 21st century. *Proc. 1st British Project Management Colloquium*, Henley-on-Thames, UK December 1993.

SECTION 3
MODELLING TECHNIQUES

This Section is the most substantial in the Workshop. Before the Workshop looked at techniques and structure to manage complex projects, it needed to look at the techniques for modelling projects required to inform this management. This Section tries to look at why traditional models are inadequate and to propose new methods of modelling.

PERT-type methods are inadequate for a number of reasons, but particularly because they do not incorporate management actions, adequately consider uncertainty or consider effects that are systemic or "soft". These papers tackle exactly these issues.

The first two papers extended PERT methods. The first is by Prof. Dmitri Golenko-Ginzburg of Ben-Gurion University of the Negev in Israel, and discusses the use of controlled networks for control problems in networks under uncertainty. The second, by Prof. Salah Elmaghraby of North Carolina State University in the U.S.A. discusses the use of diffusion networks to analyse uncertain networks, particularly where the uncertainty varies dynamically over time.

The third paper is by Prof. Chris Chapman from the University of Southampton in the U.K. This covers the field of risk analysis techniques, considering how time risk impacts on cost and what drives time risk, trying to capture important connections and recognise significant causalities without becoming too complicated. In particular, the use of operational separability is explored to deal with connectivity.

Finally, Kenneth Cooper of Pugh-Roberts Associates in the U.S.A. discusses the System Dynamics technique, which provides a simulation model of projects, capturing systemic effects, cause-effect "paths", and effects that are “soft”. As well as applications, this paper discusses the commonality often found among projects and the often counterintuitive results that can inform management practice.

PROJECT PLANNING AND CONTROL BY STOCHASTIC NETWORK MODELS

D. GOLENKO-GINZBURG and A. GONIK
Dept. Industrial Engineering and Management, Ben-Gurion University of the Negev, Beer-Sheva 84105, Israel

1. Introduction

In recent years, the problems associated with planning, controlling and scheduling projects by means of network analysis have been discussed extensively in the literature. However, the number of publications on *stochastic network projects* remains very scanty, especially for highly complicated long-term projects under random disturbances. Moreover, in this area, the application of new developments lags far behind the developments themselves.

It can be clearly recognized that in the last two decades, various control problems in project management have been the subject of lengthy debate and very sharp criticism [3, 24, 27, 31 etc.]. In our opinion, the main reason that, in practice, those projects are all usually completed late and remain uncontrolled is that they are carried out under random disturbances (new estimates of a random nature without any previous experience, random activities' durations, periodical revisions of networks over time due to random emergency situations, etc.). However, project managers usually [27] avoid probabilistic terms since they are not sufficiently trained. They are trying to control highly complicated projects with uncertainty by using deterministic techniques. This leads to biased estimates that usually underestimate the actual time needed to accomplish the project. Therefore the project's due date can rarely be met. In our opinion, there is another important reason for the numerous failures in planning and controlling stochastic network projects. This is because the models are too complicated to be effective. They are not flexible. Usually, they incorporate both scheduling and control techniques. But since it is practically impossible to develop a *proper deterministic schedule for a project under random disturbances* , such models are not adequate to the real life. Therefore both the planning and control procedures are also non-effective. *Thus the need of proper planning and control models for* stochastic *projects is very important.*

T. M. Williams (ed.), Managing and Modelling Complex Projects, 21–45.

From the broad spectrum of various planning and control problems for stochastic network projects we will consider the following ones:

1. *Alternative network projects* under random disturbances with various alternative outcomes in key nodes. The developed control model chooses an optimal outcome direction at every decision node that is reached in the course of the project's realization.
2. *On-line control models* for network projects, for which the project's progress can be evaluated only via inspection in control points. The project's due date and the chance constraint to meet the deadline are pregiven. An on-line control model determines both the control points and the control actions to be introduced at those points to reorient the progress of the project in the desired direction.
3. *Stochastic network project scheduling* with several non-consumable activity related limited resources. Each activity is operated at a random speed that depends on the resource capacities assigned to that activity. The model determines, for each activity entering the project, both starting time values and corresponding resource capacities. The model's objective is to minimize the expected project duration.
4. A *multilevel control model for several stochastic network projects* which unifies the models outlined above.

The validity of the theoretical results has been checked via intensive simulation.

2. Alternative Network Projects Under Random Disturbances

While the literature on PERT and CPM network techniques is quite vast, the number of publications on *alternative networks* remains very scanty. Various authors, e.g. Eisner [7], Elmaghraby [9], Pritsker [30], Whitehouse [32], etc., introduced the concept of a research and development (R&D) project as a complex of actions and problems towards achieving a definite goal. Several adequate network models for such projects have been considered. The first significant development in that area was the pioneering work of Eisner [7] in which a "decision box" with both random and deterministic alternative outcomes was introduced. Elmaghraby [8, 9] introduced additional logic and algebra in network techniques, while Pritsker, Happ and Whitehouse [28-30] developed the GERT techniques for alternative network models with stochastic outcomes in key nodes. Hespos and Strassman [33] introduced the concept of the stochastic decision tree, while Crowston and Thompson [4-6] and later on Hastings and Mello [23] introduced the concept of multiple choices at such alternative nodes, when decision-making is of a deterministic nature (Decision-CPM models). Lee, Moeller and Digman [25-26] developed the VERT model that enables the analyst to simulate various decision with alternative technology choices within the stochastic decision tree network. Golenko-Ginzburg [10-14] has developed a unified controlled alternative activity network (CAAN model) for projects with both random and deterministic alternative outcomes in key nodes. At each routine decision-making node, the developed algorithm singles out all the subnetworks (the so-called joint variants) that correspond to all possible outcomes from that node. Decision-making results in determining the optimal joint variant and following the optimal direction up to the next decision-making node. However, the techniques thus far developed can only be applied to fully-divisible networks that can be subdivided into non-intersecting fragments. The CAAN model does not include non-fully-divisible networks. Thus, the model is not relevant to most R&D projects, since the latter are usually structured from non-divisible subnetworks. Golenko-Ginzburg and Blokh [17] have developed a more universal alternative network—the Generalized Alternative Activity Network (GAAN model). All types of the previously developed alternative network models, namely, Eisner's model, GERT, Decision-CPM, VERT and CAAN networks, are particular cases of the GAAN model.

2.1 THE GAAN MODEL

A GAAN model is a finite, oriented, acyclic activity-on-arc network G(N,A) with the following properties:

I G(N,A) has one source node n_0 and no less than two sink nodes n'.

II Each activity $(i,j) \in A$ refers to one of three different types as follows:

Type 1: activity (i,j) is a PERT activity (PA) with the logical "must follow" emitter at node i and the "and" receiver at node j;

Type 2: activity (i,j) is an alternative stochastic activity (ASA) with the logical "exclusive or" emitter at node i. Each $(i,j) \in A$ of ASA type corresponds to probability $0<p_{ij}<1$, while node i comprises a set of not less than two probabilities p_{ij}, $\sum_j p_{ij} = 1$;

Type 3 activity (i,j) is an alternative deterministic activity (ADA) with the logical "exclusive or" emitter at node i. Node i is a decision-making node and the corresponding transfer probabilities sum to one.

III Activities of all types may leave one and the same node. Thus, unlike the CAAN model, the GAAN model is not a fully-divisible network.

IV Activities of all types may enter one and the same node.

An example of a GAAN type graph is shown in Fig.1A. Here activities (1, 2) and (1, 3), (2, 7) and (2, 8), (4, 9) and (4, 10) are of ADA type. Activities (1, 4) and (1, 5), (3, 7) and (3, 8) are of ASA type, while activities (1, 9), (2, 6), (3, 9), (5, 10), (5, 11) are of PA type.

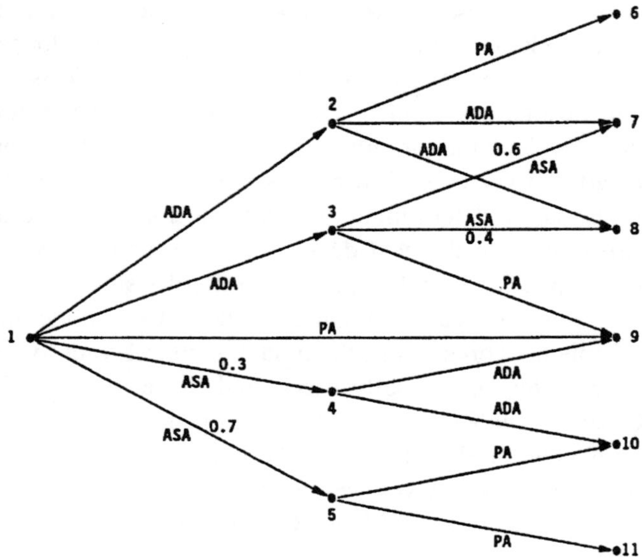

Fig. 1A. THE GAAN TYPE GRAPH

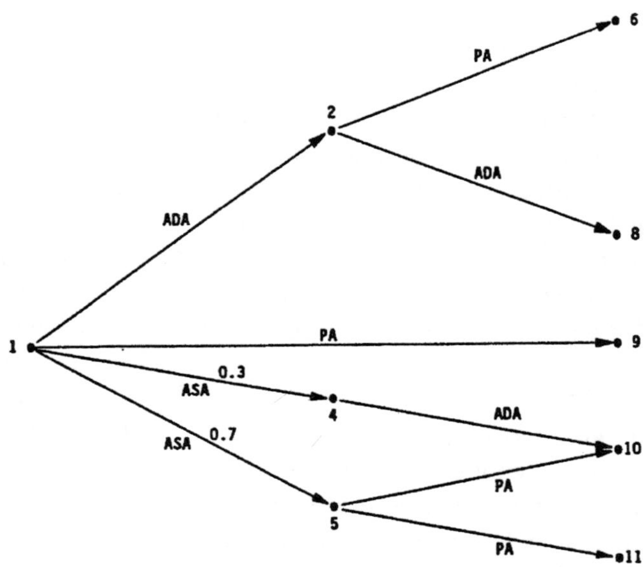

FIG. 1B. THE OPTIMAL JOINT VARIANT G_4^* FOR THE INITIAL GRAPH

We will call *a joint variant of the GAAN model* G(N,A) a subgraph (subnetwork) $G^*(N^*, A^*)$ satisfying the following conditions:

1. $G^*(N^*, A^*)$ has one source node;
2. If $G^*(N^*, A^*)$ comprises a certain node i, i.e., $i \in N^*$, then $G^*(N^*, A^*)$ comprises all activities (i, j) of types PA and ASA leaving node i;
3. If $G^*(N^*, A^*)$ comprises a certain node i which in the GAAN model G(N,A) has alternative outcomes of ADA type then $G^*(N^*, A^*)$ comprises only one activity of that type leaving that node;
4. $G^*(N^*, A^*)$ is the maximal subgraph satisfying conditions 1-3.

Call a full variant of the joint variant $G^*(N^*, A^*)$ a subnetwork of PERT type $G^{**}(N^{**}, A^{**}) \subset G^*(N^*, A^*)$ which can be extracted from the latter by simulating non-contradicting outcomes of ASA type in interconnected nodes and excluding alternative non-simulated outcomes.

Call a probability of realizing a full variant G^{**} a product of all values p_{ij} for all activities of ASA type entering the full variant.

2.2 DECISION-MAKING IN GAAN TYPE MODELS

To control a project, such as any production process, it is necessary to introduce decision-making in order to reach the goal while optimizing a given objective (the optimized value OV) subject to certain restrictions (the restrictive values RV). When the objective is the project's duration, the first restriction is usually the project's cost, and vice-versa. For a project represented by a GAAN type model decision-making means choosing the directions of the project's progress in controlled nodes (decision-making nodes) with alternative outcomes of ADA type, since alternatives of ASA type are uncontrollable. Thus the optimization problem consists of the following steps:

Step. 1. At each decision-making node which has been reached at moment t in the course of the project's realization,

- to determine and to single out all the joint variants from the remaining project G_t at moment t;
- to calculate the optimized value OV and all the restrictive values RV for each variant.

Step 2. To determine the optimal joint variant and to follow the optimal direction up to the nearest decision-making node. The problem must be repeatedly solved for the reduced network in every sequentially encountered decision-making node.

2.3 MATHEMATICAL FORMULATION

The mathematical formulation of the problem is as follows: determine the optimal joint variant $G^{*opt} \subset G(N, A)$ that optimizes the objective function

$$E\left[F\left(G^{*opt}\right)\right] = \underset{\{G^{*}\}}{Min(Max)} \sum_{\{G^{**}\} \subset G^{*}} \left[F\left(G^{**}\right) \cdot \Pr\left\{G^{**}\right\}\right] \tag{1}$$

s.t.

$$E\left[Q_v\left(G^{*opt}\right)\right] = \sum_{\{G^{**}\} \subset G^{*}} \left[Q_v\left(G^{**}\right) \cdot \Pr\left\{G^{**}\right\}\right] \le H_v, \qquad 1 \le v \le V. \tag{2}$$

Here $F(G^{**})$ is the objective function of full variant G^{**}, $\Pr\{G^{**}\}$ is the probability of realizing G^{**}, $Q_v(G^{**})$ is the value of the vth restricted criterion and H_v is the pre-set restriction level for that criterion. Note that for certain particular cases, V may be zero, i.e., the optimal problem has no restrictions, or the problem comprises only one restriction (2) without objective (1). *Since problem (1-2) is NP-complete* [17], *in order to obtain an optimal solution one has to develop a lookover algorithm to single out all the joint variants.*

The idea to enumerate the joint variants of the CAAN model [13, 14] is based on introducing lexicographical order to the set of maximal paths in the CAAN graph. The corresponding lookover algorithm is very simple in usage [13]. In the case of a GAAN network the order on the set of paths has to be substituted for the order on the set of subgraphs [17]. To develop the enumeration algorithm, we use the ideas to enumerate the so-called trajectories for assignment problems, or special matrices for traveling salesman problems [1, 2]. Note that singling out the maximal trajectory for an assignment problem is similar to determining the joint variant with the maximal objective value. Since a trajectory can be regarded as a vector and the latter, in turn, can be mapped onto a set of integer numbers, the trajectories can be enumerated. Similar ideas are used in developing a lookover algorithm to enumerate and single out all the joint variants.

As an example, consider a R&D project represented by the GAAN type network in Fig. 1A. The objective to be minimized is the project's duration while value RV restricted from above is the project's cost. The initial data of each activity is given in Table 1 (values t_{ij} and c_{ij} are constant). The mean cost of the project must not exceed \$55,000. The management has to determine an optimal decision policy, i.e., to choose an optimal joint variant together with determining optimal alternative outcomes of ADA type from every decision-making node that is reached in the course of the project's realization.

No. of (i,j)	(i,j)	ASA (p_{ij})	ADA	PA	t_{ij} (in months)	c_{ij} (in 1000$)
1	(1,2)	–	1	–	4	6
2	(1,3)	–	1	–	2	5
3	(1,4)	0.3	–	–	8	12
4	(1,5)	0.7	–	–	6	10
5	(1,9)	–	–	1	4	15
6	(2,6)	–	–	1	11	9
7	(2,7)	–	1	–	15	6
8	(2,8)	–	1	–	4	7
9	(3,7)	0.6	–	–	3	14
10	(3,8)	0.4	–	–	6	10
11	(3,9)	–	–	1	8	12
12	(4,9)	–	1	–	16	9
13	(4,10)	–	1	–	18	7
14	(5,10)	–	–	1	1	4
15	(5,11)	–	–	1	7	3

TABLE 1. The Initial Data

The GAAN model comprises six joint variants, while three of them satisfy the cost restriction. Choosing the joint variant with the minimal expected time and honoring the cost restriction results in choosing the optimal joint variant G_4* with parameters T = 18.3 months and $\overline{C}$ = $54,000. The optimal joint variant is shown in Fig. 1B. Thus, the decision-making to control the project is as follows:

a) at the beginning of the project (node 1), we choose activity (1, 2) from two alternative outcomes of ADA type: (1, 2) and (1, 3);
b) if, in the course of the project's realization, node 4 is reached, direction (4, 10) has to be chosen;
c) when node 2 is reached we choose direction (2, 8).

If the number of joint variants becomes very high, the developed lookover algorithms (both for the CAAN and GAAN models) require much computational time, especially for networks with many alternatives. We suggest an approximate method which is based on the ideas of combinatorial optimization with two parameters [1, 2]. Unfortunately, we have developed the method only for the CAAN type network models.

Note, in conclusion, that the models outlined above may be a little more costly and complicated in usage. But for modern and complex R&D projects (e.g. in defense related industries) the gain from using such models may be tremendous.

3. On-Line Control Models for Network Projects

After determining the optimal joint variant (in the case of an alternative network project) the latter is realized and controlled in order to meet the project's due date on time. A joint variant may be either a PERT, or a GERT type network, usually with random activity durations. For most R&D projects the progress of the project cannot be inspected and measured continuously, but only at preset inspection points. An on-line control has to determine both inspection points and control actions to be introduced at those points to alter the progress of the project in the desired direction. On-line control is usually carried out to minimize the number of inspection points needed to meet the target, since inspecting the project's output is usually a costly operation. In addition, an on-line control for a stochastic network project has to be carried out subject to a chance constraint. Thus, the generalized on-line control model has to be formulated as follows [15,16,18]: determine both optimal control points t_g to inspect the project and optimal control actions $CA(t_g,r_g)$ (see Notation in Appendix) to be introduced at those control points (r_g is the index of the control action), in order to minimize the number W of inspection points

$$\underset{\{t_g,r_g\}}{Min}\, W \tag{3}$$

s.t.

$$\Pr\{t_g, r_g\} \geq p^*, \tag{4}$$

$$t_0 = 0, \tag{5}$$

$$t_W = D, \tag{6}$$

$$t_{g+1} - t_g \geq \Delta. \tag{7}$$

Note that if introducing a control action $CA(t_g,r_g)$ results in determining the *project's speed* v_{t_g} to proceed with until the next control point t_{g+1} and if *several alternative speeds can be chosen*, then the optimal control action enables using the *minimal speed* while honoring chance constraint (4) [16].

Control model (3-7) is a stochastic optimization problem with a non-linear chance constraint and a random number of optimized variables. The problem is too difficult to solve in the general case. Thus, heuristic control algorithms have been developed [15,16,18] to determine the next inspection point t_{g+1}. Two algorithms are considered:

A. Using sequential statistical analysis to maximize the time span

$$\Delta t_g = t_{g+1} - t_g;$$

B. Using the idea of a risk averse decision-maker.

Algorithm A [15, 16] solves the on-line control problem as follows (see Notation): to maximize the objective (t_{g+1} - t_g) s.t. (5-7) and

$$\Pr\left\{V_t \geq V_t^*\left(t_g\right)\right\} \geq p^* \ \forall t : t_g \leq t \leq t_{g+1} \tag{8}$$

This problem can be solved by determining the maximal value T^* satisfying

$$T^* = \underset{t_g < t \leq D}{MAX}\left\{t : \psi\left(q_t\right) \geq p^*\right\}. \tag{9}$$

Here

$$\psi(x) = \frac{1}{\sqrt{2\pi}} \int_x^\infty e^{-\frac{u^2}{2}} du, \; q_t = \frac{\overline{H}_t}{S^2\left(H_t\right)}, \; H_t = V_t - V_t^*\left(t_g\right), \tag{10}$$

and $\overline{H}_t$ and $S\left(H_t\right)$ are the mean value and the standard deviation of random value H_t, correspondingly. In practice, T^* can be calculated via simulation with a constant step of length Δ. The procedure of increasing t step-by-step is followed until (9) ceases to hold. It can be recognized that $t_g + T^* = t_{g+1}$ holds.

Algorithm B is based on the concept of a risk averse decision-maker [18]. Given a routine inspection point t_g, the project's output observed at that moment V_{t_g} and the control action $CA(t_g, r_g)$ to be introduced at moment t_g up to the next inspection point, the problem is to determine that next point t_{g+1}. As for Algorithm A, the objective is to maximize the time span t_{g+1}-t_g. Value t_{g+1} is determined so that even *if the project develops most unfavourably in the interval* [t_g, t_{g+1}]*, i.e., with the minimal rate* $v'\left(t_g, r_g\right)$*, then introducing the most effective control action CA* (t_{g+1}, *r*) *at moment* t_{g+1} *enables the project to meet its target on time, subject to the chance constraints.* Here r is the index of the most effective control action, e.g. r is the index of the highest possible speed to be introduced. Value t_{g+1} is determined via a "risk averse" heuristic [18]

$$V_{t_g} + v'\left(t_g, r_g\right)\left(t_{g+1} - t_g\right) + \overline{v}\left(t_{g+1}, r\right) \cdot \left(D - t_{g+1}\right) = V^*. \tag{11}$$

Note that the minimal rate $v'(t_g, r_g)$ can be substituted for a p-quantile of the random speed $v(t_g, r_g)$ where the confidence level p is close to zero.

Both on-line control algorithms are performed in real time. However, in order to check the validity of any of them, the algorithm's work can be simulated. The comparative efficiency of Algorithms A and B has been tested on various examples of medium-size PERT projects. A conclusion can be drawn [18] that using the second algorithm rather than the first results both in essentially smaller computational time and in cheaper project realization. Both methods honor the chance constraint p^* and can be applied to various control projects of PERT type.

Note that the above outlined on-line control models can also be applied to control projects of GERT type, i.e., to network projects which comprise various random alternative outcomes. For such a project, a certain part of the activities will not be realized in the course of the project's realization. We recommend dividing the remaining project into two subgraphs at each decision-making node [19, 22]. The first is a PERT graph that is realized before meeting the nearest random alternative node, while the second subgraph is a GERT network. Determining control points is carried out only for the PERT subgraph. After the next random alternative node is reached and the random outcome is simulated, the procedure of subdividing the remaining project is carried out anew.

In order to improve the *practical* effectivity of the control techniques for stochastic network projects we come to the following conclusions:

a) Scheduling and control procedures must *not* be incorporated in one model;
b) A control model has to be based on probabilistic approaches and has to implement probabilistic terms. Such a model has to be used only at several control (inspection) points. We suggest applying the control model not to the initial network (which for some projects may comprise a large amount of activities), but to a modified one, with a medium amount of activities. For such a modified model, an activity can be a subnetwork (a fragment) of the initial network. (Note that such modified networks *for construction projects* are successfully used by Prof. V. Voropajev.)
c) Schedule procedures are applied to the initial network and are carried out *between two routine adjacent control points*. They are usually based on heuristic procedures (sometimes very doubtful) and result in biased estimates and errors. *But the latter are periodically corrected* by introducing proper control actions.
d) Thus, we *recommend developing the on-line control model as an additional tool, as a decision-making support model to help the project manager realize the project*. On the basis of such a model, the project manager may introduce any action he finds reasonable, e.g., to enhance the progress of the project.

4. Stochastic Network Project Scheduling with Non-Consumable Limited Resources

An activity-on-arc network project of PERT or GERT type with random activity durations is considered. Several non-consumable activity related resources, such as machines or manpower, are utilized to realize the project. Each activity (i,j) in such a project requires resources of various types k with variable capacities r_{ijk}. Each resource capacity r_{ijk} assigned to any activity is limited within pregiven bounds r_{ijk}^{min} and r_{ijk}^{max}. Each type of resource k is in limited supply with a resource limit R_k that is fixed at the same level throughout the project duration. It is assumed that each activity is operated at a random speed that depends on the resource capacities assigned to that activity.

The problem is to determine, for each activity (i,j) entering the project, both the starting time values S_{ij}, i.e., the timing of feeding-in resources, and the resource capacities r_{ijk} for each type of resource k assigned to that activity. The problem's goal is to minimize the expected project duration [20].

The problem's mathematical formulation is as follows (see Notation):

$$\underset{S_{ij},\vec{r}_{ijk}}{Min}\, E\left\{T\left(G/S_{ij},\vec{r}_{ijk}\right)\right\} \tag{12}$$

s.t. $$r_{ijk}^{min} \le r_{ijk} \le r_{ijk}^{max} \;\forall\, (i,j) \in G(N,A) \tag{13}$$

and $$R_k^*\left(t/S_{ij},\vec{r}_{ijk}\right) \le R_k(t) \;\forall\, t \ge 0, \quad 1 \le k \le n\,. \tag{14}$$

Model (12–14) is a stochastic optimization problem that cannot be solved in the general case; the problem only allows a heuristic solution. Decision-making, i.e., determining values S_{ij} and r_{ijk}, is carried out at decision points F_{ij} and T(i) (see Notation), either when one of the activities (i,j) is finished and additional resources become available, or when all activities (i,j) leaving node i are ready to be processed. Thus, both values S_{ij} and r_{ijk} are *not calculated beforehand* and are random variables dependent on our future decisions. If one or more activities (i_1, j_1), …, (i_m, j_m), m ≥ 1, are ready to be processed at a decision point t and *all of them can be supplied by all types of available resources of maximal capacity*, the needed resources are fed in and activities (i_q, j_q), $1 \le q \le m$, begin to be operated at moment t, i.e., $S_{i_q j_q} = t$, $r_{i_q j_q k} = r_{i_q j_q k}^{max}$, $1 \le k \le n$. Otherwise a competition has to be arranged to choose the optimal subset of activities that can be supplied by available resources.

An important auxiliary procedure precedes holding the competition, namely, calculating, for all the competitive activities (i_q, j_q), their conditional probabilities $p(i_q, j_q)$ to be on the critical path in the course of the project's realization. Calculating values $p(i_q, j_q)$ is carried out via simulation: at each decision point, all the activities that have not yet started to be operated are simulated using the corresponding probability density functions.
We will henceforth examine two cases:

a) all resource capacities r_{ijk} for each k-th type of resources are fixed and remain unchanged.
b) values r_{ijk} may vary within pregiven bounds r_{ijk}^{min} and r_{ijk}^{max}.

In the first case random values t_{ij} do not depend on values r_{ijk}, and the corresponding density functions remain unchanged in the course of the project's simulation. Later on, the critical path of the remaining graph with simulated activity durations is determined. By repeating this procedure many times, the calculated frequencies for each activity (i_q,j_q) to be on the critical path are taken as $p(i_q,j_q)$. Values $p(i_q,j_q)$ enter the zero-one integer programming model to carry out the competition [20].

For the case of variable resources rijk we do not know beforehand the resource capacity values $r_{i_q j_q k}$ that will be assigned to the activities under competition, as well as to all other activities in the remaining project. Thus, we are unable to simulate the activities' durations, that depend parametrically on values $r_{i_q j_q k}$. In order to overcome these difficulties, we suggest to use heuristics, e.g. to set $r_{ijk} = 0.5\,(r_{ijk}^{min} + r_{ijk}^{max})$. After calculating conditional probabilities p(i,j) the knapsack reallocation problem among the competitive activities has to be solved at each decision point [21].

For the case of fixed r_{ijk} a classical zero-one programming problem with a precise solution can be formulated as follows (see Notation): determine integer values $\xi_{i_q j_q}$, $1 \le q \le m$, to maximize the objective

$$\underset{\left(\xi_{i_q j_q}\right)}{\text{Max}} \left\{ \sum_{q=1}^{m} \left[\xi_{i_q j_q} \cdot p\left(i_q, j_q\right) \cdot \mu_{i_q j_q} \right] \right\} \tag{15}$$

s.t.

$$\sum_{q=1}^{m} \left(\xi_{i_q j_q} \cdot r_{i_q j_q k} \right) \le R_k(t), \quad 1 \le k \le n, \tag{16}$$

where

$$\xi_{i_q j_q} = \begin{cases} 0 & \text{if activity } (i_q, j_q) \text{ will \underline{not} obtain resources;} \\ 1 & \text{otherwise.} \end{cases}$$

In the case of variable resource capacities r_{ijk}, a heuristic model is suggested. Since the project management has to choose the subset of activities and to reallocate among them the available resources *in order to maximize the total contribution to the expected project duration*, we suggest the following resource reallocation problem (to be solved at each decision point t) [21]:

Determine optimal values $S_{i_q j_q k}$ and $r_{i_q j_q k}$, $1 \le k \le n, 1 \le q \le m$, to maximize

the objective

$$J = \underset{S_{i_q j_q}, r_{i_q j_q k}}{\mathrm{Max}} \left\{ \sum_{q=1}^{m} \xi_{i_q j_q} \cdot p(i_q, j_q) \sum_{k=1}^{n} \left(r_{i_q j_q k} \cdot \psi_{i_q j_q k} \right) \right\} \tag{17}$$

s.t.

$$r_{i_q j_q k}^{\min} \le r_{i_q j_q k} \le r_{i_q j_q k}^{\min} \;\forall (i_q, j_q) \in G(N, A), \tag{18}$$

$$\sum_{q=1}^{m} \left(r_{i_q j_q k} \cdot \xi_{i_q j_q} \right) \le R_k(t) \;\forall\, t \ge 0, \quad 1 \le k \le n, \tag{19}$$

where $\xi_{i_q j_q}$ is as before

Problem (17-19) is a NP-complete one. We suggest both a precise solution obtained via a lookover algorithm, and a heuristic solution based on an essential diminishing of feasible solutions to be examined.

The resource constrained project scheduling algorithm comprises the knapsack problem (15-16) or (17-19), together with the auxiliary problem of determining conditional values p(i,j). The algorithm to solve problem (12-14) is performed in real time: namely, *all activities can be operated only after obtaining necessary resources.* Decision moments F_{ij} and T(i) cannot be predetermined. However, if we want to evaluate the efficiency of the resource constrained project scheduling model (12-14), we can simulate the algorithm's work by random sampling of the actual duration of activities. By simulating the algorithm's work many times, the average project's duration as well as the probability of accomplishing the project by a given due date (if necessary) can be evaluated. We have undertaken [21] intensive experimentation for various medium-size PERT and GERT projects with several (3÷5) non-consumable limited resources. A conclusion can be drawn that the algorithm performs well and is easy to handle.

5. Multilevel Control Model for Several Stochastic Network Projects With Restricted Resources

A company is considered which is faced with realizing several stochastic network projects $G_\ell(N,A)$, $1 \leq \ell \leq d$. The total budget C at the company's disposal to carry out the projects is limited. We suggest a hierarchical control model [19, 22] as presented in Fig. 2. At each level the model undertakes optimal control actions as follows:

- at the *company level* the control action boils down to optimal budget reassignment among the projects (Problem I);
- at the *project level*, in case of an alternative stochastic network project, its optimal joint variant is determined (Problem II);
- at the *project level*, if the project is of PERT or GERT type, optimal control actions result either in optimal budget reallocation among the project's activities (Problem IIIA for PERT-COST projects) or in determining the optimal speed of the project's realization (Problem IIIB); thus, solving Problems IIIA and IIIB results in optimizing the progress of the project towards its goal, in order to reorient the project in the desired direction;
- at the *inspection level*, on-line control is carried out, i.e., optimal control points to inspect the progress of the project are determined (Problem IV);
- at the lowest, the *scheduling level*, resource constrained project scheduling is realized by reallocating, if necessary, non-consumable resources among the project's activities (Problem V). Note that although Problem V is an optimal one, it cannot be regarded as a control action. That is because the problem's solution is not based on the project's output V_{t_g} which is observed at control points t_g.

Optimal budget reallocation Problem I can be formulated for two alternative cases:

a) projects are of equal importance.

b) projects have different priorities η_ℓ, $1 \leq \ell \leq d$, pregiven by management.

In the case of a), the problem is as follows (see Notation):

At any moment $t \geq 0$

$$J_1 = \underset{C_{\ell t}}{\mathrm{Max}} \left\{ \sum_{\ell=1}^{d} \left[\eta_\ell \cdot P_{\ell t}(C_{\ell t}) \right] \right\} \tag{20}$$

s.t.

$$\sum_{\ell=1}^{d} C_{\ell t} \leq C_t , \tag{21}$$

$$P_{\ell t}(C_{\ell t}) \geq p^* \ \forall\ \ell : 1 \leq \ell \leq d . \tag{22}$$

In the case of b), the problem is as follows:

$$J_2 = \underset{C_{\ell t}}{\mathrm{Max}}\ \underset{\ell}{\mathrm{Min}}\, P_{\ell t}(C_{\ell t}), \quad t \geq 0, \tag{23}$$

s.t. (21-22). Both problems are solved at time $t=0$ or have to be repeatedly resolved at time $t>0$, after an emergency is called at the project level. Problems (20-22) and (21-23) are solved via simulation, under additional heuristic assumptions [14,19,22].

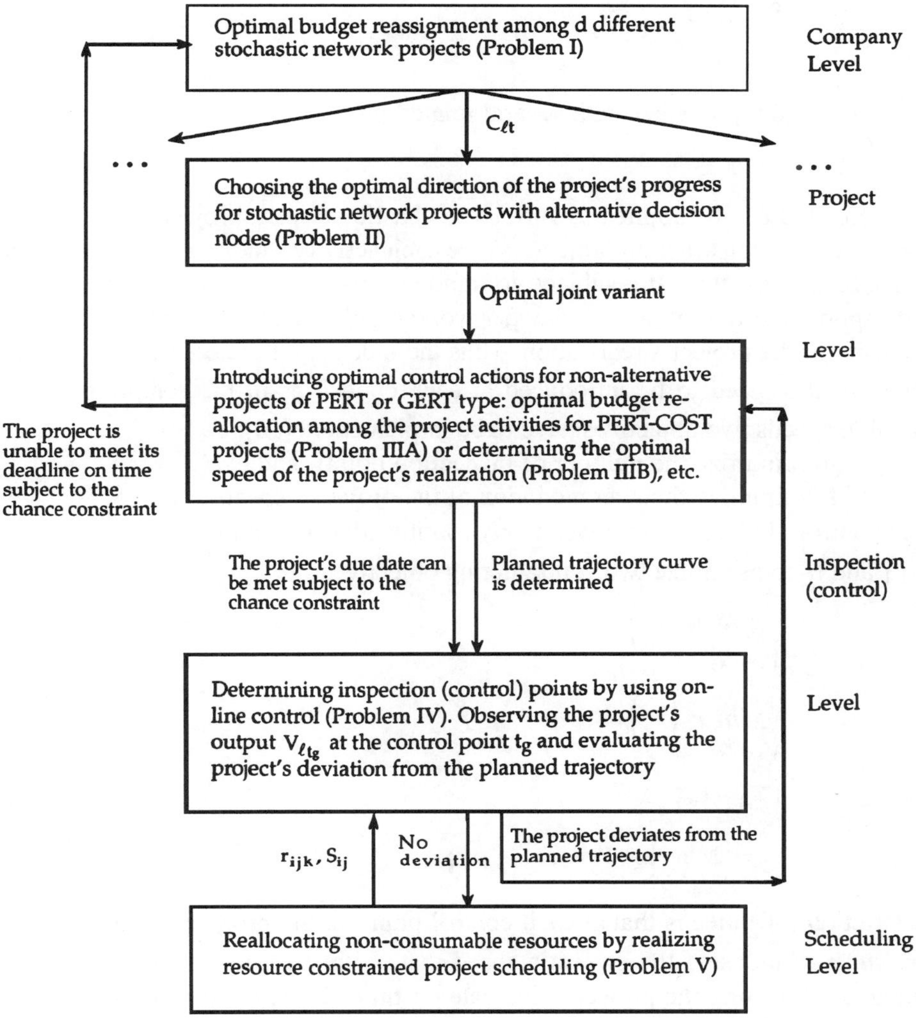

Fig. 2 Multilevel Control Model (at Moment t)

Problem II is outlined in Section 2. As to Problem IIIA, it can be formulated as follows (see Notation):

Given the budget $C_{\ell t}$ assigned to project $G_\ell(N, A)$ at moment t, determine optimal values c(i,j), (i,j)∈ A, to maximize the objective

$$\max_{\{c(i,j)\}} \{P_{\ell t}(C_{\ell t})\} \tag{24}$$

s.t.

$$c_{ij}^{min} \le c_{ij} \le c_{ij}^{max}, \tag{25}$$

$$\sum_{(i,j)} c_{ij} = C_{\ell t}. \tag{26}$$

Problem (24-26), together with several related problems, is solved via simulation [14,19,22].

Problem IIIB can be applied to various stochastic network projects with variable speeds, e.g. to construction projects where each activity can be operated at several possible speeds that are subject to random disturbances. Such speeds may correspond to different hours a day per worker and, thus, depend on the degree of intensity of the project's realization. Thus the index r_g of the control action is the index of the speed to be introduced at each control point t_g. For projects with variable speeds two objectives have to be implemented [16]:

– to minimize the number of inspection points, and
– to minimize the average index of the project's speeds.

The control model is as follows: at any routine control point t_g, determine values t_{g+1} and r_g to minimize two contradicting objectives

$$\min_{\{t_{g+1}, r_g\}} W(t_g) \tag{27}$$

and

$$\min_{\{t_{g+1}, r_g\}} \bar{r}_g(t_g) \tag{28}$$

s.t.

$$t_{g+1} - t_g \ge \Delta, \tag{29}$$

$$r_g = \min_{1 \le r_g \le r} \{r_g : \Pr(t_g, r_g) \ge p^*\}. \tag{30}$$

Restriction (30) means that at each control point t_g, the problem is to determine the *minimal* index of the project's speed that, with the given chance constraint, guarantees meeting the project's due date on time. Thus, the restriction prohibits unnecessarily high speeds. The solution of problem (27-30) is outlined in [16].

The solution of Problem IV, i.e. determining control points t_g, is outlined in [15,16,18]. Both algorithms A and B can be applied, but the second one is more efficient and offers less computational time [18]. The solution of the scheduling Problem V at the lowest hierarchical level is outlined in [20, 21].

Note that the solution of Problem I serves as the initial data for Problems II and III (at the project level). The solution of Problem III serves, in turn, as the initial data for Problem IV, which carries out on-line control, i.e., determines the optimal control points to inspect the progress of the project. This is done by determining the planned trajectories that must be repeatedly corrected in the course of the project's realization. If, at any control point, it turns out that a project deviates from the planned trajectory, an error signal is generated, and decision-making is based on resolving Problem III to reorient the progress of the project in the desired direction, i..e., to maximize the probability of meeting the deadline. If the problem's solution enables the project's deadline to be met, subject to the chance constraint, a corrected planned trajectory is determined and Problem IV is again solved to determine the next control point. Otherwise, an emergency signal is generated and decision-making is carried out at the company level. Problem I is resolved under emergency conditions to reassign the remaining resources (e.g. the remaining budget) among the non-accomplished projects. Thus, in the course of controlling a group of projects, the latter are first optimized on line from "top to bottom". In the case of an emergency, the generated "bottom-to-top" signals are converted into control actions to enable the projects' due dates to be met on time.

6. Conclusions

1. The developed planning and control models can be used for a very broad spectrum of stochastic network projects. They comprise highly complicated R&D projects with multiple alternative technology choices where decision-making has to be introduced with incomplete or inadequate information about the alternatives. Such models are especially effective for stochastic network projects with several possible alternative ways for reaching intermediate and ultimate goals. The developed control models can also be applied to a broad spectrum of PERT projects, including PERT-COST projects, projects with variable speeds, etc.
2. The developed planning and control models are easy in usage; an appropriate software to simulate the models can be implemented on a PC, mainly for projects with a medium amount of activities. Most of the models have been programmed, and have been used successfully for more than five years by students at the Ben-Gurion University of the Negev .

3. For a group of projects the developed hierarchical decision-making models indicate control actions and optimization problems for all levels. The models can be used both for controlling single projects and several projects with restricted company budget.

Acknowledgements

This research has been partially supported by the Paul Ivanier Center of Robotics and Production Management, Ben-Gurion University of the Negev.

References

1. Blokh, D. (1989), Assignment problem for three-index Jacobi matrices, *Automation and Remote Control* **50**(2), Part 2, 672--677.
2. Blokh, D. and Gutin, G. (1995), Maximizing traveling salesman problem for special matrices, *Disc. Appl. Math.* **56**, 83–86.
3. Bobrowski, P.M. (1989), Project management control problems: an information systems focus, *Project Management Journal*, **20**(2), 11-16.
4. Crowston, W.B. and Thompson, G.L. (1967), Decision CPM: a method for simultaneous planning, scheduling and control of projects, *Opns. Res.* **15**, 407–426.
5. Crowston, W.B. (1970), Decision CPM: Network reduction and solution, *Opns. Res. Quart.* **21**(4), 435–445.
6. Crowston, W.B. (1971), Models for project management, *Sloan Mngm. Rev.* **12**(3), 25–42.
7. Eisner, H. (1962), A generalized network approach to the planning and scheduling of a research project, *Opns Res.* **10**(1), 115–125.
8. Elmaghraby, S.E. (1964), An algebra for the analysis of generalized activity networks, *Mgmt. Sci.* **10**, 494–514.
9. Elmaghraby, S.E. (1977), *Activity Networks: Project Planning and Control by Network Models*, Wiley, New York.
10. Golenko-Ginzburg, D. (1972), *Statistische Methoden der Netzplantechnik, BSB, BG. Teubner Verlaggesellschaft*, Leipzig (in German, translated from Russian *Statistical Methods in Network Planning and Control,* 1966, Nauka, Moscow).
11. Golenko-Ginzburg, D., Lauenroth, H., Schultz, H. and Schultze, G. (1977), *Kybernetic zur Steuerung Ökonomischer Prozesse: Grundlage und Anwendungen*, Akademie-Verlag, Berlin (in German).

12. Golenko-Ginzburg, D., Kesler, S.S. and Livshitz, S.E. (1977), *Statistical Modeling in Project Management*, Leningrad University Press, Leningrad (in Russian).
13. Golenko-Ginzburg, D. (1988), Controlled alternative activity networks in project management, *Eur. J. Oper. Res.* **37**, 336–346.
14. Golenko-Ginzburg, D. (1993), A two-level decision-making model for controlling stochastic projects, *Int. J. Prod. Econ.***32**, 117–127.
15. Golenko-Ginzburg, D. and Gonik, A. (1996), On-line control model for COST-simulation network projects, *J. Opl. Res. Soc.* **47**, 266-283.
16. Golenko-Ginzburg, D. and Gonik, A. (1996), On-line control model for network construction projects, *J. Opl. Res. Soc.* (to appear).
17. Golenko-Ginzburg, D. and Blokh, D. (1996), A generalized activity network model (submitted paper).
18. Golenko-Ginzburg, D. and Gonik, A. (1996), Timing inspection points for stochastic network projects (submitted paper).
19. Golenko-Ginzburg, D., Gonik, A. and Kesler, Sh. (1996), Hierarchical decision-making model for planning and controlling stochastic projects, *Int. J. Prod. Econ.* (to appear).
20. Golenko-Ginzburg, D. and Gonik, A. (1996), Resource constrained project scheduling with non-consumable limited resources, *Int. J. Prod. Econ.* (to appear).
21. Golenko-Ginzburg, D. and Gonik, A. (1996), Stochastic network project scheduling with variable resource capacities (submitted paper).
22. Gonik, A. (1995), Planning and controlling multilevel stochastic projects, Ph.D. Thesis, Ben-Gurion University of the Negev, Beer-Sheva, Israel.
23. Hasting, M.A. and Mello, J.M. (1979), *Decision Networks*, Wiley, N.Y.
24. Hughes, M.W. (1986), Why projects fail: an effect of ignoring the obvious, *Ind. Engnr.* **18**(4), 14-18.
25. Lee, S.M., Moeller, G.L. and Digman, L.D. (1981), *Network Analysis for Management Decisions*, Kluwer-Nijhoff, Boston.
26. Moeller, G.L. and Digman, L.A. (1981), Operations planning with VERT, *Opns. Res.* **29**(4), 676–697.
27. Pearson, A.W. (1990), Planning and control in research and development, *Omega* **18**(6), 573-581.
28. Pritsker, A.A.B., and Happ, W.W. (1966), GERT: Graphical evaluation and review technique, Part I (Fundamental), *J. Indust. Engn.* **17**(5), 267–274.
29. Pritsker, A.A.B. and Whitehouse, G.E. (1966), GERT: Graphical evaluation and review technique, Part II (Probabilistic and industrial engineering applications) *J. Indust. Engn.* **17**(6), 293–301.

30. Pritsker, A.A.B. (1977), *Modeling and Analysis Using Q-GERT Networks,* Wiley, New York.
31. Schonberger, R.J. (1981), Why projects are "always" late: a rationale based on manual simulation of a PERT/CPM network, *Interfaces* **11**(5), 66-70.
32. Whitehouse, G.E. (1973), *Systems Analysis and Design Using Network Techniques*, Prentice-Hall, Englewood Cliffs, NJ.
33. Xespos, R.F. and Strassman, P.A. (1965), Stochastic decision trees for the analysis of investment decisions, *Mgmt. Sci.,* Ser. **B11**, 244–259.

Appendix: notation

1. The project level (to control a single project)

G(N,A)	–	stochastic network project;
$G^*(N^*,A^*) \subset G(N,A)$	–	the joint variant of project G(N,A) (for the case of an *alternative* stochastic network);
$G^{**}(N^{**},A^{**})$	–	full variant of the joint variant $G^*(N^*,A^*)$ (a realization of the joint variant);
$(i,j) \subset A$	–	the project's activity;
t_{ij}	–	the duration of activity (i,j);
c_{ij}	–	budget assigned to activity (i,j);
c_{ij}^{min}, c_{ij}^{max}	–	minimal and maximal possible budget to realize activity (i,j) (pregiven);
$t_{ij}(c_{ij})$	–	random duration of activity (i,j) on condition that budget c_{ij} is assigned to that activity;
Q_{ij}	–	the amount of activity (i,j) to be operated (pregiven); note that value Q_{ij} can be set in percentages of the total project as well as in other measures;
r_{ijk}	–	capacity of the k-th type non-consumable resource(s) allocated to activity (i,j), $1 \le k \le n$ (a decision variable);
r_{ijk}^{min}, r_{ijk}^{max}	–	lower and bound of value r_{ijk} (pregiven);
n	–	number of different resources;
R_k	–	total available non-consumable resources of type k at the disposal of the project management (pregiven and fixed throughout the planning horizon);
$R_k(t)$	–	free available non-consumable resources of type k at moment $t \ge 0$;
$v_{ij}(\vec{r}_{ijk})$	–	the speed of operating activity (i,j) in terms of Q_{ij}. Speeds $v_{ij}(\vec{r}_{ijk})$ are subject to disturbances and are random values. It is assumed that they depend on resource capacities r_{ijk} linearly, i.e., $v_{ij}(\vec{r}_{ijk}) = \sum_{k=1}^{n}(a_{ijk} \cdot r_{ijk})$ hold. Coefficients a_{ijk} are pregiven random values;

$\bar{a}_{ijk}$ – average values of a_{ijk};

$t_{ij}\left(\vec{r}_{ijk}\right) = \dfrac{Q_{ij}}{v_{ij}\left(\vec{r}_{ijk}\right)}$ – random duration of activity (i,j), on condition that non-consumable resource capacities r_{ijk} are assigned to (i,j);

$\mu_{ij}\left(\vec{r}_{ijk}\right)$ – the average value of $\mu_{ij}\left(\vec{r}_{ijk}\right)$;

ψ_{ijk} – additional value by which $\mu_{ij}\left(\vec{r}_{ijk}\right)$ can be diminished by adding $\Delta r_{ijk} = 1$, on condition that all other resource capacities $r_{ijv}, 1 \le v \le n, v \ne k$, are fixed and equal to $\bar{r}_{ijv} = \frac{1}{2}\left(r_{ijv}^{min} + r_{ijv}^{max}\right.$

S_{ij} – the time that resources are fed in and activity (i,j) starts (a decision variable);

$T\left(G/S_{ij}, \vec{r}_{ijk}\right)$ – duration of random project, on condition that feeding-in of resources r_{ijk} is carried out at moments S_{ij};

$R_k^*\left(t/S_{ij}, \vec{r}_{ijk}\right)$ – total resources of type k to be rescheduled at moment t on condition that feeding-in of resources r_{ijk} is carried out at moments S_{ij};

$F_{ij} = S_{ij} + t_{ij}\left(\vec{r}_{ijk}\right)$ – the actual moment activity (i,j) is finished;

T(i) – earliest possible time of realization of node i;

p(i,j) – conditional probability for activity (i,j) to be on the critical path in the course of the project's realization.

D – the project's due date (pregiven);

p^* – desirable probability that in practice enables completion of the project on time (pregiven);

V^* – the project's target, i.e., the planned total project volume (pregiven);

G_t – the remaining project G(N,A) at moment $t ^3 0$; $G_0 = G(N,A)$;

V_t – the project's output (project volume) observed at control point $t > 0$;

t_g – the g-th inspection (control) point, $g = 0, 1, \ldots, W$, $t_0 = 0$, $t_W = D$ (optimized variable);

W – the number of control points;

r_g – index of a control action introduced at control point t_g;

CA (t_g,r_g) – optimal control action on project G_{t_g} introduced by the control device at moment t_g to alter the project's speed in the desired direction, $1 \le r_g \le r$;

$r > 1$ – the number of possible control actions;

$\Pr(t_g,r_g)$ – confidence probability to accomplish project G_{t_g} on time after introducing control action CA (t_g,r_g);

$V_t^*(t_g)$ – trajectory control curve connecting two points $\left(t_g, V_{t_g}\right)$ and $\left(D, V^*\right)$;

$v(t_g,r_g)$ – the project's random speed set at the control point t_g after introducing control action CA (t_g,r_g);

$v'(t_g,r_g)$ – the minimal value of speed $v(t_g,r_g)$;

$\bar{r}_g(t_g)$ – the average index of the project's control actions, beginning from t_g;

$W(t_g)$ – the number of future control points, beginning from t_g;

Δ – the minimum time span between two adjacent control points t_g and t_{g+1} ito force convergence (pregiven);

2. *The company level (several controlled projects)*

C_ℓ – budget assigned by the company to the ℓ-th project, $1 \le \ell \le d$;

d – number of projects;

$C_{\ell t}$ – available remaining budget to realize the ℓ-th project at moment $t \ge 0$; $C_{\ell 0} = C_\ell$;

$P_{\ell t}(C_{\ell t})$ – probability to accomplish the ℓ-th project at the due date D_ℓ corresponding to the remaining budget $C_{\ell t}$, $t \ge 0$;

$T_{\ell t}$ – random duration of the remaining ℓ-th project corresponding to value $C_{\ell t}$ at moment $t \ge 0$;

η_ℓ – priority value of the ℓ-th project;

C – total budget at the company's disposal to realize a group of projects;

C_t – total remaining budget at moment $t \ge 0$; $C_0 = C$.

QUESTIONS on "PROJECT PLANNING AND CONTROL BY STOCHASTIC NETWORK MODELS"

Prof. Golenko-Ginzburg was asked the following question. It is true that complex models are needed for complex projects; there are a great deal of interactions and lots of interdependencies; furthermore, we need statistical parameters because there are uncertainties in real life - but how can such a complex model be presented to real-life managers in such a way that they will accept it and feel it to be useful. For example, would you suggest that managers should have "experts" who understand the probabilities and can understand the complexities, and can pass on results to the managers, or would you try to educate the managers to understand what the models are saying?

Prof. Golenko-Ginzburg made a number of points in answer. He explained his way of working for some projects in his experience. In these, managers were given probabilistic results, and this was very helpful; here, a little training can be very beneficial. He also had been using the method on not very big networks (he later suggested 40-50 activities as a good size). He pointed out that the method can give very useful indications of corrective actions - such as speeding up or slowing down a project. Current methods can show a delay, but they cannot tell the user what to do about the delay, that is, what corrective action to take. It is true that these models might be more complicated, but in this case, more complicated gives better results. Finally, Prof. Golenko-Ginzburg said that he didn't see a contradiction between practice and theory - these methods are an improvement to theory which should be useful in practice as well.

Prof. Golenko-Ginzburg also answered some technical questions on the mathematics and the optimisation in his methods.

ON THE EXPECTED COMPLETION TIME OF DIFFUSION ACTIVITY NETWORKS (DiAN)

S.E.ELMAGHRABY and M.K.AGGERWAL
Graduate Program in Operations Research, North Carolina State University, Raleigh, NC, USA

Traditionally, uncertainties in activity durations have been modelled as random variables with corresponding probability distribution functions. These latter are assumed to be known *a priori*. In reality estimates of activities' "remaining work contents" as well as their duration vary dynamically over time, with divergent variability as the planning horizon extends into the future. To capture this dynamic nature of the estimates, we model the activity as diffusion processes.

In this paper we propose a methodology to estimate the expected value of the project completion time based on such a paradigm. The results can be significantly different from those under fixed probability distribution functions.

1. Introduction

Traditionally, uncertainties in project planning have been "taken care of" within the assumed variability of the durations of the constituent activities. These durations are random variables (*rv*'s), each with known probability distribution function (pdf). Such a paradigm, while capable of modeling a wide variety of situations, fails to capture the inherent uncertainty in the estimates of the time or the resources required to accomplish an activity. It is not that an activity can take as little as t_1 or as much as t_2 time units, and so much of each of a number of resources, but it is that, at the outset, one has a rough "guesstimate" of these (and other) parameters which are subject to revision after the passage of some time *performing the activity*. Moreover, it is not practicable to foresee the myriad of issues that *may* arise and prepare a course of action for each eventuality, or combination of eventualities. Activities progress from well-defined, to the best of one's ability, albeit tentative plans to better defined plans as experience is gained in their actual performance.

T. M. Williams (ed.), Managing and Modelling Complex Projects, 47–67.

A particularly good example of this inherent uncertainty in the conduct of activities or a set of activities can be found in software development projects, where the estimate of the remaining work content changes over time both in expectation and in variance depending on internal situations that arise in the conduct of the plan *as well as on external factors* that demand changes in the very content of these plans, such as actions of competitors and market conditions.

The question that arises in such circumstances is the following: can one model such uncertainty and analyze the resulting project to secure meaningful results?

We propose to model the *remaining work content (rwc)* of an activity as a *diffusion process* [11] *with appropriate reflective and absorptive barriers*. The "work content" of an activity may be measured in any convenient unit such as man-weeks, machine-hours, horsepower or kilowatt-hours, etc. Such a process, along with its many variants (Ito process, Ornstein-Uhlenbeck process, etc.) provide us with a convenient paradigm, and a rich set of mathematical tools, which can better capture the dynamic behavior of the activity. We shall refer to the network with activities represented as diffusion processes as a "*di*ffusion *a*ctivity *n*etwork", or DiAN for short.

The *completion time* of an activity can now be modeled as the *absorption time* of its corresponding diffusion process (with an absorptive barrier at 0 *rwc*). The progress of work in the activity is measured by the *rwc*. At the start of the activity it is 1, since all (perceived) work lies ahead. If the activity experiences steady progress according to plan the *rwc* should decrease monotonically to 0, at which time it encounters the absorptive barrier. If, on the other hand, progress is uneven, the *rwc* shall meander in a random fashion and may even exceed its original estimate. The reflective barrier is devised to control the value of the *rwc*. For example, a reflective barrier at 1 *rwc* would not allow the *rwc* to exceed the previously anticipated work load. However, a reflective barrier at 1.2 *rwc*, say, would permit the *rwc* to increase to 120 of its original estimate. The absence of a reflective barrier would permit the *rwc* to exceed the previously estimated work content without bound.

The following question still persists: is modeling the activities as diffusion processes effective as a means of responding to the issues raised above?

The answer is yes, in our view. The diffusion paradigm, along with its variants and reflective/absorptive barriers, provide a powerful model for a variety of situations that are almost impossible to model with other paradigms.

We are now faced with the task of reconstructing the analysis and synthesis of activity networks under the new paradigm. Examples of analysis issues are: estimate the *pdf* of the project completion time and the *pdf* of the time of realization of intermediate "key events"; determine the net present value (*npv*) of the project under specified schedule of activities; determine the criticality indices of paths and activities. Examples of synthesis are: determine the optimal allocation of resources to accomplish the project by a given due date with a specified (high) probability; schedule the initiation of the activities to maximize the *npv* of the project.

As an example of such analysis, we shall determine the expected value of the project completion time under the diffusion paradigm. It will be seen that it is significantly different from the expected value that would have been secured under traditional assumptions.

2. The Model

We limit our discussion to activity-on-arc (AoA) representation of projects [4]. Some, or all, of the activities in the project are assumed to follow a diffusion process. The *state* of an activity is its *rwc*, represented by x or y. In this analysis we place an absorbing barrier to represent the end of the activity; no reflective barrier is used, which permits the *rwc* to expand without bound.

The activity completion process can now be written as

$$dx = a(t,x)\, dt + b(t,x)\, dz$$

Here, t is time; dz is the increment of the Wiener process; $a(t,x)$ is the instantaneous mean of the process; and $b^2(t,x)$ is its instantaneous variance. The activity is complete (i.e., its *rwc* is 0) when the process first reaches the absorbing barrier.

2.1 DIFFUSION MODELING

Let $D\,(t, x(t))$, $t \geq 0$, represent the duration of the activity with *rwc* of x at time t [1]. We make the following assumptions on D.

---

1 For the sake of brevity, we write $D\,(t, x)$ in place of $D\,(t, x(t))$.

Assumption 1. D is "regular";

$$P\,[D(t,x) > 0 \mid x > 0\,] = 1$$

In words: it is almost sure that with *rwc* the time it takes to complete the activity is strictly >0.

Assumption 2. $x(0) = 1$. That is, the initial work content (at time 0) is 1.

Assumption 3. An absorbing barrier is placed at x = 0. This denotes the completion of the project. Once state 0 is reached, the activity stays in that state thereafter.

Assumption 4. $D(t,0) = 0 \rightarrow D(t',0) = 0 \;\forall\, t' > t$. In words, if the duration of the activity is zero because it has been completed, then its "duration" stays at zero thereafter.

The last assumption permits us to define *T*, obviously a *rv* , as the completion time of the activity, given by

$$T = \min\,\{\, t \mid D(t,0) = 0 \,\}$$

Let $f(t,x)$ denote[2] the probability density function of $D(t,x)$. It can be obtained by solving the Chapman-Kolmogorov forward equation with appropriate boundary conditions [11]; which may be written as

$$\frac{\partial f}{\partial t} = -\frac{\partial[a(t,x)f]}{\partial x} + \frac{1}{2}\frac{\partial^2[b(t,x)f]}{\partial x^2} \tag{1}$$

The boundary conditions are:

$$f(t,0) = 0, \;\forall\, t > 0 \tag{2}$$

$$f(0,x) = \delta_d(x), \; x \geq 0 \tag{3}$$

where δ_d is the Dirac delta function, $\delta_d = 1$ at x = 1 and = 0 elsewhere[3].

---

2 Similarly we write $f(t,x)$ as shorthand for $f(t,x(t))$.

3 This is a loose definition for this volume. A more precise definition of the Dirac delta function is given by measure theory *(edited)*.

Using (1) we can determine the density function of the completion time of an activity. It may be difficult to solve (1) analytically for general functions *a(t,x)* and *b(t,x)*. However, numerical solutions are attainable provided three issues are resolved: the first is the polynomial or discrete approximation of functions, the second is the convolution of *rv*'s, and the third is the (perennial) issue of bounding the *pdf* of the time of realization of a node (whether it be a key event or the terminal event). We discuss these issues in the order of their mention.

3. Applicable Prerequisites

The final result we are seeking, *viz*., an estimate of the expected completion time of the project (or, for that matter, the expected realization time of any key event in the project) depends on the availability of three procedures. We have labeled them "prerequisites" because their availability renders the result attainable, from a practical point of view.

The prerequisite procedures are the following: a procedure to approximate the *pdf* of each activity duration, a procedure to evaluate the convolution of two *rv*'s, and a procedure to bound the expected value from below and from above.

As a road map to the development in this section, following is a summary of the complete procedure for the determination of the expected value of the project completion time.

1. Determine, through interpretation of each activity duration as a diffusion process, the *pdf* of the activity.
2. Approximate the *pdf*s of the activity durations with discrete probabilities as described in Section 3.1.
3. Determine bounds on the expected value of the project duration following the procedure of Section 3.3

We discuss each of these procedures next.

3.1 APPROXIMATION OF *f(t,x)*

In the mathematical analysis of activity networks one is usually faced with the problem of resolving convolutions and maxima of *rv*'s. The convolution operation is particularly difficult to accomplish, especially in the case of DiAN's because of the form of the *pdf*s involved (which are normal distributions with time-dependent variance). The problem facing the analyst is the following: can one reliably approximate the exact convolution? We discuss two approaches.

1. First, there is the use a *polynomial approximation* to *f*(t,1) (recall that we assume $x = 1$ whenever an activity is initiated). This approach has two appealing features to recommend it: polynomials are much easier to manipulate for the operations of convolution and maximum, and one can always approximate a density function with a polynomial to any desired degree of accuracy.

   The polynomial approximations we have experimented with are for *finite* and *bounded* domains of the independent variable t only. For the moment we shall drop the explicit recognition that $x = 1$. The function $f(t)$ is usually defined for $t \in [0,\infty)$, typically that is secured as a solution to (1) approaches 0 asymptotically. The range is $[0, \overline{T}]$, where $\overline{T}$ is the upper bound on the duration of the activity, which may be determined to any desired accuracy p by solving the equation $F(\overline{T}) = p$. Typically, p = 0.90 to 0.99. Normalization is achieved by dividing $f(t)$ in the range $[0, \overline{T}]$ by p. Clearly, by truncating the range of t at $\overline{T}$ we are somewhat distorting the "true" density function $f(t)$[4].

   The literature on numerical analysis is replete with polynomial approximation schemes, including orthogonal polynomials; see Dierckx and Hamming [8] for details. The choice of method depends on the situation under study and the analyst's preference; there is no "one best way". Here, we propose a rather simple approach based on the Vandermonde's determinant [8].

   Given M+1 points $(t_i, f(t_i))$; $i = 0,1,2,\ldots,M$ and $0 \le t_0 < t_1 < t_2 < \ldots < t_M = \overline{T}$ we can fit a polynomial of degree M, referred to as $P_M(t) = \sum_{k=0}^{M} a_k t^k$ to the data. The condition that the polynomial $P_M(t)$ passes through the given data points dictates that $f(t_i) = P_M(t_i)$ for all i. It can be demonstrated [8]

---

4 A problem may arise in such approximation when the polynomial approximation has negative values. Care must be taken to eliminate such occurrences and re-normalizing the polynomial approximation to yield a *bona fide* density function.

that the Vandermonde determinant V_{M+1} is well defined and is non-zero; resulting in a unique solution $P_M(t)$.

The selection of the M+1 sample points is left to the discretion of the analyst. (In the case of orthogonal polynomials, the sample points are the roots of the largest-order polynomial.) One of the popular methods of selection is to select them at uniform intervals. Another popular method is to select them as Chebychev points [2], which is what we have opted for in our experiments[5].

2. Second, there is the use of *discrete approximation* to the original *pdf*. The two most common approaches are to adopt: (i) equal intervals, which results in unequal probabilities of the intervals, or (ii) equal probabilities, which results in unequal intervals. In the experiments we have conducted we have opted for the "equal intervals" option because of its computing simplicity. In both options, the mid-point of the interval is assigned the probability of the interval.

The factor that tilted our preference towards discrete approximation is ease of computing, especially in the determination of convolutions. Specifically, we have discovered there are two strikes against approximation *via* polynomials. (i) First, when the convolution of K random variables is required, say $X_1,\ldots X_K$, one must determine $Y_1 = X_1 + X_2$, followed by $Y_2 = Y_1 + X_3$, and so on until $Y_{K-1} = Y_{K-2} + X_K$. As will be seen in the discussion of convolutions below, see section 3.2, Y_1 will have a range equal to $[0, \overline{T}_1 + \overline{T}_2]$, and it may be defined differently over *three* sub-intervals in this range. One may maintain this form of Y_1, in which case Y_2 may be defined differently over *nine* sub-intervals in its range $[0, \overline{T}_1 + \overline{T}_2 + \overline{T}_3]$ It is easy to see that Y_{K-1} may be defined over 3^{K-1} in the final range $[0, \sum_{i=1}^{K} \overline{T}_i]$ which is at best onerous and at worst impossible to compute. (ii) Second, the order of the polynomial approximation grows with the number of variables convolved. Indeed, if one convolves two polynomials each of degree n, the resulting polynomial is of degree $2n$. Convolving Y_1 with X_3 would result in a polynomial of degree $3n$ etc., and the last convolution of Y_{K-1} with X_K would result in a polynomial of degree $(K-1)n$.

---

[5] The Chebychev points for the interval $[0, \overline{T}]$ are obtained by dividing the semi-circle over the diameter of length $[t_0, \overline{T}]$ into N equal segments and then projecting the mid-points of the arc segments onto the diameter of the semi-circle.

The obvious remedy is to *re-approximate* each Y_k by a single polynomial over *its* range with a polynomial of degree *n*. This would certainly have the salutary effect of limiting the growth of the intervals that have to be maintained as well as limiting the order of the final polynomial to *2n*. But it has the drawback of extreme complexity, since the analyst must interfere at each stage of calculation to determine the sample points at which the functions are to be evaluated. Unfortunately, while it is true that "off the shelf" software exists to compute a polynomial of given order passing through a set of points, we could not locate any such software for approximating several polynomial "pieces" into a single polynomial.

3.2 THE CONVOLUTION OF RANDOM VARIABLES

As was previously mentioned, the determination of the *pdf* of the time of realization of a node depends on the repeated performance of two operations: a `maximum' operation and a `convolution' operation. The computing of maxima of independent *rv*'s whose *pdf*s are known is rather straightforward: one simply multiplies the two *pdf* 's. Such is not the case for convolutions, especially when the functions are complex in form, such as the case with the functions of interest to us here.

If one has in hand *polynomial approximations* to the density functions of the *rv*'s under study, we provide an algorithm for the determination of the desired convolution of two *rv*'s. The proposed method is an improvement over the approach proposed by Martin [13]. The algorithm has been programmed in Matlab and is available for general use on the *www*[6] [1]. It is useful in the case of two or three *rv* 's, but its accuracy deteriorates seriously for larger number of variables. This is due to the two reasons mentioned in section 3.1 above.

The *discrete approximation* of the original *pdf*s of the *rv*'s is not without its own problems either, the most important of which is the need to "collapse" the resulting convolution into a smaller number of sample points. Indeed, if two variables are approximated with *n* sample point each, their convolution may have *2n* sample points. The rapid growth of the sample points as repeated convolutions progress over the network causes most procedures to break down because of difficulty with computer memory requirements. The solution lies in re-defining the *pdf* of the convolved *rv* with only *n* sample points. This introduces an error in the estimation of the *pdf* of the final event of the project. Unfortunately, the

---

[6] The address is:
http//:www.coe.ncsu.edu/departments/or/people/faculty/elmaghraby.html

degree of error is unknown and has defied attempts to place a bound on it, since it depends crucially on the topology of the activity network.

3.3 BOUNDING THE *pdf* OF PROJECT COMPLETION TIME

Let $G(N,A)$ denote the (AoA) project network representation with $n = |N|$ nodes and $m = |A|$ arcs. We assume, without loss of generality, that the nodes are numbered from *1* to n such that an arc always leads from a lower numbered node to a higher numbered node. Let Γ_n denote the completion time of the project; a *rv*. The determination of the exact *pdf* of Γ_n is NP-Hard [10]. Many researchers over the past few decades have proposed various approximations and/or bounding schemes; see Elmaghraby [5] for an overview of such techniques.

Here we propose to use the bounding scheme proposed by Elmaghraby [7] which is based on the concept of the "control network" of the given project. It provides better lower and upper bounds than other approaches previously proposed. It also relies on computing maxima and convolutions of *rv*'s, as discussed in Section 3.2.

The complete procedure may be summarized in the following steps.

1. Determine, through interpretation of each activity duration as a diffusion process, the *pdf* of the activity.
2. Approximate the *pdf*s of the activity durations with discrete probabilities as described in Sections 3.1 and 3.2.
3. Determine bounds on the expected value of the project duration following the procedure of [7].

4. Illustrative Example

Consider the "interdictive graph" in AoA representation, Fig.1, which represents the following precedence relations among the activities (where "$\prec$" denotes "precede"): $1 \prec 3, 4$; $2 \prec 5$; $3 \prec 5$.

Here, the set of nodes is $N = \{1,2,3,4\}$; whence $n = |N| = 4$ and the set of arcs (activities) is $A = \{1,2,...,5\}$, whence $m = |A| = 5$. Assume that the instantaneous mean of activity k, $a_k(t,x)$, is a constant over time, denoted by μ_k, for $k = 1,..., 5$ [7];

---

[7] This implies that the expected value of the duration of the activity decreases linearly over time at a rate equal to μ_k.

and that the instantaneous variance $b_k^2(t,x)$ is also constant over time and =1 [8]. $\mu_k = [\frac{1}{5}, \frac{1}{10}, \frac{1}{10}, \frac{1}{8}, \frac{1}{3}]$ for k=1,2,...,5, respectively. The diffusion equation now simplifies to

$$dx_k = -\mu_k dt + dz, \;\; k = 1,2,\ldots,5$$

4.1 DETERMINATION OF $f_k(t)$

The Chapman-Kolmogorov equation (1) is now given by (we drop the explicit reference to the activity),

$$\frac{\partial f}{\partial t} = -\mu \frac{\partial f}{\partial x} + \frac{1}{2}\frac{\partial^2 f}{\partial x^2}$$

Let $f(t, y|t = 0, x)$ denote the density function at time t when the activity is initiated at time 0 in state x. The above differential equation may be solved [9] to yield

$$f(t, y|t = 0, x) = \frac{1}{\sqrt{2\pi t}} \exp\left[\frac{-1}{2t}(y - x - \mu t)^2\right]; \quad t > 0 \qquad (4)$$

The *unconditional* $f(t,y)$ is a linear combination of $f(t, y|t = 0, +x)$ and $f(t, y|t = 0, -x)$ and can be written as

$$f(t, y) = f(t, y|0, +x) - f(t, y|0, -x)\exp(-2\mu x) \qquad (5)$$

The activity completion time density function for a given initial *rwc* x is given by

$$f(t, x) = \frac{x}{\sqrt{2\pi t^3}} \exp\left[\frac{-1}{2t}(x - \mu t)^2\right]; \quad t > 0 \qquad (6)$$

Recall that we have assumed that the initial *rwc* is always x = 1, therefore (6) shall be written as

---

8 This implies that the variance of the duration of the activity increases linearly over time at rate equal to 1.

$$f(t) = \frac{1}{\sqrt{2\pi t^3}} \exp\left[\frac{-1}{2t}(1-\mu t)^2\right]; \quad t > 0 \tag{7}$$

The *pdf* of the activity duration (the so-called "cumulative distribution function") *F(t)* may be written as

$$F(t,x) = P(T \le t | x) = 1 - \Phi\left(\frac{\mu t + x}{\sqrt{t}}\right) + e^{-2\mu}\Phi\left(\frac{\mu t - x}{\sqrt{t}}\right) \quad ; t > 0 \tag{8}$$

where $\Phi \sim N(0,1)$ and the condition specifies that the activity had $rwc = \mathrm{x}$ at time $t = 0$.

4.2 APPROXIMATING *f(t)*

Each activity was approximated by a discrete probability mass function of 50 points selected according to the Chebychev scheme described above. Convolved variables were always "collapsed" to 50 sample points to satisfy the requirement that the new sample point coincides with expected value of the collapsed sample points in any interval. This automatically guarantees that the expected value of the collapsed *rv* is equal to the expected value of the convolved variable.

4.3 THE EXPECTED COMPLETION TIME

The lower and upper bounds on the expected completion time of the project of Fig.1 were estimated using the procedure outlined in Section 3.3. The result is as follows:

$$l.b.\ \underline{\mathrm{E}[T_n]} = 14.2915 < \mathrm{E}[T_4] = 15.3888 < u.b.\ \overline{\mathrm{E}[T_n]} = 15.5141$$

In these determinations, the "exact" expected value $\mathrm{E}[T_4]$ was secured by Monte Carlo sampling of the various activities of the project[9]. The estimate of the expected value based on the lower and upper bounds is equal to their mean,

9 We have used ``crude" Monte Carlo sampling, with no attempt at its variance optimization. Each experiment was based on 1600 runs to guarantee that the resulting estimate does not vary from its ``true" value by more than 5 at least 95 of the time. More refined sampling schemes can be easily devised to improve the estimate.

14.9028. The complete *pdf* of the completion time of this (minuscule) project is given in Fig.2 (dashed line); this figure is discussed again in Section 5 below.

5. Conclusions

The DiAN paradigm we are proposing demands different parameters in its specification than classical PERT models or its variants. Here, the analyst must specify the best estimate for the mean of the activity durations (which defines the "drift" in the DiAN model) as well as the changes in *variance* over time. Once these are specified, the *pdf* of the activity duration is derived. On the basis of such *pdf*s one can then proceed to estimate the various parameters of interest.

In proposing a new paradigm for the study of complex projects we are keenly aware of the need to demonstrate two propositions. The first is the difference between the results obtained via the DiAN model and the standard PERT approach or its variants. Indeed, if there is no significant difference in the results then there is no need for the new paradigm, or at least there is diminished need for it; the old would serve as well. And the second is the validity of the new paradigm as a "better model" of real life projects. Again, if it is of no greater help to the practitioner it may as well be deleted from the roster of tools available for project planning and control.

To address the first concern and demonstrate the difference between the results secured via the DiAN model and classical methodologies, we assumed two radically different distributions for the activities. The first is Poisson with parameter λ equal to the mean of the presumed activity duration. This immediately resulted in the variance of the activity duration equal to the *maximum* variance under the DiAN model (recall that we had assumed throughout that $\sigma = 1$ in the Chapman-Kolmogoroff equation, which results in the variance increasing equally with time). The second is Uniform over the (truncated) duration of the activity under the DiAN model. Naturally, this resulted in the mean (= range/2) and variance (= $(\text{range})^2/12$) of the activity duration that are quite different from those assumed under the DiAN model.

We have also used a second project network of 9 activities and 6 nodes, shown in Fig.3.

The bounds on the completion time of Project 2 following the DiAN model are as follows,

$$l.b. = 14.7733 < E[T_4] = 17.7080 < u.b. = 19.2423$$

As before, the "exact" value of the expected duration was secured via Monte Carlo sampling. The estimate of the expected duration based on the lower and upper bounds is 17.0078.

The results of the analysis are shown in Figs. 2 and 4. It is evident that the DiAN model gives *pdf*s (and, consequently, all the statistics that are based on them) that are quite different from those secured by standard PERT analysis.

The second concern, to establish the validity of the DiAN model, is more difficult to address since it requires comparing results under different paradigms in real life settings. This was not available to us, though we are still hopeful to demonstrate the validity of the DiAN model by application to software production. However, based on *a priori* reasoning we are convinced that the classical paradigm of fixed distributions is unable to reflect the inherent uncertainty in the activity duration, and that the DiAN model, as it is progressively refined, satisfies this gap in our modeling ability.

We reiterate, at the risk of being redundant, that several approximations are involved in the analysis of DiAN's: we approximate the *pdf* of each activity by a discrete probability mass function (*pmf*) over a finite range; then we "collapse" the *pmf* of convolved *rv*'s to prevent it from generating *pmf*s of ever increasing sample points; and finally we do not determine the expected time of realization of the node of interest to us but only secure lower and upper bounds on it and secure the (point) estimate of the completion time as the mean of these two values. Fortunately, this sequence of approximations do not detract from the accuracy of the final result.

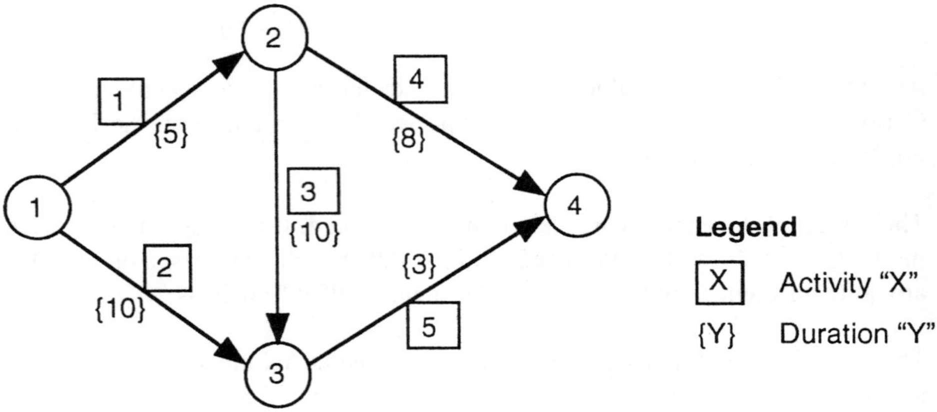

Figure 1. DiAN Project 1.

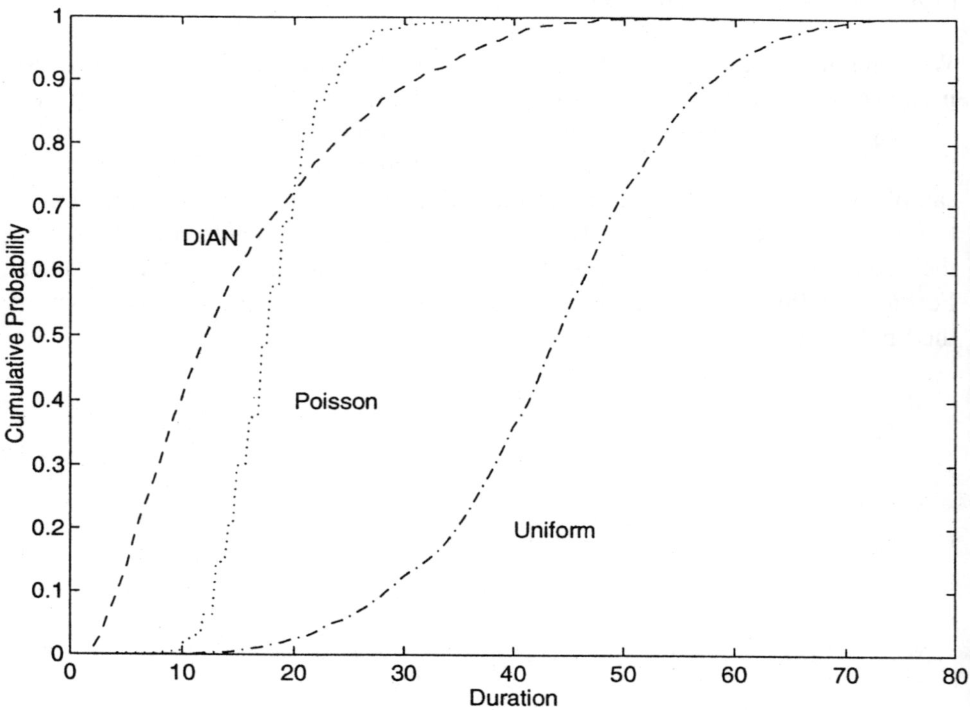

Figure 2 Results of Monte Carlo Sampling, Project 1

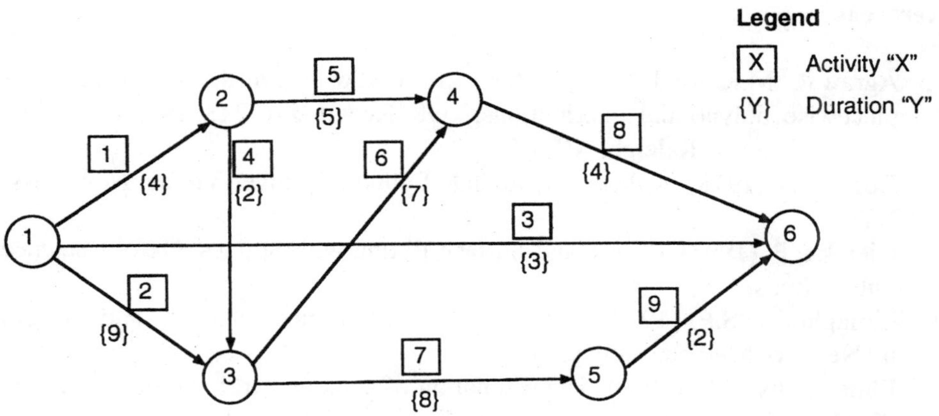

Figure 3. DiAN Project 2.

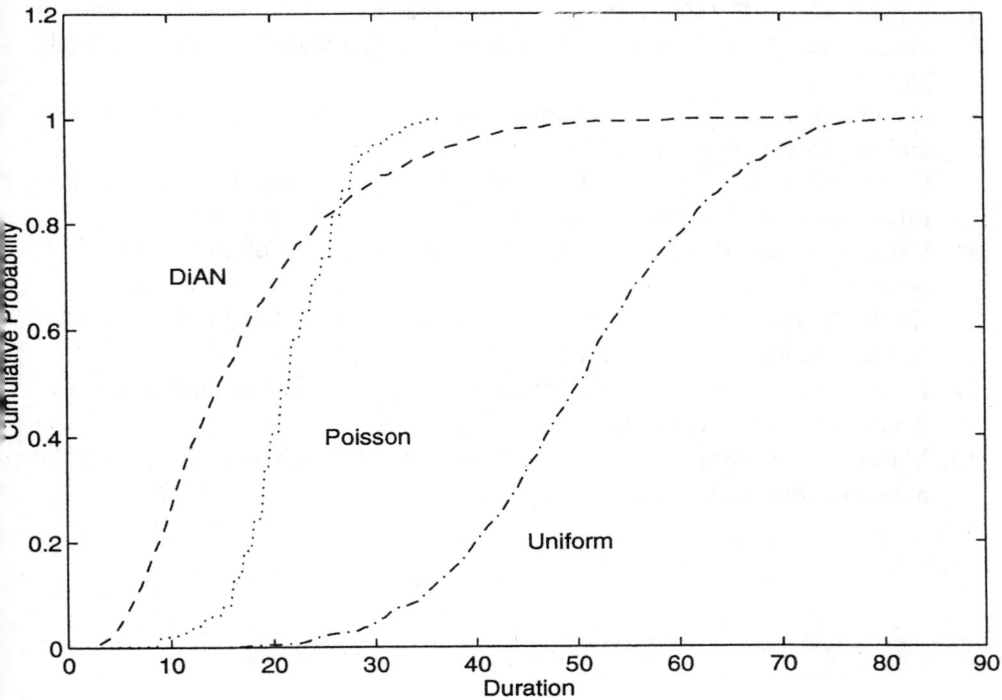

Figure 4 Results of Monte Carlo Sampling, Project 2

Rererences

1. Agrawal, M.K. and Elmaghraby, S.E. (1996). "On the convolution of piecewise polynomial distributions," OR Report No. 313, North Carolina State University, Raleigh, NC, USA..
2. Boor, C.D. (1978). A Practical Guide to Splines, Springer-Verlag, New York, NY.
3. Dierckx, P. (1993). Cu *rv*e and Surface Fitting with Splines, Oxford Science Publications.
4. Elmaghraby, S.E. (1977). Activity Networks: Project Planning and Control by Network Methods, Wiley.
5. Elmaghraby, S.E. (1989). The estimation of some network parameters in the PERT model of activity networks: review and critique, Chapter 1, Part III, in Advances in Project Scheduling, R. Sowinski and J. Weglarz, eds., Elsevier, New York, NY.
6. Elmaghraby, S.E. (1994). Activity Nets: A guided tour through some recent development Euro. J. Operl Res. 82, 348-408.
7. Elmaghraby, S.E. (1995). Bounding the expected project duration: A control net approach," OR Report No. 311, North Carolina State University, Raleigh, NC, USA.
8. Hamming, R.W. (1986). Numerical Methods for Scientists and Engineers, 2nd ed., Dover, New York, NY.
9. Grimmett, G.R. and Stirzaker, D.R. (1994). Probability and Random Processes, 2nd ed., Oxford Science Publications, New York, NY.
10. Hagstorm, Jane H. (1988). Computational complexity of PERT problems, Networks 18, 139-147.
11. Karlin, S. and Taylor, M.H. (1975). A First Course in Stochastic Processes, 2nd ed., Academic Press, New York, NY.
12. Luenberger, David G. (1989). Linear and Nonlinear Programming, 2nd ed., Addison Wesley, Menlo Park, CA.
13. Martin, J.J. (1965). Distribution of the time through a directed, acyclic network, Oper. Res. 13, 46-66.

QUESTIONS on "ON THE EXPECTED COMPLETION TIME OF DIFFUSION ACTIVITY NETWORKS (DiAN)"

Prof. Elmaghraby was asked *inter al* four questions:

- What information do you need about the activities to estimate the likely completion? Prof. Elmaghraby assumed that two very important consitutents of the activity are known: the drift (i.e. the average duration) and the instantaneous rate of variance at any time (you don't have to assume that this is constant). He went through one of the differential equations using these parameters. So while a PERT model asks for three parameters (optimistic, most likely and pessimistic), this model asks for these two parameters.
- Does this mean that the probability of a zero duration for the project is larger with this method than when using other distributions? No, for convenience the probability density functions used start at 0, giving an infinitessimal probability that you will finish in a very short time. If this is inapplicable, you can add a finite minimum duration to an activity, so it cannot finish in less than X days.
- It was suggested that with traditional PERT, although project managers often didn't give sensible optimistic, most likely and pessimistic durations, at least they understood the question being asked of them. Would they understand the two parameters required in this paper? Prof. Elmaghraby said that it wasn't necessary to ask the question in the terms in which they have been expressed in this paper. You could ask, say, "in two weeks time, can you give me a confidence interval (upper and lower bounds) for what the work content will be?", then go back and deduce what the implicit instantaneous variance is.
- On the advantages of this method, Prof. Elmaghraby pointed to the fact that shorter activities are generally felt to have less variability than longer activities, and thaat this model shows this - although this is a feeling that is being represented, and feelings are very difficult to measure.

OPEN DISCUSSION: NETWORK MODELLING METHODS

Discussion chaired by:
P. RAWLINGS
EuroLog Ltd, 3 High Street,
Glasgow,
Teddington, UK

This summary by:
T.M. WILLIAMS
University of Strathclyde,
UK

Mr. Rawlings firstly stated in introduction why he felt that such models were needed: in order of importance:

- to aid understanding of the venture being undertaken;
- to communicate with all the people involved what is going on;
- to be able to make some predictions about the outcome of the venture, in terms of its time, its cost, resource and its revenue, and its profitability;
- to be able to control the project.

He also felt that we should question one of Dr. Williams' assumptions earlier, namely that complex projects require complex solutions. Rather, "complex projects are run by simplex people, if not simple people" - we each as individuals need to be able to communicate to project managers what is going on. Although we should always try to find more appropriate modelling techniques and solutions, he believes that the modelling methods that exist today are by no means exhausted, and we need to find ways of using those methods more effectively as well as finding new methods.

Dr. Williams said that his point was not that we need more complex methods but that current methods were inadequate; we might indeed need simpler methods - we probably don't know how to use our current methods properly. For example, does anyone really know what a "pessimistic" or "maximum" duration is? and do they exceed it anyway?; many project managers don't even know what the "duration" of an activity is: is it the mean, the mode, a target to aim for if they are *machismo*, something safe? We need to increase our understanding of networks rather than necessarily to make our methods more complex.

Prof. Elmaghraby felt that it was not necessary for a project manager to understand the modellers' methods (as a patient does not understand the doctor's pill). So our fault is not in the complexity of the models (less complex than the medical theory underlying the pill), it is in the distilling of the complexity into our equivalent of the "pill" so that the project manager can use it. Prof. Golenko-Ginzburg agreed that the complexity was not the problem; rather, the problem was that people do not understand what network models are, what controlling a project means and how to estimate the parameters.

It was pointed out by a Russian participant that the best-selling software was MicroSoft Project, which uses very simple heuristics and calculations. The methods in all the current software he uses are from the 1960s/70s: none of it is using the new methods. So we should continue the research, not being frightened of complexity, but then technologists must turn it into simple technologies: project managers don't have time to learn very sophisticated methods. However, a practitioner participant pointed out that sometimes models can explain what has occurred.

The idea that projects were getting more complex (as claimed in Dr. Williams' paper) was disagreed with by one participant - as he looked back to previous projects, previously unrecognised complexity becomes clear. What is the evidence that project failure has been more due to complexity than to the inappropriate application of traditional project management techniques? There are several studies which show that project failure has been because the basic tools have not been used in the proper way - not that the tools in themselves were inadequate. Although there was no consensus over the first of these points (are projects becoming more complex?), the latter point became a theme of the conference - so while it might or might not be appropriate to look for better tools, it is essential that we seek to make the current, basic, tools, better understood and applied. Education and training in fundamental project management is important.

It was suggested that many of the plans actually found in practise are often complete fiction: network-logic and durations might be fictitious, designed to agree with a pre-determined end-date. Furthermore, resource-constrained methods are rarely used, and this technology was introduced into the literature 30 years ago. Even the tools currently available are not being used - will we need to wait another 30 years for the idea of controlled networks to be taken up in practice?

The conclusion from the point claimed that in fact projects were *not* getting more complex was the idea that perhaps we should be looking at de-complexing projects - that is, rather than looking for more complex methods, we look for ways to decompose complex problems into ones that lend themselves more easily to easily-understood solutions. A number of speakers called for ways to simplify projects. It was pointed out that what project managers need is a system to help them to manage the remaining work, to maximise the probability of bringing the project to a successful conclusion - and this might need quite simple methods, although it does require that you sub-divide the project into manageable spreads of decision-making for each of the project managers in the hierarchy.

In summary, Mr. Rawlings pointed out that the issue of "network models" hadn't really been addressed, rather more fundamental issues about complexity.

- There might be better methods than what we are using to model projects, but we need to convince project managers that these are valid, that they are useful, and that they are practical.
- There is a need for much more training for project managers, since the use of even basic tools is often very poor.
- Network models - indeed any models - are only a small part of the problem of managing complex projects; there is not just one solution to this problem.

As a particular issue in this last point, the question of decomplexing both projects and the management of those projects is taken up partly in this Workshop, but needs to be more fully explored elsewhere

TIME RISK: WHAT IT DRIVES AND HOW IT IS DRIVEN

CHRIS CHAPMAN AND STEPHEN WARD
University of Southampton, UK

This paper addresses two complementary issues: how time risk impacts on cost and other measures of performance, and what drives time risk. The focus is how to structure modelling and management processes in a way which captures important connections and recognises the significant causalities involved without becoming too complicated. Formal risk management processes will be used as the basis for exploring appropriate operational separability as an effective and efficient way of dealing with connectivity.

1. Introduction

Most people who are knowledgeable about project management understand in considerable depth the complex concepts lying behind the simple time-cost-performance (measures other than time and cost, quality measures for example) often depicted by a triangle, with an interval point representing relative priorities for a particular project [1]. A simple triangle is a useful device for reminding everybody concerned with a project that low cost or short duration or high performance may be feasible, but attempting to achieve all three usually means achieving none of them. However, this simple triad is limited as a structure for operational risk management.

Cooper [2] and Williams [3] have explored the underlying complexities of time-cost-performance relationships using influence diagrams and systems dynamics in terms of forensic risk analysis (explaining why a project's cost was so high after the fact). This kind of modelling needs further development in the context of risk analysis as part of the project management process. It captures the feed back and feed forward loops that drive risk connections in a very useful manner. For example, if a high technology project like channel tunnel rolling stock is behind schedule because of government safety inspector induced design changes, more staff and more concurrent engineering introduced to cope with these and other sources of delay can make the situation worse, not better.

T. M. Williams (ed.), Managing and Modelling Complex Projects, 69–88.

Chapman [4] provides a comprehensive risk management process referred to as SCERT (Synergistic Contingency Evaluation and Review Technique), designed for project duration and cost risk analysis of offshore North Sea oil and gas project, which takes duration risk as a key driver of cost risk, assuming performance is not significantly variable. Generalised to 'Risk Engineering' [5], alternative forms allow a focus on more general time-cost-performance structure, considering capital cost/operating cost and revenue trade off in a reliability/availability analysis context for example [6]. However, the models used by these processes do not address the feed back and feed forward loops referred to above in a direct explicit manner.

Further generalisation of the 'Risk Engineering' methodology addressed in the APM (Association for Project Management) PRAM (Project Risk Analysis and Management) approach [7] provides the structure for a possible integration of all the above and other modelling approaches. Chapman and Ward [8] develop some aspects of this in a comprehensive project risk management framework. This paper provides some focus in terms of 'what drives time risk and how it is driven'.

Three frameworks are outlined in three sections:

1. six aspects of a project,
2. eight stages in a project's life cycle,
3. nine phases in a risk management process.

A further section explores examples of linkages suggested by this structure. A concluding section suggests how this structure might be developed.

2. Six aspects of a project

Figure 1 (from Chapman and Ward [8]) portrays six aspects of a project (the six W's) in terms of initial influences as a project is conceived and the main feed forward and feed back loops as a project evolves through the early parts of its life cycle.

As a simple example of the 'who' and the 'why' for an illustrative project, assume a man buys a dog to deter potential intruders into his home and provide companionship for his family. Assume his wife objects to the idea of dog hair on the carpets and furniture in the house, and they jointly decide to put a kennel (dog

Figure 1 The six Ws

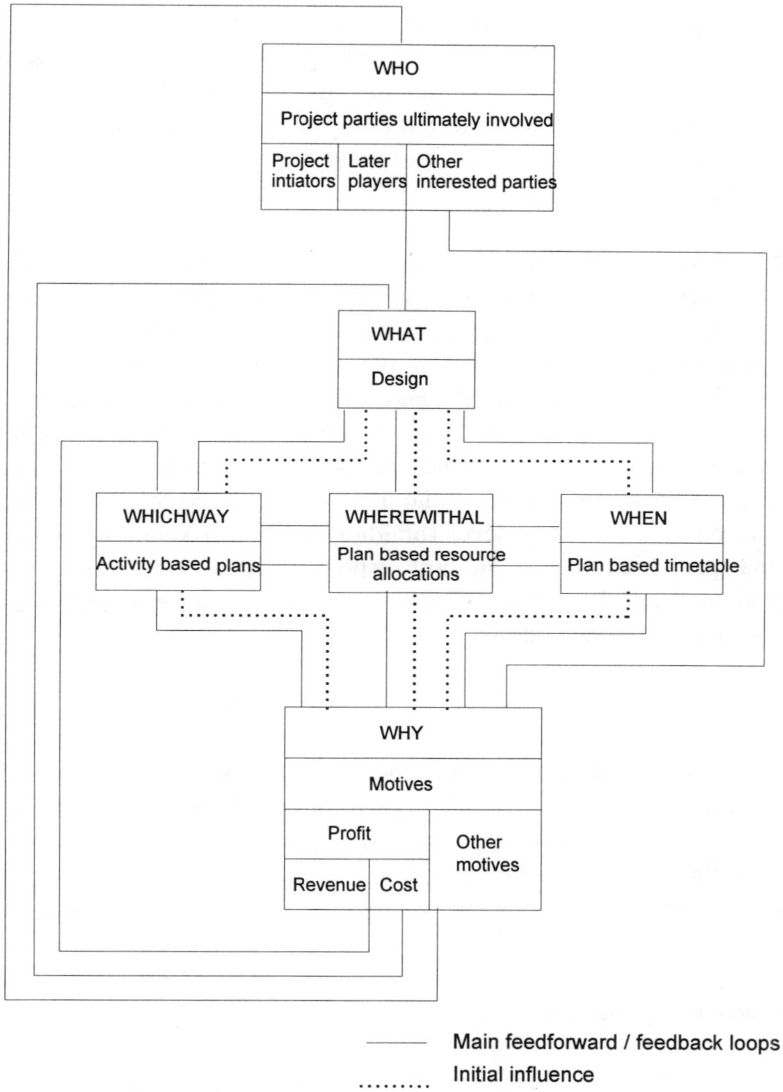

house) in the garden. Assume 'the project' is thus initiated and defined as 'acquire a kennel in the garden'.

Assume they now consider the 'whichway, wherewithall and when', deciding the man will build it, primarily in wood, next weekend.

Assume the man then designs it, readdressing plans, resource allocation and timetable in more detail, and comparing the outcome to the original motives. At this point, or later when he perhaps realises he has overestimated his skills as a builder, he may decide to buy a kennel instead. Alternatively, if this would damage his pride or wallet too much, he may appeal to his children to persuade his wife it would be a good idea to keep the dog in the kitchen. If the dog decides it doesn't like being outside, and barks all night, this may lead to any even later adjustment to the project.

Major projects with a complex nature can make this diagram worth careful study in terms of sources of uncertainty and risk for reasons explored in Chapman and Ward [8], just touched on here. The key issue here is the broad nature of the who and why, the interconnected nature of these six W's, the need to maintain consistency with respect to all six W's, and the risks associated with failure to do so. For example, it is very easy for those involved in the whichway to develop detailed plans which loose sight of the priorities of the issues addressed in design, and for designs to lose sight of the initial motives which triggered the project. If risk is managed at the activity level (building a wooden kennel), higher level strategic changes which may be the best response may be overlooked (buying a ready built kennel) and doing something altogether different (letting the dog sleep in the house) may not be considered.

Consider the relationship between Figure 1 and the simple cost-time-quality triad.

If the Figure 1 'when' equates to 'time' and 'wherewithal' equates to 'cost', then 'who, why, what and whichway' equate to performance. That is, Figure 1 provides another way to look at time-cost-performance triad. Its purpose is more detail because the additional detail is useful at a conceptual and operational level. It is a formalisation, a process model, to help us structure and discuss the issues.

3. Eight stages in a project's life cycle

Figure 2 (from Ward and Chapman [9] and Chapman and Ward [8]) portrays eight stages in a project life cycle (PLC). It decomposes the more usual four phases as indicated in Table 1, for reasons just touched on here. The key issue here is the mapping between the six Ws and the eight stages. The usual 'planning' phase is decomposed into 'design, plan, and allocate' stages. These three stages generate different kind of risks in later stages if not properly managed. The roots of time risk in the execute stage lie in earlier stages, and time risk in the execute stage can generate other risks in later stages, through the six Ws feed forward and feed back loops. To use our simple example, if the man recognised in advance that the dog might not accept sleeping outside, his wife might not allow the dog inside, and his children might not allow him to give the dog away once they became friends, he might forget about buying a dog and buy a burglar alarm and a cat instead.

Figure 2 Inter-stage progression and feedback structure

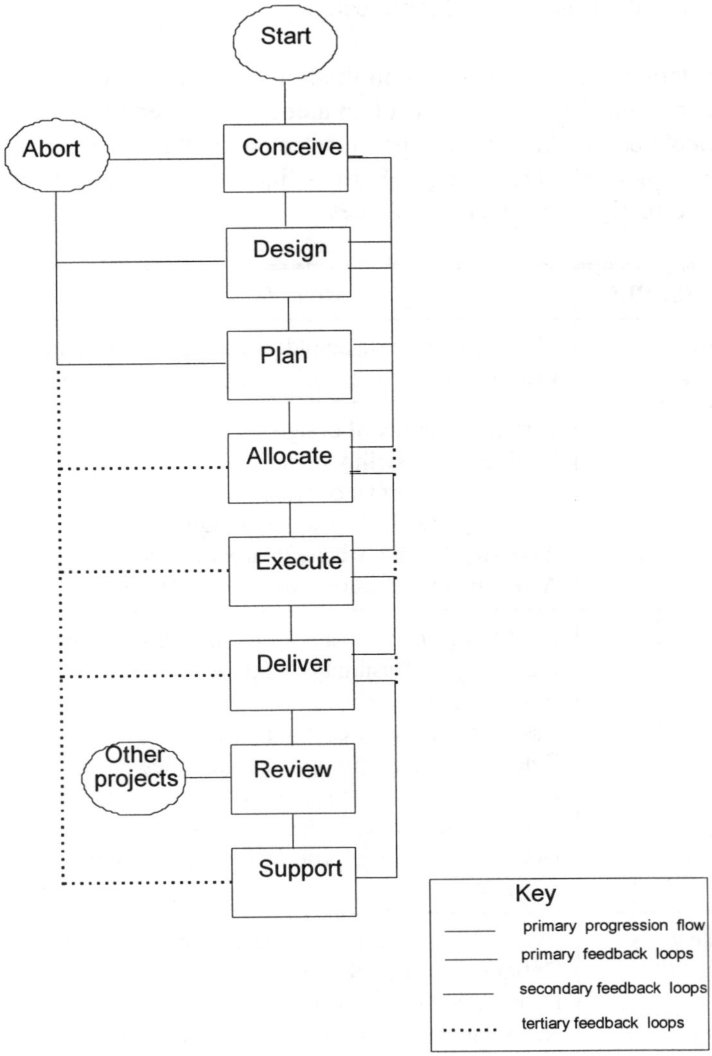

Complex projects can make this diagram a useful tool for exploring risks that become evident (are realised) at one stage in the PLC but owe their origins to earlier stages.

In particular, time risk in the execute stage drives and is driven by cost and performance risk in that stage. It is also driven by a range of factors in all earlier stages. And it drives risk in all later stages. The structure provided by Figures 1 and 2 can provide the basis of a checklist or search structure for linkages appropriate to a particular project, in the sense that they are important enough to address formally for that particular project.

We cannot look at all possible links all the time. In each particular case we must choose the key relationships. This is often a decision linked to the project priority issue as depicted by the interior point selected in the time-cost-performance triangle. It is also linked to a range of issues like forms of contact, which can be designed to directly shape the risk linkages.

Stages of the PLC	Roles for risk analysis
Conceive	Identification of stakeholders and their expectations Identification of appropriate performance objectives
Design	Testing reliability of design Testing the feasibility of design Setting of performance criteria Assessing the likely cost of a design Assessing the likely benefits from a design Assessing the effect of changes to a design
Plan	Identifying and allowing for regulatory constraints Assessing the feasibility of a plan Assesssing the likely duration of a plan Assessing the likely cost of a plan Determination of appropriate milestones Estimation of resources required Assessing the effect of changes to the plan Determination of appropriate levels of contingency funds and resources
Allocate	Evaluation of alternative procurement strategies Defining contractual terms and conditions Determining appropriate risk sharing arrangements Assessing the implications of contract conditions Assessment and comparison of competitive tenders

	Determining appropriate target costs and bid prices for contracts Estimation of likely profit following project termination
Execute	Identification of remaining execution risks Assessing implications of changes to design or plan Revising estimates of cost on completion Revising estimates of completion time of execution stage.
Deliver	Identification of risks to delivery Assessing feasibility of delivery schedule Assessing feasibility of meeting performance criteria Assessing reliability of testing equipment Assessing requirement for resources to modify project deliverable Assessing availability of commissisoning facilities
Review	Assessing effectiveness of risk management strategies Identification of realised risks and effective responses
Support	Identifying extent of future liabilities Assessment of appropriate level of resources required Assessment of profitability of the project.

Table 1: Phases, stages and steps in the project lifecycle

4. Nine phases in a risk management process

Figures 3 and 4 and Table 2 (from APM [7] and Chapman and Ward [8]) portrays nine phases in a risk management process which can be applied at each and every stage in the PLC, considering all six Ws and their relationships, as discussed elsewhere (Chapman and Ward [8]). The process is iterative, as illustrated by three complete passes in Figure 4, each with a single loop or subcycle between the 'estimate' and 'evaluate' phases. It decomposes the more usual four or five phases as indicated in Table 3 for reasons not developed here. The key issue here is different kinds of risk generated in each phase by a failure to address that phase properly.

The 'define' phase is concerned with defining the project for risk management purposes, including comprehensive treatment of the linkages between the six Ws. Failures in the project management process in early phases should be detected in this phase. If we are still in the plan stage of the PLC when this analysis is undertaken recovery from ineffective earlier project management may be costly, but it is much worse if we are in the execute stage, and a disaster after execution. In the context of our kennel building exercise, a key question is "is the project building a kennel or providing security and companionship?" The answer is central to defining what we mean by "time risk".

Figure 3 Risk Management Process phase structure flowchart

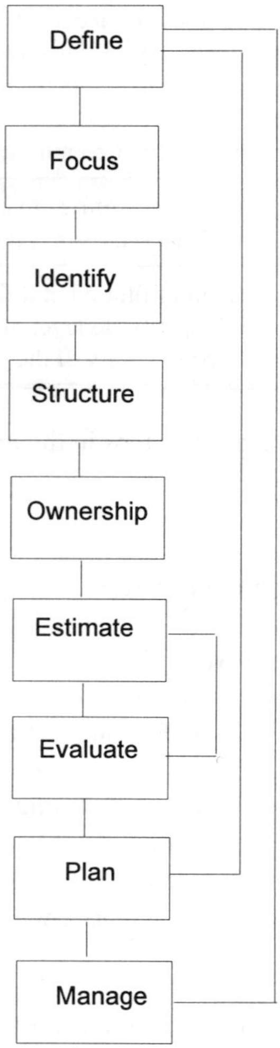

The 'focus' phases is concerned with defining the project risk management process for this particular project, developing appropriate models and processes from the generic sets available, including SCERT, systems dynamics, and others. Failure in the project management process in earlier phases, and other sources of risk, will only be rectified if an appropriate process detects them and facilitates their control.

The 'identity' phase is concerned with implementing the risk management process (RMP) defined in the focus phase to identify all significant risks and all appropriate responses. Failure in this part of the RMP means risks are not managed and opportunities (provided by responses which may bring benefits beyond dealing with the triggering risk) are lost.

The 'structure' phase is concerned with understanding relationships and assumptions, like statistical independence. It is in this phase that systems dynamics ideas could be embedded. Failure in this part of the RMP means the whole project management process is flawed at a fundamental level, perhaps to a catastrophic extent.

The 'ownership' phase is concerned with ensuring who owns which risks and responses is clearly understood by all relevant parties. Failure in this part of the process mean inefficient and ineffective external and internal contracts, which can drive a series of knock on risks. For example, if a contractor thinks a contract is unfair, and starts 'claims engineering', disputes can impact time risks directly and indirectly in a wide range of ways.

The 'estimate' and 'evaluate' phases and their subloops are concerned with using time effectively and efficiently to understand the implications of risks and responses when and where quantitative analysis is useful (it may not be required). Failure in this part of the process can just mean inefficient analysis. Ineffective analysis is a more serious concern, with obvious knock on effects.

The 'plan' and 'manage' phases rest on and depend on the earlier phase, earlier errors of omission and commission beginning to be realised in more expensive terms once the project begins the execute stage.

5. Linkage examples

The dog and kennel example served to illustrate the six aspects of a project, some of the eight stages of a project PLC and the first phase in the RMP, but more realistic examples are needed to take the discussion forward.

The SCERT approach [4] to offshore oil and gas projects lends itself to an initial focus on time risk in terms of complex drivers for time risk like weather risk associated with "weather windows", risks (like buckles in a pipe during offshore pipelaying) which are dependent upon weather, and responses which affect more than one risk (a larger pipelaying barge reduces the chance of a buckle and increases pipelaying speed whether or not buckles occur). Modelling this time risk can be very sophisticated [5]. This time risk, once understood, can then be related to cost risk in a reasonably straight forward manner [11]. Performance risk was not an issue in terms of the projects SCERT was first developed for. In terms of the earlier kennel example, this is comparable to very careful risk management of wooden dog house construction over the planned week-end construction period. Sometimes this is what is needed. Simplified versions of this approach [5] were applied to combined cycle gas turbine electric power generation projects in the late 1980s, applying approaches to simplifications discussed elsewhere [8, 11]. For reasons elaborated elsewhere [8], the key benefits of the process were widely perceived to be comparable to borrowing a dog house to try out the idea of keeping the dog in the garden before building a kennel. The project changed in two key respects. First, a separate gas supply was laid on from British Gas to allow extensive testing of the new equipment before the main gas supply (with a take-or-pay contract) became available, providing a high level of confidence that delay and take-or-pay penalties would not be an issue. Second, completion was timed to coincide with the start of the winter peak, to improve cash flow. Links within "dog house construction" were simpler, but links between the take-or-pay gas contract, the new equipment testing regime and generation of electricity cash flow in relation to peak seasonal demand was understood and managed effectively.

Simplified versions of this approach [11] have also been applied more recently to a project concerned with the disposal of intermediate level nuclear waste. In this context my perception is that the key benefits came from starting discussions comparable to "what size dog do we want", "do we want to wait until summer to build it", and "how many helpers do we need". Links between the volume of waste to be disposed of and the unit cost of disposal, economics of delay, and the need for a prime contracting strategy driven by limiting the need to manage risks over contractual boundaries were amongst the key issues understood in depth as a consequence of the RMP.

Ongoing is the application of this evolving thinking [8] to managing a major project concerned with computerisation of branch banking. The process is known as "benefit management" (not risk management), to emphasise that the project is "providing security and comfort" (not a dog house). The benefits used to sell the idea of the project are linked to deliverables, deliverables are linked to tasks, and risks are defined in terms of this linked structure. Risks address customer perceptions of benefits and employee ability to respond to training as well as the more technical information systems risks.

Figure 4. An example Risk Management Process (RMP) over time

Time scale: days, weeks or months, depending on the nature of the project

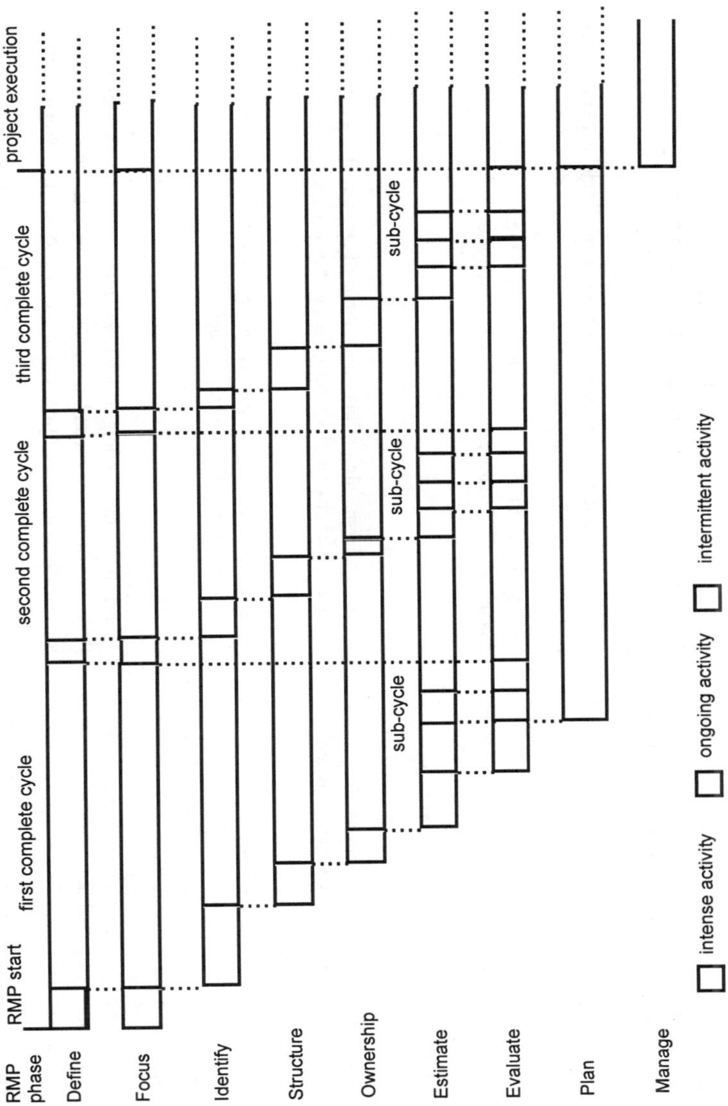

Table 2 A generic risk management process structure (Client perspective/Plan stage initiation)

Phases	Purposes	Deliverables (may be targets not achieved initially)
Define	Consolidate relevant existing information about the project. Fill in any gaps uncovered in the consolidation process.	A clear, unambiguous, shared understanding of all relevant key aspects of the project and the associated RMP, documented, verified and reported.
Focus	Scope and provide a strategic plan for the RMP. Plan the RMP at an operational level.	A clear, unambiguous shared understanding of all relevant key aspects of the RMP, documented, verified and reported.
Identify	Identify where risk might arise. Identify what we might do about this risk, in proactive and reactive responses terms. Identify what might go wrong with our responses.	All key risks and responses identified, both threats and opportunities, classified, characterised, documented, verified and reported.
Structure	Testing simplifying assumptions. Providing more complex structure when appropriate.	A clear understanding of the implications of any important simplifying assumptions about relationships between risks, responses and base plan activities.
Ownership	Client/contractor allocation of ownership and management of risks and responses. Allocations of client risks to named individuals. Approval of contractor allocations.	Clear ownership and management allocations, effectively and efficiently defined, legally enforceable in practice where appropriate.
Estimate	Identify areas of clear significant uncertainty. Identify areas of possible significant uncertainty.	A basis for understanding which risks and responses are important. Estimates of likelihood and impact in scenario or numeric terms, the latter including identification of assumptions or conditions, sometimes with a focus on "showstoppers".
Evaluate	Synthesis and evaluation of the results of the estimate phase.	Diagnosis of all important difficulties and comparative analysis of the implications of responses to these difficulties, with specific deliverables like a prioritised list of risks, or a comparison of base plan and contingency plans with possible difficulties and revised plans.

Plan	Project plan ready for implementation and associated risk management plan.	1. Base plans in activity terms at the detailed level required for implementation, with timing, precedence, ownership and associated resource usage/contractual terms where appropriate clearly specified, including milestones initiating payments, other events or processes defining expenditure, and an associated base plan expenditure profile. 2. Risk assessment in terms of threats and opportunities, prioritised, assessed in terms of impact given no response is feasible and potentially desirable, along with assessment of alternative potential reactive and proactive responses. 3. Recommended proactive and reactive contingency plans in activity terms, with timing, precedence, ownership and associated resource usage/contractual terms where appropriate clearly specified, including trigger points initiating reactive contingency responses and impact assessment.
Manage	Monitoring. Control. Developing plans for immediate implementation.	Diagnosis of a need to revisit earlier plans, and initiation of replanning as appropriate, including on a regular basis specific deliverables like the monitoring of achieved performance in relation to planned progress, and prioritised lists of risk/response issues. Exception (change) reporting after significant events, and associated replanning.

APM (used here)	UK MoD [9]	SCERT (Chapman 1979)
Define	Initiation	Scope
Focus		
Identify	Identification	
Structure	Analysis	Structure
Ownership		
Estimate		Parameter
Evaluate		Manipulation and interpretation
Plan	Planning	
Manage	Management	

Table 3 Risk management process (RMP) phase structure comparisons

6. Conclusion

The sequence adopted in the last section is not intended to suggest that complex time risk of the kind addressed by SCERT is not relevant. It may be the key. However, there is a very rich range of drivers for time risk, and time risk drives other risks in complex ways. The range is too wide for any single simple approach. We need to select an approach which simplifies in a manner appropriate to the particular context. And a lot depends upon whether we view "the project" as "building a wooden dog house" or "providing security and companionship".

References

1. Barnes, N.M.L. (1984) Construction project management, *International Journal of Project Management* 6(2), 60-79.
2. Cooper, K.G. (1980) Naval ship production: a claim settled and a framework built, *Interfaces* 10, 20-36.
3. Williams, T. Eden C., Ackerman, F and Tain, A. (1995), The effects of design changes and delays on project costs, *Journal of the Operational Research Society* 46, ( ) pp 809-818; and Vicious circles of parallelism, *International Journal of Project Management* 13, 151-155.
4. Chapman, C.B. (1979) Large engineering project risk analysis, *IEEE Transactions on Engineering Management*, EM-26, 78-86.
5. Chapman C.B. (1990) A risk engineering approach to project risk management, *International Journal of Project Management*, 8, pp 5-16.
6. Chapman C.B., Cooper D.F. and Cammaert A.B. (1987), Model and situation specific OR methods: risk engineering reliability analysis of an LNG facility, *Journal of the Operational Research Society*, 35, 27-35.
7. Association of Project Managers (*APM*) (forthcoming) *Project Risk Analysis and Management (PRAM) Guide*
8. Chapman, C.B. and Ward, S.C. (1997) *Project Risk Management: Processes, Techniques and Insights,* John Wiley and Sons, Chichester.
9. Ward S.C. and Chapman C.B. (1995) Risk management perspective on the project lifecycle, *International Journal of Project Management*, 13, 145-149.
10. MOD(PE)-DPP(PM) (1991) *Risk Management in Defense Procurement*, Reference D/PPP (PM) 2/1/12, published by and obtainable from the Ministry of Defence, Procurement Executive, Directorate of Procurement Policy (Project Management) Room 6302, Main Building, Whitehall, London, SW1A 2HB.
11. Chapman C.B., Phillips E.D., Cooper D.F. and Lightfoot L. (1983) Selecting an approach to project time and cost planning, *International Journal of Project Management* 3 (1), 19-26.

QUESTIONS on "TIME RISK: WHAT IT DRIVES AND HOW IT IS DRIVEN"

The following Open Discussion deals with Risk analysis methods in general, but the following immediate questions were put to Prof. Chapman.

Prof. Chapman was asked whether there was a risk that the risk management process as described by him was becoming like the network analysis process in the early days - practised by specialists; shouldn't we be looking at how it impacts upon the line managers? Prof. Chapman pointed out that the UK Association of Project Managers version of the above account defined the process as "Risk *Analysis and* Management". He had used the term "Risk Management" because he saw all the earlier phases described in the paper as being done solely to help with and under-pin the management. The actual making of decisions might be separate from the analysis to some extent, the job of the analyst being to help whoever has to make the decisions understand the issues, but it is essential to build this process into the project management process: it is not separate, "it's not an add-on, it's an add-in".

In answer to questions about practical experience, Prof. Chapman described some work he was currently carrying out in the computerisation of a major UK High-Street Bank.

In further discussions about the complexity of models, Prof. Chapman made two points:

- previously, project risk had been driven by the risk to time, and this is why complex time-models had been developed and used;
- formal mathematical models are only useful if you understand the other complexities involved in the project.

OPEN DISCUSSION: RISK ANALYSIS METHODS

Discussion chaired by:
D. HILLSON
HVR-CSL Ltd, Selborne House,
Alton, Hants, UK

This summary by:
T.M. WILLIAMS
University of Strathclyde,
Glasgow, UK

Dr. Hillson introduced the discussion session by saying that complex projects are inherently risky. The previous sessions had concluded that complex methods may be necessary, but that they must have simple implementations, because project managers are busy people. This is certainly also true for risk analysis and management. The purpose of risk analysis and management is to focus attention on those areas of the project which are most at risk. Risk analysis should help to tell the busy project manager where to spend his time, money, effort and resources, particularly important in complex projects. There are already many effective techniques for risk analysis, and the UK experience is that we do not need better risk analysis, we need better risk management - the weakness is in the implementation. So the challenge to this Workshop is: how can we use risk analysis methods in a pragmatic and practical way to ensure that risk is effectively managed.

Prof. Elmaghraby raised the question about how risk can be measured: are they crisp or fuzzy measures, and if the latter, what sort of technology can be used to come up with recommendations? In answer, it was generally felt that the data was fuzzy and subjective; there might be precise units (cost, time etc), but the levels of confidence around the measures is low (and we never know whether the estimates were right or not).

Dr. Hillson felt that the important issues were flexibility in process, and clarity in communication. The analysis is important, but is subservient to recognising, dealing with, and shrinking the risks themselves; this latter is not a difficult or numerate issue. However, Prof. Elmaghraby had posed some problems for which it was necessary to come up with specific, precise recommendations.

It is important to be able to communicate the results of such analyses to senior management. Esoteric findings will be ignored. Senior managers have to evaluate several factors before deciding on whether to carry out projects, including many broad business issues, as well as the risk analysis.

In innovation projects, it was suggested that the focus should be on the gaps between acquired and required knowledge and skills. Concentrating here will give a reduction of complexity in focussing on risks. It was also pointed out that the manager needs to discriminate between risks that can and those that cannot be influenced, and between complex issues that do and those that do not affect the strategic aims of the project.

Three further points were made in discussion:

- These techniques must stimulate the thinking, or enhance the quality of the thought-processes. Any techniques that do not do this are not useful.
- Problem formulation is a very important phase in risk management: clarifying the needs and the methods; the latter is relatively easy, but the first is essential also, otherwise you end up with "white elephant" projects.
- The stages of maximum sensitivity need to be looked at.

In summary, Dr. Hillson pointed out the order of the words in the title of the Workshop: "Managing and modelling complex projects". The management of the risks is the key issue, and the modelling must be subservient to this.

SYSTEM DYNAMICS METHODS IN COMPLEX PROJECT MANAGEMENT

KENNETH G. COOPER,
PA Consulting,
Cambridge MA, USA

A brief introduction to system dynamics modeling as related to the conditions, problems and phenomena of complex project performance and management. Increasing numbers of applications of system dynamics modeling and analysis methods to different kinds of project are reviewed; they provide the basis for an experience-based diagnosis of common project problems

Application of this concept-rich methodology, as documented in case studies, results in improved understanding of how projects work, better management and more insightful decisions, and significantly improved project cost and schedule performance.

Differences in characteristics and performance usually obscure the degree of commonality among projects. Use of system dynamics model structures yield insight into the common problem sources and the cause-effect "paths" by which they affect projects. From this view we identify often-counterintuitive management practices that strongly affect the performance of projects and the organizations dependent upon those projects for success.

1. Introduction

CASE A: The project manager watched as his customer required several changes to the system in development. Virtually no adverse impact on project performance was visible. More changes followed, but these now produced big delays and extra costs. Why the difference?

CASE B: Another manager, faced with a project behind schedule, got her staff to work overtime. Progress improved, so she continued using overtime. Within a few months, however, the pace of work had slowed worse than ever. What happened?

T. M. Williams (ed.), Managing and Modelling Complex Projects, 89–108.

CASE C: This project manager hired new staff when he saw the project falling behind schedule. Progress did not improve quickly enough, and more new people were brought in. Work progress remained slow. Why?

These three managers experienced non-linear (A), time-delayed (B), causal feedback (C) on their projects; they are like most other managers in their experiences running complex projects. Precisely because of the commonality of these kinds of phenomena and characteristics, system dynamics methods are increasingly used to help the management of complex projects.

System dynamics is a methodology used for the description, diagnosis, and simulation of complex systems. The method is particularly well-suited for analyzing systems (businesses, industries, markets, ecosystems...) for which the performance over time is driven by the interdependence of many factors working through cause-effect relationships (e.g., worker skill levels *affect* productivity, and productivity *affects* work progress). The characteristics of such systems are that they exhibit *feedback*, with *non-linear*, *time-lagged* effects. It is clear to see why complex projects qualify...

1) Complex projects exhibit many feedback processes.
   For instance, a system being redesigned to incorporate changes may need additional staffing. Less experienced new hires make more mistakes, leading to... more redesign. (This closed loop of cause-effect relationships constitutes feedback.)
2) Complex projects are driven by non-linear effects.
   But one example is the effect of changes experienced by Manager A above. Each of many (non-linear) effects on staff productivity also qualify, such as schedule pressure, overtime levels, organizational change,...
3) Complex projects are subjected to time-delayed effects.
   As much as we managers would like to see the instantaneous effects of each action we take, that is simply never the case in project management. Time delays in monitoring, reporting, responding, and implementing are just part of the time lags in project systems.

After a review of the range of system dynamics applications to projects, we will examine some of the resulting analyses of problem sources and attempted remedies.

2. System dynamics applications to projects

An excellent review of system dynamics work applied to project management is provided by Rodrigues and Bowers [1] Over a dozen works from the past two decades are described. Most of these models have as a central structure a version of the "rework cycle," which we first designed and built in 1976 and published in 1980 [2].

2.1 THE REWORK CYCLE

As described in Cooper [3], the Rework Cycle structure portrays *flows* of project work, be it feet of cable, tons of steel, lines of code, design drawings The structure is the core of a simulation model developed for and applied to dozens of complex design and construction projects.

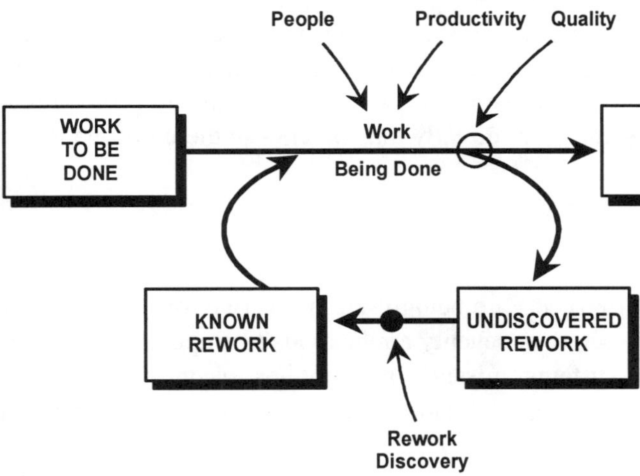

Figure 1. The Rework Cycle

At the start of a project or project stage, all work resides in the pool of **Work To Be Done** (see Figure 1.) As the project begins and progresses, changing levels of **People** working at varying **Productivity** determine the pace of **Work Being Done**. But unlike all other program/project analysis tools and systems, the Rework Cycle portrays the real-world phenomenon that work is executed at varying, but usually less than perfect, **"Quality"**. A fraction that potentially ranges from 0 to 1.0, the value of Quality (as well as that of Productivity) depends on many variable conditions in the project and company. The fractional value of Quality determines the portion of the work being done that will enter the pool of **Work Really Done**, which will never again need re-doing. The *rest* will subsequently need some rework, but for a (sometimes substantial) period of time the rework remains in a pool of **Undiscovered Rework** - work that contains as-yet-undetected errors, and is therefore *perceived* as being done. Errors are detected by "downstream" efforts or testing; this **Rework Discovery** may occur months or even years later, during which time dependent work has incorporated these errors, or technical derivations thereof. Once discovered, the **Known Rework** demands the application of People, beyond those needed for completing the original work. Executed rework enters the flow of Work Being Done, subject to similar Productivity and Quality variations. Even some of the re-worked items may then flow through the rework cycle one or more subsequent times.

The full simulation models of these development projects employ thousands of equations. They explicitly portray the time-varying conditions which cause changes in productivity, quality, staffing levels, rework detection, and work execution, as well as the interdependencies among multiple project stages. All of the dynamic conditions at work in these projects and their models (e.g., staff experience levels, work sequence, supervisory adequacy, "spec" stability, worker morale, task feasibility, vendor timeliness, overtime, schedule pressure, hiring and attrition, progress monitoring, organization and process changes, prototyping, testing ...) cause changes - some more directly than others - in the performance of the rework cycle.

2.2 RANGE OF USE

System dynamics models of projects are being used in a variety of industries for several purposes. We have seen over seventy applications to shipbuilding, power plants, software, air traffic systems, missiles, automobiles, electronics systems, radars, telecom networks, submarines, finance systems, aircraft, command and control systems, and civil construction.

In each industry there has been a wide range of uses of system dynamics models of projects; they fall into four categories.

<u>Pre-project bidding and risk analysis</u>

These applications are typically focused on competitive and high-stakes projects that justify the analysis investment. In this setting, a validated model of one or more prior projects in the organization is used as a baseline. Explicit changes in numerical factors represent differences in the setting, work scope, schedule and other programmatics. New practices and processes are “overlaid” (via model inputs) on the (simulated) new project in order to foresee legitimate achievable improvements over prior project performance. Then the improved prospective project can be subjected to (simulated) risk conditions that might emerge during the project - technical problems, labor difficulties, scope change, vendor performance... Here the intent is not only to assess the degree of cost/schedule performance variability on the prospective project, but also to formulate and test contingency plans to mitigate any adverse impacts foreseen in the risk tests. Managers can then be prepared to enact the pre-tested plans if the tested risk item(s) should emerge.

The combination of these model-based analyses has provided project managers and executives with a stronger, more rigorous basis for determining and submitting responsible cost bids and schedule estimates.

Mid-project management support

When even the best-laid plans go awry, project managers face a variety of "what if" choices. *What if I bring on new staff to help catch up? Use more overtime? Proceed without the vendor information? Accept this new customer change? How much? When? What combination?...* For each such condition and question, the consequences for the staff, the project and the business need to be evaluated. Most such decisions are based purely on hunches and experience, influenced by business and external pressures from several sources. Typically the immediate effects of the decision are clear and become the determinants of the decision, but the full effects ripple far "downstream" in the project, widely separated in space and time from the instant source. The effects are less clear, seem less certain, and often run counter to the intuition of managers (more on this later). As system dynamics models consist of chains and loops of cause-effect relationships, they have proven valuable in identifying and quantifying the "distant" (and often larger) counter-intuitive effects of actions and decisions. The models and analyses are used by managers to identify (and explain to others needing convincing) the longer-term effects. Well-calibrated models are even capable of quantifying the magnitude and timing of effects on schedule, cost, staffing needs, productivity, rework, and more.

Post-project diagnosis and dispute resolution

Contracted projects that exceed cost and schedule targets often lead to some degree of dispute between contractor and customer. If the disputed amount is large enough, enormous efforts may be mounted by both parties in order to obtain or defend against a formal or even litigated dispute award. Most prevalent in the United States, the litigation capital of the world, such disputes have now surfaced in many parts of Europe as well.

By far the most contentious segment of project cost dispute is the "delay and disruption" cost - secondary impacts on other parts of the project, originally caused (or alleged so) by faulty specifications, customer design changes, delayed information, and a host of other sources. Fact-finders (judges, arbitrators...), after determining contractual entitlement, seek to answer one basic question in quantifying any cost damages owed the contractor: *What would have been the cost of the project but for, and only but for, the impacts of customer-responsible actions?*

A variety of methods have been used to estimate the answer to this question; system dynamics models have proven especially useful in doing so. Their cause-effect structures provide causal "traceability" from incident to impacts; the basic what-if capability provides the means of directly answering the fact-finders' question. Since most such disputed projects have reached completion, the system dynamics model is set up first to re-create accurately the documented historical performance of the project, including the incidents in dispute. The model is then exercised to simulate the performance of the project *excluding* those incidents. All other conditions remain the same, so the *difference* in the simulated costs in the two cases is the full cost attributable to the incidents. Some of the cost difference is the "direct impact" of the incidents, and the rest is "delay and disruption" - secondary impacts on cost as a result of lowered productivity, increased rework, or extended performance time.

The strong logic of this approach has been used over two dozen times on large complex project cost disputes. Virtually all cases using system dynamics project models have been resolved in less than typical time, on better terms, and with less long-term damage to customer-contractor relationships.

Training and Improvement

Much of the use of system dynamics project models has been directed toward training managers or improving management processes. Again, the causal structures provide clearer explanations of how projects work and how managerial decisions affect the project.

In this setting different formats are used. Sometimes the models are nothing more than causal "maps" used for instruction. Operating computer simulations may be conducted in conventional "what if" form or as an interactive simulator - presenting situations to the trainee, eliciting decisions injected into the model, updating the situation in the model, and so on - stepping through time to project completion. "Post-mortem" diagnoses then help to explain lessons to trainees. With multiple individuals (or teams) participating, cross-comparisons of results enliven the diagnostic discussion and hasten the learning process.

Be it through advance planning, what-if analyses, retrospective diagnosis, or pure training, the role of system dynamics models in project management always goes beyond providing a tool or an "answer." System dynamics models aim to provide a greater understanding of complex systems and improve our ability as managers to achieve better performance. In the remainder of this paper, we draw upon dozens of real case studies to explore some of the managerial questions that arise so commonly on different projects. We employ a system dynamics model to explain the counterintuitive consequences that typically reduce our effectiveness as project managers.

3. Project management lessons from the use of system dynamics

While we as managers have little absolute control, we have enormous *influence* on project performance. Our decisions and actions work through multiple *cause-effect paths* to affect the rework cycle and the project outcome. All too often, the countermeasures we use in response to typical problems along the way are counterproductive. The use of system dynamics modeling can help us anticipate the many other cause-effect paths by which the decisions influence the project. In the discussions that follow, we will employ system dynamics to explain some of the most common failings of project resource management, and their many consequences.

Staff Overtime

Figure 2 adds to the rework cycle structure a "path" showing a chain of cause-effect relationships, indicated by the thin arrows. Using information from the "pools" of work in the rework cycle (but without knowing, of course, the size or content of as-yet undiscovered rework), managers or their aides periodically estimate the progress made to date in the project, or in a stage of the project. Based on this, they assess the extent to which additional staffing is needed, to try to finish the remaining work on the prevailing schedule. A common and reasonable response to falling behind schedule is to supplement the effective staffing by the temporary use of overtime. This avoids the cost, hassle, and long-term commitment of bringing in additional people through hiring or transfer.

The intended direct effect (increasing the people effort) is achieved, and the overtime is easily extended. However, the *secondary* effect (through fatigue) on worker productivity and work quality, though often acknowledged in spirit, is consistently under-estimated (this path of unintended secondary influence is shown in the diagram with dashed-line arrows -- this is what causes the counterintuitive effect.)

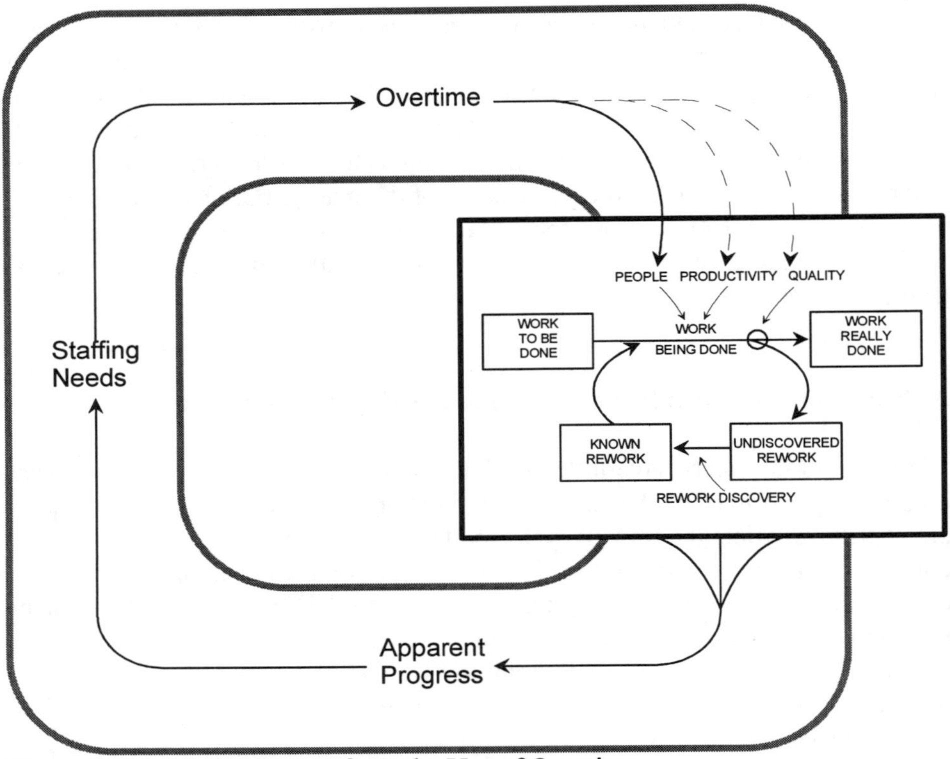

Figure 2. Paths of Influence from the Use of Overtime

We used our system dynamics project simulation model to generate the performance of a typical four-year, million-person-hour (planned) project under two different conditions. In the first simulation was typical overtime usage, but only with the intended direct effect on staffing - no unintended secondary effects on productivity or work quality. In the second simulation we activated the secondary influences of overtime, at the strength of effect that we have observed operating in dozens of development projects. In this (more realistic) case, the per-person productivity drops and rework increases (by 25%), as a result of overtime-induced fatigue. Overall, the same project scope requires 15% more effort to complete when using extensive and sustained overtime.

Staff Hiring

Unavoidable at the start or in moderation, bringing new people into a project organization (by hiring or transfer) has a degree of counterproductive impact that is, again, widely underestimated or ignored. But the impacts - potentially harsh ones - will occur nonetheless.

The diagram in Figure 3 shows another set of cause-effect paths of influence that work over the course of time, when managers decide to hire in response to a perceived need for additional staff. The obvious intent is to bring in new people who will supplement the staff available to work the project.

With any substantial hiring or even transferring of people unfamiliar with the project, the new people enter with less experience or skill than those already on board (despite individual exceptions, this holds true, especially for technically demanding work or specialized skill needs). So the average skill level of the growing staff drops, at least long enough for new people to "get up to speed" (which can take years). Worse, the more hiring occurs in any constrained labor market, the lower the entry skill level.

As if that were not enough, newly hired individuals tend to have a higher attrition rate than longer-time experienced people. A higher attrition rate requires still more new hiring, just to remain even - let alone grow the staff more. So a vicious circle develops, a "churning" with little forward progress: new people are hired, departure rates increase, and still more new people are hired to sustain the organization.

It gets worse still. Any problem response that adversely affects productivity or quality, such as lowered skill levels from hiring (or sustained overtime fatigue, for that matter) can trigger self-reinforcing impacts on a circular path of cause and effect. As the same diagram indicates, the hiring, which yields skill-worsened productivity and quality, slows the pace of progress relative to the schedule. With less than planned progress, we clearly need... more people! (Thus more hires, new people, skill dilution, productivity and quality reduction, slowed progress...) Further, there is the extra coaching and training time invested by experienced staff, and the disruption of a rapidly growing organization. We have observed these exact phenomena in many organizations; they start innocently and build insidiously, becoming a trap into which one can easily fall.

When we activate all of these secondary effects in our typical project model, total hours climb another 45%, and more rework accounts for nearly 2/3 of that increase. Indeed, the rework effort grows to be nearly half of the total hours spent on the design. The added rework cycling also extends the time of performance by another three months. In a particularly vicious circle, the additional delay induces

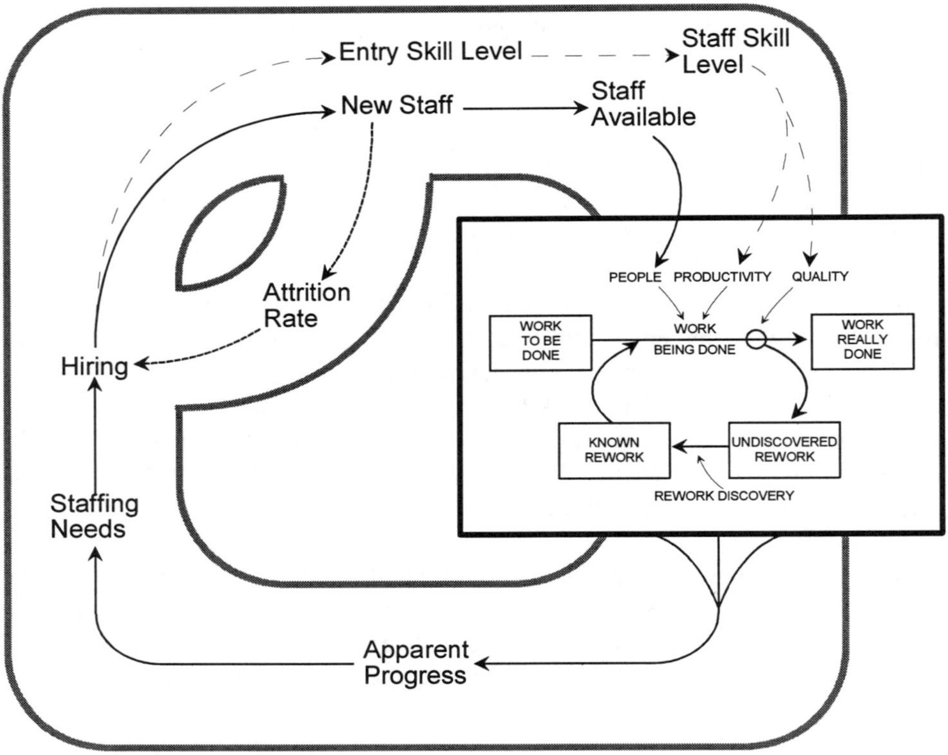

Figure 3. Paths of Influence from New Employee Hiring

the managers to employ more overtime, too - thus incurring even more secondary impacts on productivity and work quality (and cost and schedule).

<u>Staffing and Scheduling Work Phases</u>
True progress in development efforts is inherently difficult to assess, so, understandably, we have devised surrogate measures which are easier to monitor during the conduct of projects. While they certainly are *precise* (793 drawings, 45.2%, ...), they are not *accurate*.

In deciding how and when to staff, we too often depend on the “precise” measure when we assess the readiness of prerequisite work products to support the execution of the next dependent tasks. We need to take better account - even if approximate - of the quality of the work logged as "done" when making staffing decisions.

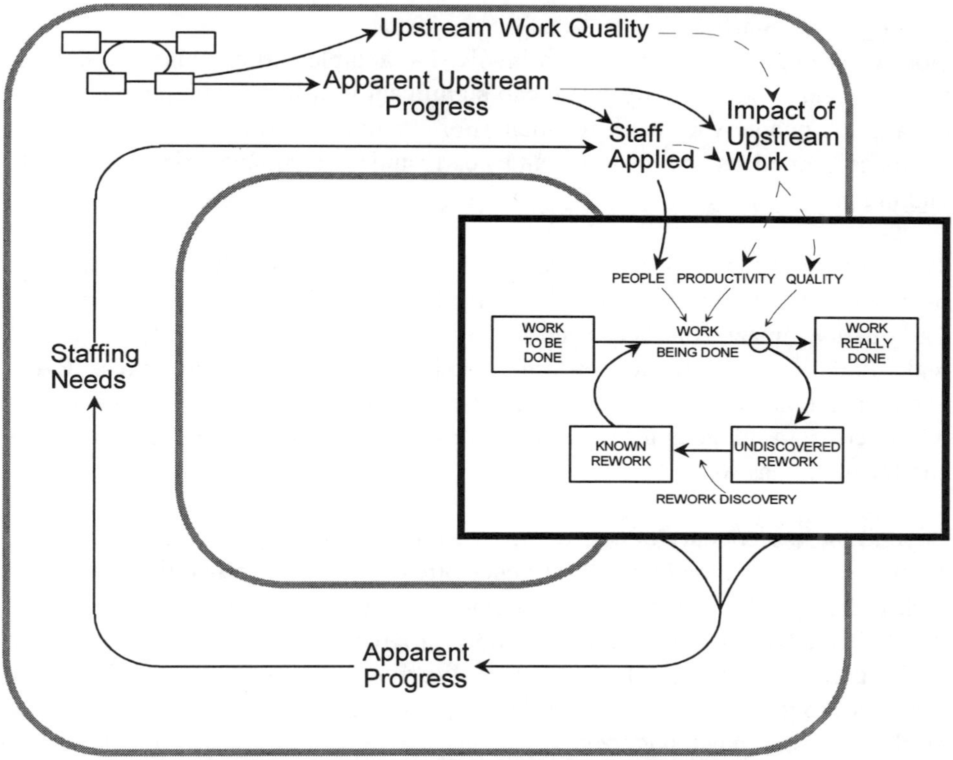

Figure 4. Paths of Influence from Staffing on Low-Quality Work

The diagram in Figure 4 displays this additional pathway through which management decisions affect projects. In determining appropriate staffing levels, managers consider the apparent availability of prerequisite tasks' work products. But poor *quality* of that prerequisite work will cause unintended reductions in subsequent productivity and work quality. The more aggressively the project (or stage) is staffed in this condition, the more people and work that are subject to these adverse impacts.

When we add to the simulation of the same project these unintended but real productivity and quality effects caused by staffing dependent work while "quality" problems remain in upstream work products, project costs climb an additional 40% and completion time is four months later still. Rework is up disproportionately (it is nearly ¾ of the total increase in hours - the rest from reduced productivity), now accounting for over 50% of the design effort.

Pressuring the Staff

Not everything we do as managers involves a tangible action. What we say and the "incidental" gestures we make can significantly influence the people around us, and the people who work for them. For example, exerting schedule pressure on those around us is a natural, and nearly universal, managerial response to lagging progress.

The diagram in Figure 5 demonstrates the intended path of influence of schedule pressure - an increase in productivity. It also shows the unwanted side-effects. The quality impact, (the later need to do it again) is far less visible, but no less real. Equally real are longer-term impacts of sustained pressures. Every experienced manager has seen the considerable morale impacts when the staff is pressured from all sides to improve not only on the schedule, but also the costs and the quality as well.

Beyond these secondary effects on productivity and quality, Figure 5 depicts yet another cause-effect pathway of influence often activated by schedule pressure. In such a highly-pressured condition, engineers (or programmers or...) are induced to work on anything that will demonstrate "progress". This usually translates into working more and more out-of-sequence. We may achieve the desired short-term apparent progress gains, but we set the stage for even more trouble later when the resulting rework gets recognized.

Once more we'll add these effects, at the strengths we have seen them operate in many development projects and programs, to the simulation model. The results should be no surprise. The poorer performance generally shows up in the latter half of the effort, as the pressures mount well into the project. Rework effort now accounts for about 60% of the total hours spent on design, and causes the vast majority of the 100,000-hour increase.

How far we've come. It was such an innocent-looking project at the start. Each step along the way added some more troubles and realism. The "normal" managerial responses, all aimed to exert control, brought unintended side-effects and penalties through paths of influence generally not well-understood, or at least under-estimated.

Figure 6 helps illustrate just how far we "progressed." We added 600,000 hours to the design effort (6a, the curves labeled "X" show the original conditions, "P" the final) through productivity losses (6b) and more rework (6c). Rework alone increased by 500,000 hours. A plan-beating staff profile that had peaked in mid-

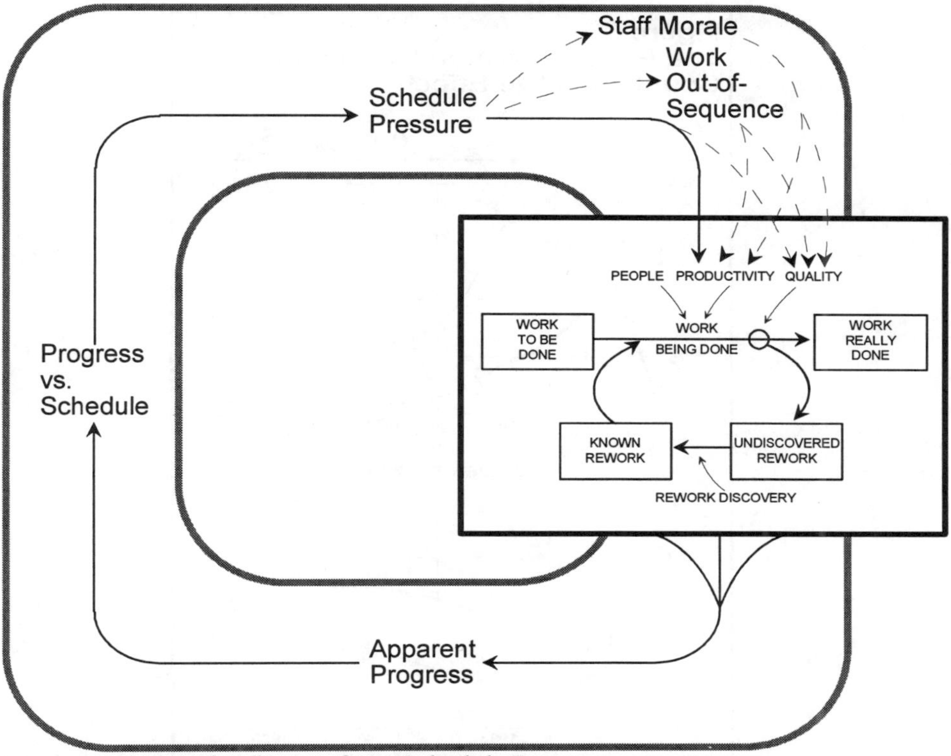

Figure 5. Paths of Influence from Schedule Pressure

1992 at *150* people became a serious overrun, with a second peak at *250* people, a year later.

By mid-1993, when the build effort that was using the design product was under way, the original simulation displayed a perception of design readiness (7a) near 95% (and accounting for the undiscovered rework, really was at 75% - see 7b). The final, real project simulation displayed a perceived design readiness of 85% at the same time (7a), but much more undiscovered rework meant the design product was really only 50% complete (7b).

While we've focused on the design effort, the build effort also is impacted by the higher and later design changes - enough to cause more than a doubling in build labor costs (7c) - *and* a full year's slippage.

Figure 6.
The Project, With and Without the Counterintuitive Effects

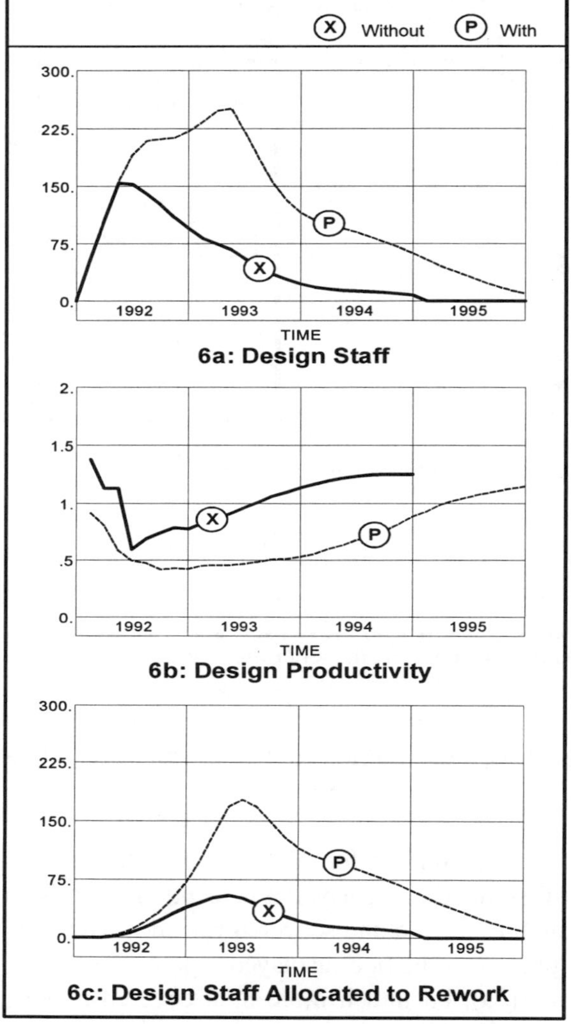

Figure 7.
More About the Project, With and Without the Counterintuitive Effects

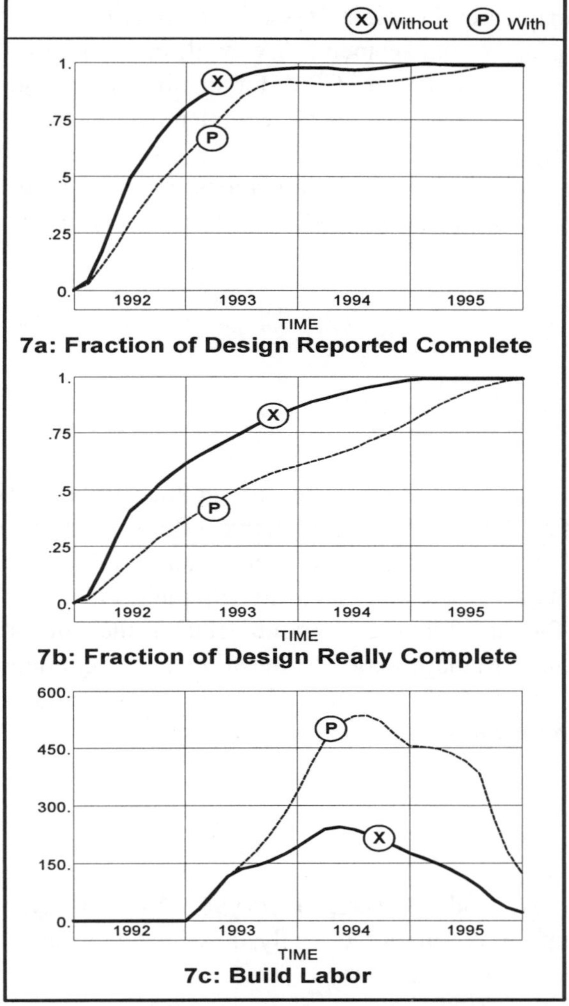

7a: Fraction of Design Reported Complete

7b: Fraction of Design Really Complete

7c: Build Labor

4. Conclusion

The appropriately increasing attention to the art and skill of project management is being accompanied by increasing use of system dynamics modeling to support and improve project management. We will continue to need and to use "traditional" techniques such as CPM for detailed planning; however, system dynamics offers a valuable strategic complement to the collection of tools available to project managers. Be they in the form of causal diagrams, simple models, complex highly-validated simulations, or interactive trainer-simulators, these models provide a valuable additional capability for improving projects' performance.

Acknowledgments

Much credit must go to all of my colleagues at Pugh-Roberts Associates, division of PA Consulting Group. Their superlative efforts over the years have added consistently valuable new insights; it is an honor and delight to work with them daily. The clients who have allowed us to help their projects and companies have been the most forward-thinking and courageous in their industries. Great thanks also to the Project Management Institute (USA) for continued interest in this work, and for publication of many of the concepts underlying this paper in *PMNETwork* and *Project Management Journal.*

References

1. Rodrigues, A. and Bowers, J. (1996) System dynamics in project management: a comparative analysis with traditional methods. *System Dynamics Review*, 12, 2
2. Cooper, K.G. (1980) Naval ship production: a claim settled and a framework build. *INTERFACES* 10, 6, 30-36.
3. Cooper, K.G. (1993) The re-work cycle. *PMNETwork*, February 1993.

QUESTIONS on "SYSTEM DYNAMICS METHODS IN COMPLEX PROJECT MANAGEMENT"

The following Open Discussion deals with Systemic modelling methods in general, but the following immediate questions were put to Mr. Cooper.

- Have managers felt confused at trying to quantify these parameters that are difficult to quantify, or has the management decision-making been improved by this process? Mr. Cooper replied that PA's experience has always been the latter: they spend a lot of time with managers designing and improving each model. By the time analysis results come, the project manager already understands the basic structure of the model generating those results, and there's usually a great sense of relief at having some better quantification of the actions being contemplated.
- Projects are clearly very different, with many varying factors, so more explanation was requested of his concept of a "common model" for System Dynamics project-models. Mr. Cooper said that at the very basic level of the re-work cycle, he saw a near-universal model of projects; that is a very simple structure. Going beyond that to look at the many factors, there are some that are nearly always working on all kinds of projects (experience levels of staff doing the work, testing and re-work discovery processes....), while others exist in some projects and not in others. So it isn't a perfect common model, but one that goes some way towards being common. Mr. Cooper felt that if we can make understandable the basic processes at work in projects, we will be able to transfer learning from one type of project to another much more effectively.
- It was suggested that perhaps re-work shouldn't happen at all - the work should be "right first time". Mr. Cooper replied that every complex project has re-work, and undiscovered re-work. It is desirable to aim for "right the first time", so long as the manager recognises that that will never be achieved; the manager must take not action which discourages the detection of re-work that is already there, otherwise it will have to be fixed in more expensive stages later.

OPEN DISCUSSION: SYSTEMIC MODELLING METHODS

Discussion chaired by:
D. LONGSTREET
Longstreet Consulting,
MO, USA

This summary by:
T.M. WILLIAMS
University of Strathclyde,
Glasgow, UK

Mr. Longstreet discussed the need for modelling the systemic nature of projects. He used the example of water: at room temperature it has the property of being "wet", but neither of its constituent parts - oxygen and hydrogen - have this property in themselves. So there are many interactions within projects - between tasks, between different people in the project and those within the rest of the organisation, between people and their personal life - and it is only by considering these that we understand the impact of schedules and the impact on morale. (The importance of the effect of schedule pressure appeared to be an emerging theme from the conference). Similarly, if you look at Quality, the total level of quality is *not* the sum of the individual quality tasks (and so on for Productivity and similar factors). It is necessary to look at the global picture, not just the small parts.

The question of how to quantify "soft" factors such as quality, morale and stress was raised, both from Mr. Longstreet's introduction and Mr. Cooper's paper. Mr. Longstreet pointed to indicators for factors such as morale and stress (turnover, sickness etc), and for quality, the need for management to specify their requirements precisely. Mr. Cooper quantified quality by the number of revisions to work.

This led to a discussion by one CP participant about the aggregation problem: a quality level is achieved by each person at each stage, and system quality depends on the combination of the individual indicators. He exampled his work (Burkov, Russia) estimating the level of tension in a part of the country: the mayor realises the overall indicator, but this depends on narrower indicators, e.g. people's standard of living, the level of criminality etc. This gives the important problem of formulating the relationship between the lower-level data and the overall measures.

Another CP delegate discussed TECHINTEL's experiments of modelling large electricity and gas plants, and goods-transport systems, and deriving the characteristics of a complex system from the parameters of sub-systems; this is complicated when the system is dynamic, and even more so when sub-system objectives can change over time; again work was quoted.

Mr. Longstreet agreed that to solve a problem at one level, often it fixes the system at many different levels, and attempting to solve at a lower level, or at the wrong level, may not have any impact at all. There may not be models to quantify these effects exactly, he suggested, but "common sense goes a long way".

SECTION 4
DOMAIN SPECIFIC EXAMPLES

The Workshop so far has been looking at projects in general. This Section looks at aspects of managing complex projects in particular domains. The first domain considered is that of projects in Ukraine, particularly High-Technology projects, in a discussion paper prepared at very short notice by Dr. Leonid Romanenko of the Center of Strategic Investment and Project Management in the Ukraine Ministry of Economy. The second domain considered is software projects, which need very specific considerations and have very specific techniques, with a paper given by Tom Conrad of the Architecture and Engineering Division in the US Naval Undersea Warfare Center. Finally, the domain of infrastructure projects is considered by Prof. Khim-Teck Yeo, of Nanyang Technological University in Singapore, concentrating in particular on the human factors involved in infrastructure projects. Before these papers, in order to put these differing domains in context, this Section is introduced by a short talk by Dr. Peter Dunton of IBM UK Ltd.

THE MUSHROOM PROJECT MODEL: A NOTE

Dr. Peter Dunton
IBM UK Ltd, Bedfont Lakes, UK *(edited)*

The concept and scope of a project appear to be different depending upon the area of industry or commerce. The following model can be used to explain the differences of approach to project management by different groups and puts into context the key expectations of a business project. The objective is to recognise that the concept of a project varies depending upon familiarity with a particular industry sector and that all are usually correct for that environment. In this narrative several views are expressed by the author that are deliberately polarised to exaggerate differences in the view; in reality there are many shades of grey.

The Solution Model

The traditional concept of a project is demonstrated in construction projects where something is built, starting from some specification and normally ends with some form of acceptance. In Prof. Chapman's example of the dog kennel project [1] this would correspond to his specification - a place outside for the dog, and acceptance (sometimes forgotten) - the dog is *able* to live in the kennel. A project is required to build the kennel. In the model this can be referred to as the **solution**, which forms the emerging cup of a mushroom and where traditional design, build, test, implement processes can be observed. In this model it is important to recognise the symmetry of the mushroom (see Figure 1).

Requirements and Benefits

In some project environments such as manufacturing and engineering we see that the solution is driven by requirements, leading to the specifications and also some expectation of the solution meeting these requirements to fulfil some purpose and provide some form of benefit. The model can be expanded to cover these aspects by growing the cup of the mushroom symmetrically to include requirements on the right hand side and benefits on the left, as in Figure 2. In the dog kennel example we may find that the requirement for a dog living in the garden is to have somewhere to live, sleep and be sheltered from adverse weather. The kennel then forms the solution. The solution delivery remains the same, that is, the

T. M. Williams (ed.), Managing and Modelling Complex Projects, 111–115.

construction of the kennel and the subsequent well-being of the dog, living in the kennel, then forms the benefit. In considering these two additions the scope of the project has expanded as the project is now not a straight forward construction project since the dog's requirements need to be considered.

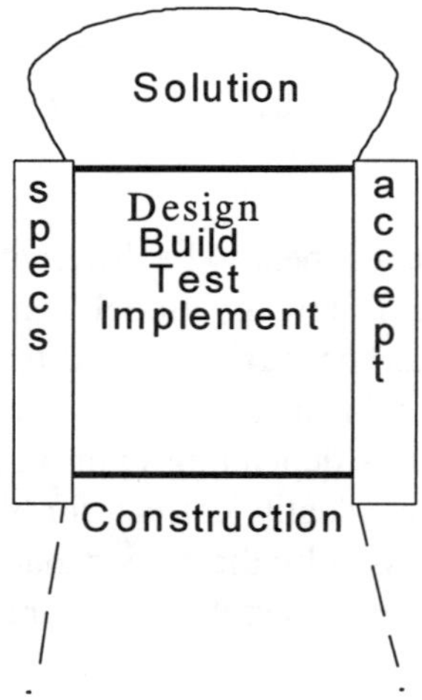

Figure 1.

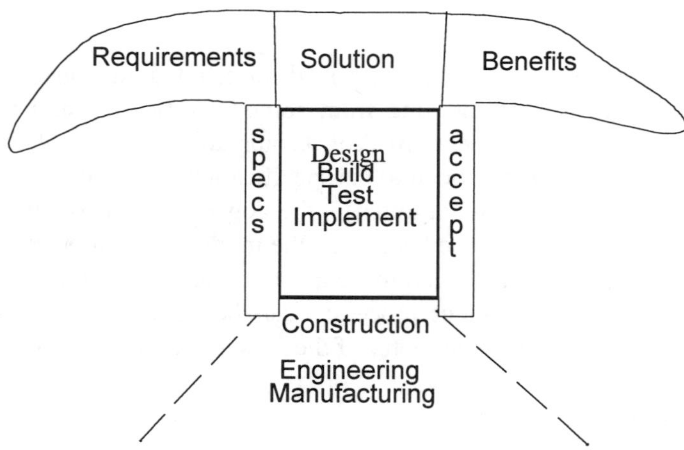

Figure 2

Needs and Opportunities

Projects in the commercial and financial sector expand the mushroom model further with the requirements often driven by commercial needs and opportunities and realisation of the benefits linked to these being deemed the success criteria. In our example: "the dog must sleep in the garden, therefore a kennel must be built in the garden to keep it warm and dry" might be voiced as needs statement and we might possibly hear "then it will be more comfortable and happier" (in the garden) as a benefit statement. Alternatively we might hear "as the dog is expecting babies she needs her own place, this is an opportunity to move her outside", or "now that we have installed a fence around our property there is an opportunity to move the dog outside". All of these statements are linked to the requirements and also define benefits that are expected to be fulfilled (following construction of the kennel outside). For the project to be successful, in the eyes of all those who might judge, it is as important, if not more to manage these aspects of the project as that of the construction of the kennel. Thus the mushroom cup in the model now begins to grow with needs, opportunities and requirements driving the solution and providing the basis for the measurement of benefits.

Benefits

Benefits are often difficult to express explicitly in project terms (objectives) since "more comfortable" and "happier" cannot be quantified, and when included in the project raise the question of when should the project end? - a point addressed in the discussion of 'The Sponsor's Dilemma' [2]. Business projects are very much about achievement of benefits and projects in these environments generally need to include a measurement system for the benefits provided by the solution.

Effects of Marketing Pressures

In today's environment marketing and social pressures can often influence the needs aspect. For example: "All our friends' dogs have kennels to live in, we must have one also", or numerous advertisements, could well instigate the kennel construction project. These can cause the project to be undertaken without a good basis for the decision making. In the model Marketing Pressure can be shown driving the left hand side of the mushroom balanced by satisfaction falling from the right (Figure 3). The widening root system shows the increasing scope and lifecycle depending upon the type of project environment.

If the project is to be successful then the solution resulting from the needs should provide benefits that give satisfaction to those to whom the need is being addressed. Thus the puppies are born with the mother having the benefits of a secure comfortable environment, or in the former example the dog should be satisfied with a more comfortable, happier life. This very statement begins then to widen the discussion about **needs**. Is it the just the needs of the dog?

For Whom is the Project Intended ?

It is possible that the needs statement could be something like "we *need* the dog to live in the garden because we have just brought a new carpet in the house" in which case, following the solution we would expect that the new carpet is saved from the dog thus providing a benefit to the family finances. Examination of the needs and opportunity aspect lead to the question *WHY?* This is fundamental to business projects, since if time and effort is to be expended on the solution, only answers to this question can justify the cost, dictate the expected benefit and be the foundation of the decision to proceed. Similarly if *WHY* is the driver for the left hand side of the mushroom then *IMPROVEMENT* must be the driver on the right. Thus to complete the mushroom model reasons *why* need to added to the left hand side of the model and *improvement* to the right. (Figure 4).

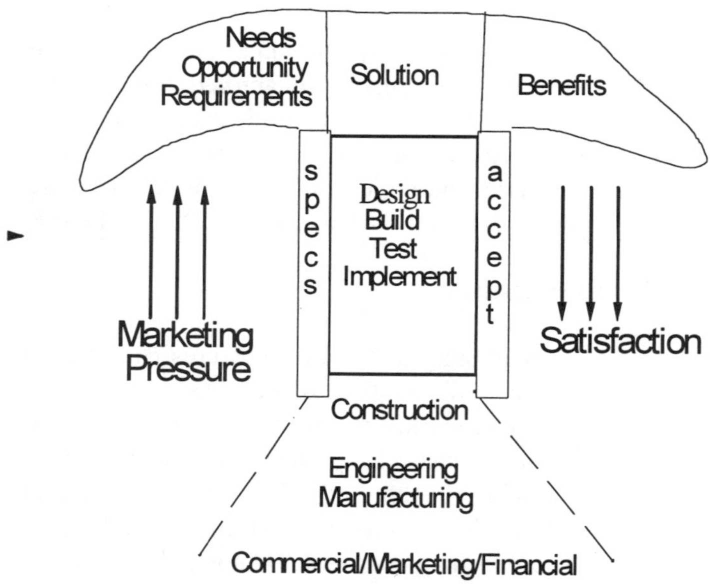

Fig.3

Summary

When the *why* (referred to as goals or aims) is developed into what must be achieved (objectives that can be measured) and the accepting body explicitly defines where the improvements are to be expected and how satisfaction will be measured (benefits), the project scope can be considered to be fully defined. The solution can then be integrated into this for a full project definition.

More could be made from this model but there is not space here. It is important to realise that the concept of a project lifecycle and scope can be different for different types of project, or industry environment, and that all are correct assuming that everyone uses the same model. Unfortunately often the left and right hand sides of the mushroom are neglected in projects where they are of fundamental importance.

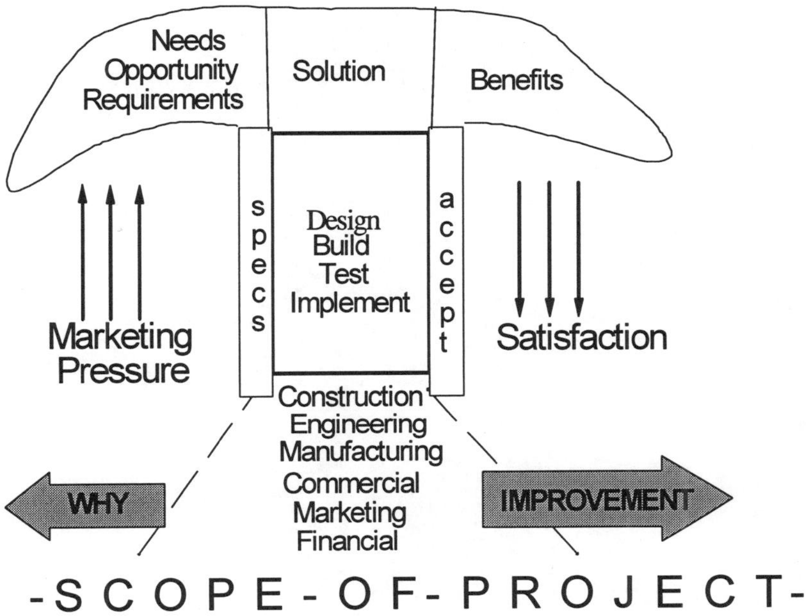

Figure 4

Human Factors and Complex Projects

This example can be made progressively more complex by introducing human and situation factors. For example the family may say "we need a dog as a security deterrent", or "we need a pet to consolidate our family group", or "our friends have a puppy that needs a loving family". All of these could result in a kennel being built, where the traditional construction project management processes can be applied. However each of these statements varies the scope of the project, although they all *could* involve the kennel. However it is possible that with everything associated with the acquisition of a dog into the family, the numerous views of the members of the family this simple project could become quite complex and the best decision might be to buy the kennel from the pet store. At which point we might introduce supplier management ....

References

1. Chapman C. (1996), Paper in Section 3 of this Volume.
2. Dunton, P.C.(1996), Sir Monty Finniston Award. UK APM, May 1996

MANAGING COMPLEX PROJECTS IN UKRAINE

Dr. Leonid I. Romanenko
Director of Center of Strategic Investment and Project Management
Ministry of Economy, Kiev, Ukraine

Editor's comment: Dr. Romanenko gave this discussion paper at very short notice, due to the late withdrawal of another CP-country speaker. There was not time to prepare a written paper. What is given below is the Editor's summary of the paper.

Dr. Romanenko is the Director of a Centre for foreign investments, within the Ministry of Economy in Ukraine. For 2 years, he has been working on problems of forming management systems in Ukraine. The title of the Workshop is "Management of Complex Projects" but this paper is more about the systematic approach of PM. It looks in detail at the question of creating a state system of management of projects, because this is not solved in any of the CP countries. Dr Romanenko spoke on this topic also at the SOVNET conference (St. Petersburg, September 1995), and there was much support from CP delegates.

Dr. Romanenko started with the thought with which Mr. Schpek finished his talk. The Ukrainian economy is very young, and that is why it tries to get as much knowledge from outside that it can. In particular, Ukraine knows it has problems with PM, and that is why it looks for new approaches, new abilities and particularly new professional contacts.

There are three main problems in Ukrainian PM.

- Foreign investments in the Ukraine is limited due to the limited investments of Ukraine itself, because of its limited budget. As one figure: only 10% of demands of Ukrainian enterprises for investment is satisfied, and only 1-2% of demands are met by foreign capital that can be relied upon. Most Ukrainian projects fall through even before their start because of lack of financing.
- The second problem is the problem of social construction, or the investment economy. By its nature, it fails to transfer to a project-oriented approach for the development of the economy. Ukraine is as if it was still imprisoned by the

T. M. Williams (ed.), Managing and Modelling Complex Projects, 117–122.

old structures and models of development, when the resources were directed from the centre (characterised by long lists of titles in organisations). The project-oriented approach has up till now not been accepted willingly by the governmental level. The Ukrainian PM association is working hard at this level to change views. This is a very important task and it can solve a lot of the problems in the many multi-projects.

- The third problem is that on the governmental level there are many departments which claim interest but do nothing, or act but not in the spheres in which the PM professionals are interested.

Dr. Romanenko described the structure of the governmental levels which are supposed to take part in projects. This begins with the cabinet of ministers, which has a special council on foreign investments. Then there is the Ministry of Economics, and this has a department of foreign investment credits (in which Dr. Romanenko works). The Ministry of External Economics has a trade department, and also a department for investment politics. The Ministry of Finance is connected with internal debts. There is the Ukrainian State Investment company, which has the Ukrainian Agency of Co-operation with Foreign Investors. Mr. Schpek is the manager of the Ukrainian Agency of Reconstruction and Development. And this only touches the tip of the iceberg of what the government does. Then there are the regional levels which should start looking into PM as well. The problem is that no-one co-ordinates this work. So Dr. Romanenko and his colleagues in the Ministry of Economics are trying to some extent to do this work systematically.

First of all, they are looking at how to attract foreign investors, and have developed some methods to formulate problems to make them more attractive internationally. Secondly, they have opened within the Ministry of Economics a branch to monitor free capital abroad: currently on the basis of the Central European Representation Council in Budapest they are developing a network, with the help of which they hope to obtain systematic information about the movement of capital, first of all in Europe, and then hopefully in Canada, in US, in the Eastern countries and Asia.

These are the organisational steps. But of course these are not enough, so they have been considering how to organise in the Ukraine a mechanism by which Ukrainian businessmen could find a foreign investor, and so a foreign investor wouldn't have to come to the corridors of the ministry to try to find a project in which he would be interested. He gave examples of foreign companies who had not received a good reception in the Ukraine.

So they have begun to develop a system. It is still only a skeleton, but the main principles have been established. It is called the State System of Project Management. It has four modules: an information analytical module, a module of development of investment projects, a module of preparation management, and a module of realisation of investment. If we look into the structure for example of the first module, it has three blocks: a data-base, information/analytical block and the block of co-operation of partners. The data-base has two branches: first the people who want foreign investors and have projects and the second data about foreign investors who want to take part in Ukrainian projects. Soon this system will hopefully go onto Internet. The second block should provide all the information materials: at the moment they are working on all the regions of Ukraine, to accumulate all the information about the enterprises they have there, about active businessmen, and about the projects that will suit these regions. The third block is very important as it provides the opportunity to find partners for business. It's already working in the form of conferences, seminars, symposiums etc. There are more effective forms, but they are still not developed, e.g. auctions, employment agencies, computer centres; it is hoped that these will be developed. Each module mentioned provides the opportunity to develop the system more effectively. Particular attention is paid to the projects and question of expertise which have been lost in the process of economic change in Ukraine.

OPEN DISCUSSION: MANAGING HI-TECHNOLOGY PROJECTS

Discussion chaired by:
V. BYKOV
Kiev, Ukraine,

This summary by:
T.M. WILLIAMS
Strathclyde University
Glasgow, UK

Prof. Bykov discussed the problems posed by the management of high-technology projects. Human/machine systems, which combine complex computer programmes and human intellect certainly come into this category. Both the context and the structure of such projects have badly-defined characteristics, both static and dynamic. Also, products are often defined by a system of criteria such as a functional specification, which can result in a multi-objective nature to the project. On the other hand, he included systems which are often described by systems of differential equations, stochastic models, and other models and forms of formal analysis which are well-known, many of which haven't been solved analytically. Consideration also has to be given to the adequacy of models in the real project system, a particular area of interest for Prof. Bykov.

Mr. Wakeling, speaking for the UK Ministry of Defence, suggested that the phrase "high technology" not be used, since it has many different meanings. He proposed three different types of such project:

- projects whose requirement requires a new technology for its realisation; these are high-risk projects for which the emphasis should be on early demonstration of feasibility, and for which there should be a structured process of risk reduction;
- projects for which the requirement is uncertain, because it is using a commercial technology and you are looking for a use for the technology;
- projects for which the technology is developing very fast, so that during the life of the project the technology will change; the risk is that a solution will be implemented which is obsolete by the time it is in place. This requires a PM approach focussing on reducing the project timescale to the minimum, and realising the benefits in shorter increments rather than in one big project.

(Some CP participants expressed particular interest in the third type of projects, where the aim and demands of the project changes, so providing a new class of problems for PM.) Mr. Wakeling also referred to the UK Private Finance

Initiative, in which the (government) customer specifies the service he wishes provided, and the service provider is rewarded by a revenue stream over the whole life-time of the provision of the service.

Prof. Bykov agreed that the problem lay in projects whose technology was changing quickly, and being developed and used outside of the project so that the development is not under the project's control. He suggested the use of open systems, which give a chance for open structures and characteristics, and features that reflect the dynamics of natural development and regulation of relationships when working on projects.

Following Mr. Wakeling's definition, Dr. Barnes said that the projects which have special problems are the new-technology not high-technology projects. High-technology is not a real concept, since everything now seems high-technology to people 50 years ago. The difficulty comes from the novelty of the technology, which creates the uncertainty in the tasks, and the risk that the client will make changes as his perception of what is possible changes with the advances in technology. Prof. Bykov had said that this led to the need for open systems, but Dr. Barnes challenged this: it creates the need for systems that are carefully arranged to cope with this situation, which are not open systems. What are needed are systems of information flow which are about what we *now* know about the system and not about what we *thought* we were going to have to do when we started. It leads to a system of management which is systematic but concentrated on reacting to new information to finish the project and less to keeping to the original project plan. Prof. Bykov responded that the discussion was about flexible systems which should take into consideration the environment and the all the elements that are used in the systems (and the systems should be made readily understandable).

Prof. Cleland quoted a benchmarking study of US companies considered "good" at managing technology. They found a number of common things that these companies did: a close link between the management of technology and long-range strategic planning; a highly participative company culture; teams of people interested in a particular technology area were given seed-money to pursue the technology; an excitement about technology; a lot of time spent studying the market and studying the technology. Thus it is the attitudes and philosophies on the part of the senior personnel that is important.

It was suggested that it is important to reduce a highly complex new-technology project into smaller projects, producing a new-technology line. The Japanese automobile industry was pointed to as an example of this.

MANAGING COMPLEX SOFTWARE PROJECTS
Experiences With Real-Time Mission-Critical Systems

T. P. CONRAD
U. S. Naval Undersea Warfare Center
Newport, Rhode Island 02841
United States of America
email: tconrad@C223.npt.nuwc.navy.mil

1. Introduction

Managing any large development project is a challenge. When the system under development incorporates vast amounts of software, the management challenge is made much more difficult. When that software must perform at rates dictated by external events, when its successful operation involves human-machine collaboration to make decisions in the presence of uncertain data, and when system failure cannot be tolerated, then the difficulty in managing the development to a successful conclusion is beyond most managers' ability.

Failure to deliver a useable product on time and within budget is, unfortunately, the norm for such development efforts. There is no lack of data to support this claim, either in the software literature, in the newspaper accounts, or in the assessment reports of expert panels commissioned to study the matter. Indeed, one has only to consider the mortality rate of software engineering standards to conclude that we are far from a complete understanding of how to manage complex software engineering projects effectively. For the past decade, we have been averaging approximately twenty new software engineering standards per year. The September 1994 IEEE Forum on Software Engineering Standards noted the existence of some 334 such standards deriving from 61 standards organizations. There were forty standards just for the management aspect alone!

The situation is far from hopeless, however. As more approaches are tried, the anecdotal evidence accumulates as to what works well in various circumstances. Similarly, we become more certain of which practices lead to failure more often than not. Since by their very nature complex software development projects are

T. M. Williams (ed.), Managing and Modelling Complex Projects, 123–146.

costly and lengthy, the number of such projects is relatively small. As a result, it is imperative that lessons learned on all projects of this scope be reported and shared. Thus, the motivation for this paper is clear.

In the paragraphs below will be found a characterization of a most difficult application domain, submarine command and control. An elaboration of the factors that contribute to the difficulty of successfully developing such systems is made, and observations are drawn from the author's experiences on several such projects. Finally, suggestions are developed for fundamentally beneficial practices, including approaches for assuring reliability, measuring status and cost, optimizing operability of the resulting system, avoiding early technological obsolescence, and achieving effective integration of the system under development.

2. The Nature of the Challenge

The observations made here derive from many years of experience with systems that are among the largest and most complex ever built, namely submarine combat systems. An appreciation for the nature of the application domain is essential to full understanding of the observations made and the management practices recommended.

2.1 THE TECHNICAL CHALLENGE

The technical challenges are numerous. The problem that the system is built to solve is not precisely known. The system requirements exhibit continual change. The critical nature of the system places severe reliability demands upon it. The most important time for the system to operate effectively is when it is most highly stressed. The system must be operable and maintainable by the ship's crew, and it must function continuously around the clock. Further, the system is of vast size, incorporating millions of lines of software instructions.

The combat system constitutes the principal inboard electronics suite for the submarine. It handles all the signal and data processing, taking data from the thousands of sensors, converting, sorting, and analyzing that data to produce decision-coordinating information that is subsequently used by the attack center personnel in coordinating ship maneuver and weapon deployment actions. There are several attributes of the application that pose significant challenges and impact the way that the system must be designed, built, and tested.

2.1.1 The Uncertain Environment

The hardware and software alone cannot drive the ship nor perform its combat mission. The attack center crew, on the order of 20 officers and specialists, are an integral part of the submarine combat system. Thus the computing subsystem that is deployed is in one sense a tool set for a highly trained crew. It does not represent a preprogrammed solution to a known "combat system problem". Indeed, for the submarine, the mission elements cannot be precisely predicted in advance. For one thing, the mission is on the order of months and the combat system is active twenty-four hours a day. For another, the very nature of the warfare business is that the opposition is trying very hard to confuse you about what he is doing, including trying to corrupt the data that your sensors are providing. Thus, the combat system must be very flexible in its ability to deal with unpredicted scenarios and with data of unknown accuracy, much of it subject to the vagaries of acoustic transmission through water.

2.1.2 Resilience Demands

As one might expect, this application domain is very demanding in terms of the requirements for resilience. The hardware and software is in continuous use over a long mission time. There is no good time for the system to fail. Consider the ramifications of a software failure causing a restart of the system while you were tracking the progress of an incoming torpedo. Or losing all sensor data as you transit a narrow channel beneath the polar ice cap. The reliability and fault tolerance demands are extremely high, so deployment of the system requires high confidence in the quality of the product. A related characteristic of the application domain is that whatever does fail at sea should be correctable at sea, or else the mission must be aborted.

2.1.3 Real Time Performance

Another key aspect of the application domain is that it requires real-time performance. Although the case can be made that, for submarines, time moves relatively more slowly than for surface combatants or aircraft, it is also the case that localization of subsurface contacts is substantially chancier and involves the fusion of data from thousands of sensors. This means that more time is needed to identify and deal with threats. Thus, the submarine combat system is indeed a demanding real-time system, with processing timelines driven by sensor data streams and where the value of information decreases rapidly with time.

2.1.4 Vast Scope

It is also the case that submarine combat systems are large, complex, embedded systems. Millions of lines of deployed software are involved. That is millions of lines of high reliability, hard real-time embedded code operating on vast databases updated through thousands of sensors. There are several reasons for the huge scale of these systems. As already mentioned, they have to be very flexible so that the human-computer team can deal successfully with whatever challenges are posed over a multiple month mission. And as weapon technology and the submarine mission have both evolved rapidly in recent years, the scope of the combat system has also risen dramatically. The combat system must deal with more weapon types, more sensor types, more threat types, more operational areas, and more mission types than ever before. For example, where a few years ago the submarine typically operated alone, it must now also support joint operations from the sea. Also, the combat system must provide for on-line diagnosis and fault location, graceful degradation and reconfiguration in the event of casualty, and an on-board training capability that provides realistic scenarios.

2.1.5 Long Lifetime

One consequence of the nature of these systems is that they are naturally very expensive to build and acquire. The unit cost of submarine combat systems is also high because of the very limited market for such systems. Submarines are numbered in the tens, not the thousands, so there aren't any volume discounts to big customers. It is not difficult to understand that consequently, one does not just throw combat systems away every couple years and buy a new model. No, submarine combat systems have a long lifetime, 20 to 30 years. And during that lifetime we can be sure that hardware technology will turn over perhaps ten times, and mission requirements will change, and threats will change, and weapon systems will evolve. Thus, the submarine combat system must change as well. The import of this is that it must be built with change in mind from the outset. Otherwise, we will likely not be able to afford the cost of owning such systems. The salient point here is that the architecture of the combat system is of fundamental importance.

2.1.6 Architectural Approach

The system must be partitioned in such a way as to permit growth and change in an incremental, modular fashion. The topology must facilitate the garnering of intellectual control over a huge development that is technically challenging in every dimension. It must support concurrent development by multiple development teams. And in the end an effective integration process must result in a system fit for operational use.

2.2 THE ACQUISITION ENVIRONMENT

In addition to the technical challenges, there are relevant attributes of the acquisition environment to consider. In the first place, the buyer of the system is the government, in particular, the Navy. A consequence of this is that the acquisition process is highly regulated with the goal of assuring that the public's money is wisely and fairly spent. Typically, the process has involved preparation of a requirements specification by the government acquisition team. Then competitive bids are invited from industry, with a winner being selected based on an assessment of the bidders' technical and cost proposals.

Affordability is always a prime concern. While acquisition approaches continue to evolve, various mechanisms have been tried to contain costs, including the use of fixed price contracts. Also, the Navy typically employs a Technical Direction Agent (TDA) to oversee the contract. The TDA is established to assure that the Navy has sufficient insight into the development to make wise decisions about progress of the contract which often spans seven to ten years. The TDA reviews interim deliveries of documentation products, participates in regular reviews of progress, performs independent modeling and testing activities, and makes recommendations to the Navy's Program Management Office.

Large software projects tend to be delivered late, over budget, and with less functionality than originally planned. It is understood that the customer and contractor must work closely together to detect and resolve problems at the earliest possible date, and keep the project moving toward a successful on-time delivery. The Navy approach has been to incorporate in the contract provisions for support for an on-site government engineering team at the contractor's plant to provide day-to-day insight into progress. In addition, the contract is written to provide government independent test time in the contractor's integration test facilities. Both of these have proved to be invaluable activities even though they seem at odds with fixed price contracting schemes.

An additional strategy that has been employed on some combat system acquisitions is implementation of a formal Independent Validation and Verification (IV&V) program. This approach requires significant investment by the customer. It calls for an independent government team to assess both the engineering processes being applied by the contractor and also the quality of the software products as they are generated.

As a final note on the acquisition environment, consider the recent consolidation in the defense industry. It is not unusual to award a contract to one company, only to see that company bought by another which is subsequently purchased by a third firm, with all of these changes taking place during the period of performance of the contract. Changes of this sort are bound to have some impact on employee retention, morale, and productivity, and thus on contract performance and program risk.

This same consolidation in industry, coupled with rapid technological advances poses other problems for the developer as well. There is a risk that suppliers of important software and hardware components will disappear from the marketplace or no longer support products (or versions of products) planned for inclusion in the system. The current trend toward maximal use of off-the-shelf commercial components in building large systems exacerbates this problem.

2.3 DECISIONS AND CHALLENGES

In light of the technical domain characteristics described above, and the constraints of the acquisition environment, it is evident that a concerted management effort is required to keep such projects on a path to success. Consider the challenges facing the management team. On the developer's side, there are dozens of technical decisions to be made:

- Shall we build the system as an all new development or shall we incorporate off-the-shelf components available in the marketplace and legacy components from previous generation systems?
- Can we reuse components without significant revision? Will it cost more money than it saves?
- What computer language(s) should be employed in software development? What are the implications of the choice for the required software development environment? What are the costs of retooling and retraining?
- How can we anticipate failure of the system to meet performance timelines?
- How should the software be partitioned into functional components and how should the development of those components be scheduled?

At the same time, the customer's management team faces tough decisions as to what oversight will be most effective. Again, the issues are diverse:

- What data should be collected to judge progress by the developer?
- Should there be significant investment in an Independent Validation and Verification effort?
- What steps are needed to assure that the product will be maintainable after its delivery? What strategy for life cycle management of the product is best?
- How should technical risks be managed?
- On what basis should the product be judged acceptable? What is the minimum acceptable testedness of the system?

Despite (or perhaps, as a result of) the many challenges posed by the application domain and the acquisition environment, there has been no lack of boldness in dealing with the situation. Consider a recent case in which the development contract called for the use of a new standard software development process, the use of the Ada programming language, the collection and periodic delivery of software metrics data, the employment of Computer-Aided Software Engineering (CASE) tools, the insertion of commercial off-the-shelf components as a cost containment measure, and the incorporation of software reuse as a significant objective. The combat system to be developed would have new computers, a new operating system, a new operator interface, a new distributed architecture, and a new database technology. The risk in attempting this major software development with so many new aspects is evident. The effectiveness of the technical and management approaches applied in this and similar cases is discussed below.

3. Lessons Learned and Fundamental Issues

Almost all the problems encountered in managing complex software developments derive from one or more of five root causes. These are: the problem of scale, the inherent nature of software, lack of attention to architecture, substitution of technology for engineering, and unenlightened contracting. We will consider each of these in turn.

First, there is the problem of scale. The size of a software system bears strongly on how long it takes to develop, how many people need to be involved, how much and what sort of documentation is required, how many latent errors will exist upon delivery of the software, and how adaptable or disposable the software is, to name just a few important things. For the multi-million line of code system that is our concern here, it means we can expect tens of thousands of staff-months of effort for development, and that effort will require many teams of people, perhaps enough workers to fill a football stadium. When they are finished and deliver the product we can expect it to contain thousands of defects, some of which may never be discovered. Also, there are limits to the size of software systems that can feasibly be built with the technology at hand. The marketplace of software development tools is not oriented to systems of grand scale since there are too few such systems to justify the investment required by tool builders.

The second basic cause of problems derives from the inherently different nature of software. It is dangerous to think of software as analogous to hardware if you are managing a major development project. That view leads one to plan a software development as one would a hardware development, and that is not the path to success. Managers who dismiss the software development effort associated with their project as a matter of secondary, or even negligible, importance are certain to fail. In the end they are apt to claim that the software was overly expensive and a waste of resources, and that they would have been better served by building the system entirely in hardware. They are correct in only one matter. Software is expensive. However, software *should be expensive.* It provides virtually all the useful functionality in most systems. As such, it is the specialized component of the system. At the same time, software is intended to be variable, to provide the flexibility that a fixed hardware solution cannot. For discriminating between the nature of hardware and that of software it is useful to consider the following definitions: Hardware is something that, if you use it long enough, it stops working. Software is something that, if you use it long enough, it starts working. Software does not wear out. Neither is it possible to exhaustively test software systems of significant size. When you accept a software product, it has to be with the understanding that you only know what you've tested, that you haven't tested most possibilities, and that there is some acceptable set of remaining problems with the software as delivered.

The third issue is lack of attention to architecture. For large, costly, long-lived systems the architectural decisions are the most important. They are also often the most overlooked in the rush to get some demonstrable capability implemented. Consider that for large systems, partitioning is required in order to permit their effective construction by multiple teams of developers, with subsequent system integration. The partitioning is also necessary for another reason. It is necessary if we are to produce a maintainable system. This includes the corrective and perfective maintenance that is a reality for all software systems. It also includes adaptive maintenance. We want the software system to be able to respond to changes in requirements over time. We also want to be able to take advantage of technological advances that impact some portion of our large system. If the software does not exhibit a modular structure that anticipates change in both mission and technology, then we will be faced with a heavy, and perhaps unaffordable, burden of maintenance costs. Of course, architectural decisions will also drive the performance of the system, so careful attention to these is necessary from the start.

Fourth is the tendency to substitute technology for engineering. Especially during times of rapid growth, being first to deliver some new capability seems to be held in higher esteem than demonstrating a capability over the long haul. The technological leap, even if short-lived, is viewed as intrinsically good. One needs to be wary of the technological imperative, that drive to do things because we can do them, to use technology because it is there. The danger is that in the headlong rush to apply technology we lose track of the real program objectives. When this loss of perspective occurs, the technology tends to become the objective. Not only does the real problem not get solved, but worse, the energy that might have been applied to doing so is spent instead on assuring the technological lamp is shining with sufficient brilliance. It is important to find the discipline to assert that the technology is not the mission. Unfortunately, there is a tendency to try to substitute technology for engineering, to declare that it is good, and move on. This mind set leads one to presume the solution and try to build the product on it, instead of figuring out the solution and trying to engineer the system. And the larger the system, and the more demanding its performance and reliability requirements, the more essential that engineering discipline is to delivering a quality product.

Finally, there is the matter of unenlightened contracting. The myth of fixed price contracting for complex software systems seems at last to be understood as an unworkable idea. However, there are many more subtle issues that remain. Control of subcontracts is almost always a problem, especially as regards control of development processes and management of requirements. The use of COTS software components as preferable to new development raises new issues of rights in data for the third party software. The trend toward customer/developer teaming to get the system built through shared participation on Integrated Process and Product Development teams institutionalizes the risk-sharing that was a reality anyway. However, contract management paradigms are not keeping up with such trends.

It is also the case that outmoded product support mechanisms need to be replaced with ones that reflect the use of off-the-shelf software components. Off-the-shelf products tend to have a very short lifetime in terms of support from the original producer. It is quite possible for a large complex software system to incorporate a few small third party software products on which the larger system is strongly dependent. If those products disappear from the market during the twenty-five year life of the large system, what is to be done to assure its continuing maintainability? In fact, the problem is deeper than that. If even the version of the product produced for inclusion in the larger system is no longer supported by its manufacturer, then there are hard choices to be made by the large system developer. It is clear that we ought not proceed so quickly with new development approaches that we outpace the availability of acquisition and support practices necessary to get the job done.

With this discussion of underlying causes in hand, we turn now to some lessons learned on past submarine combat system development projects. Sadly, many of the lessons have had to be learned numerous times despite their having been previously reported. We shall recite here briefly some of the most significant lessons learned and map them to the fundamental causes cited.

3.1 FIXED PRICE CONTRACTS

First, there is the area of contracts. Firm fixed price contracts are appropriate for many acquisitions. In particular, they are appropriate when the cost of the product to be acquired is reasonably straight-forward to estimate with good confidence. This includes cases where the marketplace is mature and potential vendors have long experience in producing like systems. Also, they are appropriate where requirements are not likely to change significantly during the development period and are known with good certainty at the time the invitation for bids is advertised. They are much less satisfactory on projects of huge scale and complexity with long development cycles that span several technology turnovers. There just aren't enough of such projects for a good database of applicable experiences to exist on which to predicate a credible cost estimate. Recent trends show an appreciation for the superiority of "cost plus" type contracts for large complex software systems.

3.2 SUBCONTRACTING

Also, most contracts of the scope of the systems under discussion will involve subcontracting relationships between the prime contractor and subordinate companies hired to supply subsystems or specialized equipment or experience. It is important to review subcontractor agreements against the prime contract for contractual consistency. Although you cannot control how your prime contractor writes subcontracts, you can review, and be prepared for, the potential problems arising from the subcontracting relationships. It is likewise important to maximize the prime contractor's control of subcontractors. Insist that subcontractor representatives attend major meetings, and require them to provide briefings. This provides visibility into subcontractor progress, but it also emphasizes the need for close coordination of all participating developers.

3.3 PLANNING FOR COTS PRODUCTS

The current acquisition paradigm favors the incorporation of commercial off-the-shelf (COTS) products over new development. Unless all licenses and other agreements concerning such products are identified, the program budget will almost certainly not account for them. Consequently, it is essential to identify all commercial products and licensing agreements early.

3.4 SYSTEM ENGINEERING

With multiple development teams proceeding in parallel on various components, it is easy for everyone to lose the big picture. A lesson learned on many large developments is that the system is more than the sum of its components. The contract should encourage a continuing system perspective throughout the development. Insist that the contractor model the system performance up front and maintain the model throughout the development. We have also found it useful for the government Technical Direction Agent (TDA) to run an independent model to validate the results.

3.5 QUALITY ASSURANCE

In the realm of quality assurance (QA), combat system acquisition programs have imposed a requirement for a separate reporting chain for the contractor's QA organization. We found this to be beneficial and important to keeping programmatic considerations like cost and schedule from affecting the QA group's ability to do its job and to protect its budget, especially under the strains of a fixed cost contract. It is also useful for the customer to establish a relationship with the head of the QA organization and promote the importance of the QA role. It is highly beneficial for the traditional QA focus on products to be broadened to include consideration of processes as well.

3.6 SOFTWARE METRICS

If you don't ask contractually for software metrics data, you won't get it. The one thing worse than this is asking for data that you don't have a plan to use. It is important to have a metrics approach based on program issues that identifies what metrics are required to effectively assess the contractor's progress against his plan. Include in the contract only the delivery of metrics data that you need for planned analysis and reporting. Also, control the data that you receive, and restrict its visibility to program representatives only.

3.7 DOCUMENTATION

On large systems, it is especially important to require standardized "Style Guides" for documentation produced across multiple developer sites. Our experience has been good with style guides for code (assembly and high order language) and also for requirements and design documents. Review of the products and maintenance of the delivered system is greatly facilitated by consistent standardized styles. Related to this is the issue of electronic delivery of documentation products on large systems. It is not unusual for a combat system's Interface Requirements Specification to exceed 77,000 pages. Electronic format is a good idea as an optional delivery medium to specify in the contract.

3.8 SOFTWARE PLANS

There are also process-oriented lessons. Early baselining and continual update of plans is crucial, particularly the Software Development Plan and System Integration Plan. While changes in these plans are to be expected as progress reveals weaknesses or inefficiencies, the important point is to view the plans as living documents to be used during development rather than as deliverables per se. In other words, having a plan is not nearly as important as using it.

3.9 INFRASTRUCTURE SOFTWARE

Another key to success is to identify high risk requirements up front, invest in them early, and closely manage critical path software items. In a large, real-time, distributed system, it is essential for the infrastructure software, or Combat System Services as we call it, to be stabilized at the earliest possible date. Those resource management, operating system, database management, network communications, graphics generation, and like functions are a key to progress on the mission application functions. Without them, no realistic testing can be done. Also, parallel development of infrastructure and applications software runs the risk that different application development teams may be building to different versions of the infrastructure software, with a result that rework may be necessary before integration begins.

3.10 SIMULATION/STIMULATION

A related issue is the development of test harnesses and simulation/stimulation systems to support system testing. For much of the functional and performance testing these must be in place, and not just built but validated. Investment in this support structure must not be either underestimated or delayed.

3.11 SOFTWARE DEVELOPMENT ENVIRONMENT

Despite all the arguments for greater efficiency when using a support environment and development tools that a subcontractor is most familiar with, experience shows that you're better off mandating that the developer identify a standard set of support tools for use across the whole development team. The penalty for ignoring this is a miserable integration experience with much rework and finger-pointing. Make a single standard development environment a contractual requirement and enforce it. This means everybody uses not just the same tool, but the same version of the same tool!

3.12 PLANNING FOR COTS COMPONENT SUPPORT

There are hidden risks associated with the use of COTS tools especially on long developments. It is no less fatal to the program to have a vendor stop supporting your COTS development and integration environment than it is to have a computer chip manufacturer go out of business. Rolling over to a new version of a COTS product will eat up money and time that probably hasn't been budgeted, and may not even be possible. It's also the case that for very large applications, the tools may need some customization to cope with the size of the software being manipulated. The solution is to find out how much a support agreement will cost up front (it probably won't be cheap), budget for it, and make the deal.

3.13 AUTOMATED TOOLS

Finally, it pays to invest in automated tools for assessing software quality. Much of the assessment necessary to assure the maintainability of the product is repetitive in nature and well suited to automation, including, for example, tools to check adherence to standards and style guides. The larger the application, the bigger the payoff in using automated tools.

All of the lessons learned on recent combat system development projects (whether the experience was positive or negative in nature) are traceable to one or more of the five primary issues identified above. Table 1 maps the observations made here to the five primary underlying issues.

4. Best Practices

For the past three years there has been increased attention to the challenges of producing large complex software systems. Within the U. S. Department of Defense there have been several notable developments. A Software Best Practices Initiative was launched in 1994 to identify, document, and promulgate those practices that characterize successful software acquisition programs. In a separate development initiated in the same time frame, the Joint Logistics Commanders' Joint Group on Systems Engineering began work on a practical software measurement approach that would provide an effective framework for measuring software projects. Finally, in recognition of the superiority of standardizing on

Table 1. Mapping of Observations to Primary Issues

AREA OF OBSERVATION	UNDERLYING ISSUES
Fixed Price Contracts	Contracting
Subcontracting	Contracting
Planning for COTS Products	Contracting
System Engineering	Scale, Nature, Architecture
Quality Assurance	Scale, Nature
Software Metrics	Contracting, Scale
Documentation	Scale, Nature
Software Plans	Nature, Contracting
Infrastructure Software	Architecture
Simulation/Stimulation	Nature
Software Development Environment	Contracting, Technology, Scale
Planning for COTS Support	Technology
Automated Tools	Scale

interfaces rather than on implementations, several important steps were taken to advance the open system architecture approach to system acquisition. These included the Navy's Next Generation Computer Resources (NGCR) program, and a Department of Defense effort to define a Technical Architecture for Information Management (TAFIM). We will consider next how these efforts have contributed to the establishment of an improved approach for managing the risks associated with large complex software system development projects.

4.1 SOFTWARE ACQUISITION BEST PRACTICES INITIATIVE

The Software Acquisition Best Practices Initiative was established in July 1994 by the Department of Defense (DoD) to improve and restructure the DoD software acquisition management process based on identification of practices used by successful software projects in both government and industry. Dozens of experts from industry and government participated in the initiative, which held workshops to assess and refine proposed best practices. Seven separate Issue Panels addressed practices suggested for:

- Risk Management
- Planning and Baselining
- Program Visibility
- Program Control
- Engineering Practices and Culture
- Process Improvement
- Solicitation and Contracting

Details of the practices selected can be found in The Program Manager's Guide to Software Acquisition Best Practices [1]. For our purposes here it will suffice to note the nine Principal Best Practices identified by the so-called Airlie Software Council of software experts and industry leaders which was convened under the initiative. It is believed that these practices, if implemented, will improve software development and maintenance productivity and quality, reduce cost, and improve user satisfaction.

4.1.1 *Formal Risk Management*

A formal risk management process requires corporate acceptance of risk as a major consideration for software program management, commitment of program resources, and formal methods for identifying, monitoring, and managing risk.

4.1.2 *Agreement on Interfaces*

A baseline interface must be agreed upon before the beginning of implementation activities, and the user interface must be made and maintained as an integral part of the system specification. For those projects developing both hardware and software, a separate software specification must be written with an explicit and complete interface description.

4.1.3 *Peer Reviews (Inspections, Walkthroughs, etc.)*

Reviews should be conducted on requirements, architecture, designs at all levels, code prior to unit test, and on test plans.

4.1.4 *Metric-Based Scheduling and Management*

Cost and schedule estimates should be based on empirical data. Metric-based planning requires early calculation of size, projection of cost and schedules from empirical patterns, and tracking of project status through the use of captured result metrics.

4.1.5 *Binary Quality Gates*

Completion of each task in the lowest level activity network needs to be defined by an objective binary indication. These completion events should be in the form of gates that assess either the quality of the products produced, or the adequacy and completeness of the finished process.

4.1.6 *Program-Wide Visibility of Project Plan and Progress Vs. Plan*

The core indicators of project health or dysfunction should be made readily available to all project participants. Anonymous channel feedback should be encouraged to enable bad news to move up and down the project hierarchy.

4.1.7 *Defect Tracking Against Quality Targets*
Defects should be tracked formally at each project phase. Configuration Management (CM) enables each defect to be recorded and traced through to removal. In this approach there is no such thing as a private defect, that is, one detected and removed without being recorded.

4.1.8 *Configuration Management*
Configuration management is an integrated process for identifying, documenting, monitoring, evaluating, controlling, and approving all changes made during the life cycle of the program for information that is shared by more than one individual. The discipline of CM is vital to the success of any software effort.

4.1.9 *People-Aware Management Accountability*
Management must be accountable for staffing qualified people (those with domain knowledge and similar experience in previous successful projects) as well as for fostering an environment conducive to high morale and low voluntary staff turnover.

4.2 PRACTICAL SOFTWARE MEASUREMENT (PSM)

The Department of Defense (DoD) Joint Logistics Commanders Joint Group on Systems Engineering sponsored the PSM project in response to the need for more effective techniques for managing DoD software intensive programs. The result has been the publication of a guidebook for program managers, Practical Software Measurement: A Guide to Objective Program Insight [2], that has gained wide acceptance in government and industry.

PSM is based on actual software measurement experience on DoD and industry programs. The PSM guidebook represents the best measurement practices used within the software acquisition and engineering communities. It differs from many other measurement approaches in that it treats measurement as a flexible process, not a predefined list of graphs or reports. The process is adapted to address the specific software issues, objectives, and information requirements unique to a particular project.

The PSM process is defined by a set of nine best practices, called measurement principles. Experience has shown that a measurement process that adheres to these principles is more likely to succeed. The Measurement Principles are:
(1) Program issues and objectives drive the measurement requirements.
(2) The Developer's process defines how the software is actually measured.
(3) Collect and analyze low level data.

(4) Implement an independent analysis capability.
(5) Use a structured analysis process to trace the measures to the decisions.
(6) Interpret the measurement results in the context of other program information.
(7) Integrate software measurement into the program management process throughout the software life cycle.
(8) Use the measurement process as a basis for objective communications.
(9) Focus initially on single program analysis.

The basic premise underlying the PSM approach is that measures should be selected and organized to track program-specific issues. An issue is anything that might affect the achievement of program objectives. Issues include risks, constraints, and any other concerns. Experience has shown that some issues are common to almost all programs, and the PSM guidebook elaborates six of these as a starting point from which related issues specific to a given program can be recognized. The six common software issues are:

(1) *Schedule and Progress* - This issue relates to the completion of major milestones and individual work units. A program that falls behind schedule can usually only make it up by eliminating functionality or sacrificing quality.
(2) *Resources and Cost* - This issue relates to the balance between the work to be performed and personnel resources assigned to the program. A program that exceeds the budgeted effort usually can recover only by reducing software functionality or sacrificing quality.
(3) *Growth and Stability* - This issue relates to the stability of the functionality or capability required of the software. It also relates to the volume of software delivered to provide the required capability. Stability includes changes in scope or quantity. An increase in software size usually requires increasing the applied resources or extending the program schedule.
(4) *Product Quality* - This issue relates to the ability of the delivered product to support the user's needs without failure. Once a poor quality product is delivered, the burden of making it work usually falls on the software support organization.
(5) *Development Performance* - This issue relates to the capability of the developer relative to the program needs. A developer with a poor software development process or low productivity may have difficulty meeting aggressive schedule and cost plans.
(6) *Technical Adequacy* - This issue relates to the viability of the proposed technical approach. It includes features such as software reuse, use of COTS software and components, and reliance on advanced software development processes. Cost increases and schedule delays may result if key elements of the proposed technical approach are not achieved.

If you are not tracking these common issues then you are probably not managing all of your program risks.

The PSM guidebook also provides assistance in selecting from a proven set of measures appropriate ones based on a prioritized list of program issues and questions. It further provides help in specifying associated data and implementation requirements and provides examples of how measurement indicators can be defined, analyzed, and visually depicted.

4.3 OPEN SYSTEMS ARCHITECTURE

The emergence of open architectures as a pre-eminent concern in recent years has had the beneficial effect of focusing attention on architecture in general. Architecture's significance has long been misunderstood and under-appreciated.

The architecture embodies the compatibility standards that permit successful assembly of the system from its parts. It is the case that architecture impacts, indeed constrains, design. This means that architecture must be defined first in order for it to be useful.

Of course, *any* system implementation will reflect *some* architecture. But it is far superior to choose architecture with deliberation at the outset than to accept whatever architecture evolves during development. The useful architecture lays down the principles on which development will be based. It thus constrains subsequent design activity. This is not bad. In fact, it is necessary if you intend to achieve a system that exhibits desired levels of such qualities as modularity, openness, resilience, scalability, operability, affordability, insensitivity to (tolerance for) technological or mission changes, understandability, and performance, to name a few important characteristics. When a system is finally delivered it will, per force, exhibit some amount of success in delivering the desired qualities. But at that point (i.e. at delivery) the degree of success is a mostly untested promise. The bottom line is that architecture is something that should be intended and prescribed, and not something that should be discovered and lived with. The role of the architect is not to unearth and repair bad design decisions. It is to elaborate goal-directed design principles that result in good design decisions that yield an effective system.

Thus, the interest in open architectures provides welcome attention to an important area. The motivation for the interest is the need for improved interoperability of major systems and the desire for portable, reusable software systems as an aid in containing the cost of large software development efforts.

The key to open systems is interface standardization which is broadly accepted by industry. In the U. S. Navy the first serious moves in this direction were made by the Next Generation Computer Resources (NGCR) program over the past five years. The NGCR program brought together industry and government experts to seek consensus on robust, forward-looking interface standards for local area networks, backplane buses, operating systems, database management systems, graphical user interfaces, and project support environments.

Subsequent to that activity, a separate DoD effort established a standard framework, or model, to serve as a basis on which to construct program specific architectures. The Technical Architecture for Information Management (TAFIM) seeks to promote open systems through a common set of adopted standards. It reduces the likelihood of building vendor-specific proprietary interfaces between systems. Details on the TAFIM are available in [3].

While these efforts, and others like them, have made a significant positive impact on our ability to construct large software systems in an effective way, there is more left to do. There remains much misinformation about open architectures. Many managers still believe that the mere incorporation of commercially available software products in a system automatically results in an open system. Others have the misguided view that if all systems are required to have a set of common modules or components, then you will have achieved an open system and the interoperability sought from having such a system. While the interoperability might well be achieved through this approach, at least for a while, there is nothing open about an architecture that requires all cooperating systems to be built from the same components. The essence of openness is support for diversity. This means independence from reliance on any given vendor, any given product, or any given technology.

The first key to successful use of off-the-shelf software components is being able to integrate them easily into your existing system. The second key to success is being able to extract and dispose of them when something better comes along. Open architecture based on adherence to widely accepted commercial interface standards is what makes this work.

4.4 OTHER EFFECTIVE PRACTICES

For systems with high reliability requirements, there is no such thing as too much testing. However, exhaustive testing is also not a feasible option for large complex systems. A set of principles regarding product validation has evolved from years of working with submarine combat systems. These apply to the government as customer and technical oversight agent for complex software projects. The product validation principles are:

(1) Develop and maintain a Product Evolution Plan. The Evolution Plan documents the build sequence for each component product for both interim and final delivery, and defines the interrelationships among products.
(2) Provide a critical path process. Prepare and maintain detailed schedules and documentation that describe system requirements, milestones, critical paths, interdependencies, decision points, and metrics.
(3) Develop and employ a standard set of repeatable regression tests. These tests are used to confirm that hardware and software operate according to requirements and provide quantitative pass/fail criteria for making such a determination.
(4) Conduct pre-certification testing with identical software, hardware, and procedures for all certifications that are to be performed by external organizations. No important external test event should occur without having been previously tried successfully internal to the project.
(5) All equipment and software must be validated within project test facilities prior to delivery to external sites, including the end user sites.
(6) Provide an independent validation process. Employ "red team" reviews prior to all critical reviews for the project and as a part of all major program milestones.
(7) Ensure that all requirements have been successfully validated.

Related to the testing issue is the matter of operability. The resulting system will not be fit for its intended purpose if those who are to use it cannot operate it effectively. The answer is early and continuing end user involvement in the development. Conduct a series of early operational assessments by the end users. A "concept of operation exercise" involving the end users and employing scripted scenarios, mockups of equipments, and conceptual displays is essential before any real development is begun.

5. Conclusion

There is no easy solution to complex problems, and the challenges posed by large software intensive projects are immense. Many observations have been reported here, both as regards the nature of the problems and also the best practices for addressing them.

Summing up in an effective manner is difficult in its own way. Perhaps it is best to reflect on the main issues once more. Software systems are essentially different from hardware ones and the wise project manager will plan with the differences in mind. The vast size of large projects demands that special attention be paid to tooling for the development and to effective partitioning of the work for parallel development, and to the consequent integration task at the end. The integration is almost always more challenging than anticipated. The scheduling of the development should be such as to focus on architecture first, with special attention to development of interface profiles that support an open architecture. The infrastructure should be built before the application modules that depend on it. There is no "silver bullet" substitute for engineering. Avoid the temptation to add technology that you don't need. Recognize that the end product will not be the 100% solution and plan for the product to evolve over time. Finally, avoid the checklist mentality in customer/developer relationships. The truth is that on large software projects the customer and the developer will be sharing the risks. Early recognition of this fact will provide the opportunity for an effective measurement program to be defined that tracks the issues relevant to the project and gives the managers on both sides the best opportunity for success.

References

1. Software Program Managers Network (1995) The Program Manager's Guide to Software Acquisition Best Practices, available by email request to best@spmn.com or via World Wide Web Home Page at URL http://spmn.com/.
2. McGarry, J. et al (1996) Practical Software Measurement: A Guide to Objective Program Insight, Joint Logistics Commanders Joint Group on Systems Engineering, available by email request to mcgarry@ada.npt.nuwc.navy.mil or via World Wide Web Home Page at URL http://www.cards.com/psm/.
3. Defense Information Systems Agency Center for Standards (1995) Department of Defense Technical Architecture Framework For Information Management, available via World Wide Web Home Page at URL http://www.itsi.disa.mil/.

OPEN DISCUSSION: SOFTWARE (and) PROJECTS

Discussion chaired by:
A. TSVIRCUN
Institute of Problems of Control
Russia

This summary by:
T.M. WILLIAMS
University of Strathclyde,
Glasgow, UK

This brief part of the Workshop revealed a difference in interest between the two sets of participants. Although much interest was shown by NATO-country participants in *methods for managing large software projects*, there was more interest from the CP participants in *software for managing large projects*. This summary of the discussion deals only with the latter aspect, the former being dealt with in smaller discussions outside of the formal Workshop. The reasons for this difference can only be speculated upon, but it may be that on the one hand PM software is much less available in CP countries (reference is made below to the difficulty in production of Russian-language PM software), and on the other hand there is possibly less experience in CP countries of the type of complex software project described by Mr. Conrad above.

Dr. Tvirscun introduced the discussion by looking at PM in CP countries. The experience of managing big projects has existed for a long time, and there has been little new in the last few years. What needs to be taken into account in the new situation? Previously, project managers didn't have experience of calculating profits. Even widely-used financial terms are often not understood by managers. In the West such terminology as "auctions" and "joint ventures" have been known for a long time, but they have been known in CP countries only in the last 4-5 years. Such ideas change the way projects are chosen and carried out. Another difficulty in CP countries is that people who make decisions now who occupy high positions were educated 10-15 years ago; they are new to these problems, and don't know how to solve them and what to do (note the two papers above by Voropayev (Russia) and Romenko (Ukraine)). So 4 years ago Dr. Tvirscun and his colleagues started work on understanding and applying the new techniques. The Institute of Management and Problematics in the Academy of Science of Russia has developed a suite of programs allowing the user quickly to calculate the main project characteristics, taking into account factors such a inflation and different forms of ownership, and to calculate the sensitivity of the project to various parameters. These programs have been used for a number of years on projects from $1M-$50M; they are also used by banks and consulting firms (this is itself a new phrase in Russia - they need more information about how such consultancy works, what contracts are made, who is responsible for decisions, and how effective these activities are). They have found mistakes that are often made by managers inexperienced in free-market financial concepts. Other areas which need exploring are complex programmes of many interacting projects; uniting bank capital and industrial property; and finance-industrial groups.

Prof. Bushuyev said he had been working on software since 1965, and there has been a lot of software produced. Although a competitive group was organised to develop software, 5 years ago the competition in software in the market destroyed the Russian and Ukrainian markets completely, so he had decided to integrate into the world system, worked within IPMA (the International Project Management Association) and finally made contact with Primavera Inc. However, programs are always at risk in CP countries, since the demand is great but no-one actually wants to pay (he quoted his work producing a Russian-language version of SureTrak). Such projects will always fail while there is no world-wide system of software copyright protection. Furthermore, for PM software to be successful, it is necessary to have mutual understanding and commonality of culture between the countries, which does not exist between East and West (or even between America and Britain!); this is exacerbated by the lack of a mutual programming language. Finally, technology today provides a unique capability for integration of packages and different products; this gives us new programmes and new software, and often project managers don't take full advantage of such packages since they do not see all the abilities of IT.

HUMAN FACTORS IN PLANNING AND MANAGING COMPLEX INFRASTRUCTURE PROJECTS-A Singapore Experience

K.T. YEO
Division of Systems and Engineering Management
School of Mechanical and Production Engineering
Nanyang Technological University
Nanyang Avenue, Singapore 639798

This paper gives focus to human factors in the planning and managing of large and complex infrastructure projects. Singapore's experience and her long term ambition to create a world class transport system for the island state will be of relevant interest. Three levels of human factors are discussed relating to the role of political leadership and environment, strategic planning and organisational learning, and enlightened project management. Concepts and ideas of systemic thinking, learning organisation, and TQM are explored as potentially powerful sources of ideas and mental frames of reference to deal with problems of complexity in large-scale system planning and implementation. The challenges and lessons for future project management of complex projects will hinge on human learning in both hard systematic approach and soft human factors. Special lessons, challenges and problems faced by project management in a newly industrialised economy can be also of interest to other recently reformed economies.

1. Introduction

This paper discusses some relevant experience in major infrastructure construction in Singapore which will be used to illustrate aspects of human factors in planning and managing complex projects. Three levels of human factors and their environment will be discussed namely human factors relating to: the political leadership and environment, project planning and learning, and enlightened project management.

In the course of discussion, relevant systemic concepts and perspectives will be incorporated to reinforce the ideas of human factors and its contribution to major

T. M. Williams (ed.), Managing and Modelling Complex Projects, 147–163.

project successes. Singapore's experience to date to build a world class land transport system will be highlighted for illustrative purposes.

2. Toward a World Class Transport System

The mission to build a world class land transport system is under the charge of a newly established government agency, Land Transport Authority (LTA). The vision is to provide such a world transport class system that meets the needs and demands of a dynamic and growing global city with a population that will increasingly expect high standards in service and infrastructure. The success of the overall transport system hinges on providing a high quality service that is convenient, accessible, comfortable, safe, speedy, and affordable to the majority of the community [1].

The master plan is a comprehensive integrated public transport system to provide commuters with a 'seamless' journey. The public will be provided with the choice of one or more of the following modes of transport:

- A Mass Rapid Transit (MRT) system to serve as heavy transit corridors,
- A Light Rail Transit (LRT) system to serve as feeders to the MRT network
- Fleets of large capacity buses to serve the less heavy routes to complement the MRT-LRT integrated network,
- Premier buses to provide higher grader services, mainly serving the business districts, and
- Fleets of taxis to provide car-like services.

In addition, an electronic road pricing system will be implemented to regulate the road usage. The successful implementation of the integrated system depends on the Authority's success in tackling a number of key issues pertaining to Singapore's geographic and land space conditions:

- Integrating transport and land use in a land-scarce island state
- Expanding the existing network of roads and expressways, and
- Managing proactively the increased demand of road usage, and
- Providing a range of complementary and cost effective public transport choices.

The initial phases of a world class MRT system have been a great success; and the system has already won international acclaim for its efficiency and performance. Experiences learned from the construction of the MRT system will be used to illustrate the importance of human factors in project success [2].

The Singapore MRT system's construction cost was one of the lowest in the world. The initial phases of the system constructed towards the end of 1980s were built at 41M US$/km of track. The cost for Taipei's MRT system was 210M US$/km, and the cost of Hong Kong's planned airport link was projected at 123M US$/km. Similar systems in Tokyo, and Washington DC, cost 186M US$/km and 116M US$/km, respectively [3].

3. Human Factors as in Political Leadership & Environment

3.1 A POLITICAL WILL TO SUCCEED

For the success of major infrastructure construction, government commitment and support at the highest level are vital. A combined strong political leadership and a political will to succeed have proved to be one of most critical success factors in the infrastructure development in Singapore. For all the infrastructure projects, the Singapore government has been the sponsor, champion, owner and project manager. Any potential bureaucratic inefficiency was overcome by such 'political will' within the government leadership who had the authority to decide and capability to effect the project implementation. The government in turn empowered the project planning and implementation teams with full project authority to efficiently plan, design, and manage all technical, financial, legal and contractual matters.

3.2 POLITICAL STABILITY

The strong political leadership was also reinforced by a stable government and finely tuned civil service. The continuity of the political leadership and government provides a conducive environment for major infrastructure construction. The project team can plan and implement the project with greater efficiency and confidence. The effect of a strong political leadership and a stable government, and backed by strong national reserves had filtered down and led to a concerted effort by all who were involved. And all wanted the project to succeed. History showed that in some parts of the world, major infrastructure construction were in jeopardy whenever there is a change of government, political agenda and development priority, or due to a lack of funds. The project became a so called "political football".

3.3 CLEAR AND CREDIBLE PROJECT OBJECTIVES

The objective to solve a long-term transport problem in the land-scarce island of a mere 630 km^2, and with a growing population of 3.1 million, was clear and high on the government's political agenda. Without a highly efficient public transport system in place, the government's policy and effort to control private car ownership would be in jeopardy.

The rate of car ownership has grown tremendously in the recent decades, resulting from the rapid economic growth and rising income of its population. The present car to population ratio is 1 to 10, even with the enforcement of a strict demand control policy over private car ownership through a Certificate of Entitlement (COE) bidding scheme. The bidding price of the certificate can be as high as US$ 40-50K for top range models. By 2010, the forecast car-to-population ratio is 1 to 7. There is a sense of urgency to build a highly efficient world class transport system to satisfy its people's demands and long-term economic development needs.

4. Human Factors in Planning, Organising and Learning

A world class transport system will not come easy. However, the planning authority, LTA, strongly believes that achieving it is not impossible if they get the basics right, dare to try out bold and imaginative solutions and have the political will to carry them out [1]. Some of the 'basics' are in recognition of the constraints imposed by the scarcity of usable land space, and a clear objective of providing a viable and cost effective solution of national significance, and a comprehensive understanding of the long term requirements.

The role of human factors is closely linked to the value, perspectives and experience of the human actors. A systemic worldview and the practice of systems approach are advocated as potentially powerful to deal with the problems of complexity and uncertainty in a large project or system setting.

4.1 SYSTEMS THINKING

The strategic project planners must be capable of systemic thinking. A world class land transport system to support a human-vehicle traffic of 10 million trips per day, has to be perceived as a large-scale system. It will have many subsystems and components to be fully integrated as described in Section 2. It is truly a large and complex human activity centred system as described by Yeo [2].

For such a system, the metaphor of a living 'organism' is appropriate. The living and dynamic people-mover system is capable of growth and expansion, and requires continual rejuvenation to keep its vitality throughout its long and useful service life span. It is imperative that the planners take a large-scale living systems perspective when formulating the master plan.

The implementation of a large-scale infrastructure system is never an one-off project assignment. It involves continual, stage-by-stage, phase-by-phase, sustained planning, learning and implementation efforts. The planning and leaning intensity will be the highest during the pre-project planning, pre-commitment stages. Effective and dynamic systemic thinking inevitably involves a great deal of concentrated learning. The learning process is particularly necessary for Singapore as a newly industrialising economy, due to the lack of prior experience, the absence of a relevant and adequate mental model on large-scale construction of a modern and integrated land transport system. The availability of a relevant and adequate mental model will determine the intensity of learning. Such learning can take place at national, industry, organisational and individual levels.

4.2 CREATING A CONTROL STRUCTURE AND TEAM

The planning and control processes for the mass transit system took place within a global project organisation which included the relevant government agencies, and a specially established MRT Corporation. The latter acted as both the project manager and operator, and worked closely with a large number of international consultants and prime contractors during the peak period of implementation. It was a highly co-ordinated, motivated and committed global team. All constituent organisations, including the international construction firms, wanted the project to succeed since they all had a stake in the successful implementation of a highly prestigious world class project. Control of the project development was done in terms of tight but adequate budget allocation, stringent work schedule and performance standards. In the MRT project, the cost, time and quality objectives were all met and indeed, the final costs were well within the original budgets, which was a rare phenomenon for a large complex of this nature.

For the further implementation of a world class transport system, the Land Transport Authority (LTA) was created in September 1995, after a major re-organisation of several existing government agencies, the MRT Corporation, and part of the Public Work Department, to create a single entity to oversee the planning and implementation of the entire system. The Authority was granted broad authority and had the power to acquire land, divert utilities, and award multi-million dollar contracts.

4.3 PLANNING AS ORGANISATIONAL LEARNING

Purposive planning leading to final decision on a large-scale infrastructure development can be an arduous, protracted and iterative process. The process depends on the ability of the pre-project planning team in formulating viable alternatives and schemes. The planning team must be able to absorb information, learn about the project delivery process, and analyse relevant issues over the broad sphere of economic, technological, social and political environment; and make the right recommendations.

During the planning phase, the pre-project team was faced with the choice between a rail-based MRT system and an all-bus system. The ensuing fact finding, debating, and learning process has been a protracted and arduous one. Indeed, the decision on a MRT system has taken more than ten years of several phases of study. All the studies involved multi-disciplinary teams consisting of professional officers seconded by government and from international consultants. Similarly, in the recent planning and design of a LRT system, the project planning and design team has visited some of the best light rail systems in the developed countries to learn about their experience.

4.4 LEARNING TO BUILD MENTAL MODELS

The mental model is an idealised conceptual model constructed on the basis of one's past experience and knowledge, and currently available information and ideas. Such mental model shapes one's view of the world around him and his understanding of the issues in hand such as those concerning the tasks, the processes, the technology, the community, the environment, the culture and value when planning for major infrastructure development. When making a recommendation, a planner would simulate the potential outcome(s) using his implicit mental model.

Based on the soft system methodology (SSM), mental models can be continually revised, refined and improved in its predictive capability over the real-world situations, as the project planners or 'actors' continually learn from new information gathered, from insights generated through purposive debates and concentrated thinking and rethinking processes over critical issues and rearrangement of key relationships and priorities over a protracted period of time.

The mental model is a highly intellectual construct, which is necessarily bounded by the aggregate of past experience and knowledge [4]. Such mental models may be found to be inadequate due to incomplete information, a lack of prior experience, or if there is no genuine learning taking place on the part of the actors. The development of a set of adequate and relevant mental models is therefore extremely important to ensure the quality and accuracy of the plans.

The Singapore experiences have shown that its planners have invested tremendous amount of time and resources to the building up of such relevant and adequate mental models needed for the construction of a major infrastructure project. These actors are relatively young professionals in comparison to their counterparts in the advanced countries. However, they are able, determined and above all they have the capacity to learn and absorb quickly, and to construct the necessary mental maps that guide subsequent problem-solving and project implementation.

4.5 THE CATWOE ANALYSIS

The SSM uses a CATWOE analysis, relates and compares the real world situations or requirements with the available mental models, encourages purposeful debates and learning, and thus provides a potentially powerful framework for dealing with the problem of complexity and ambiguity.

The six C, A, T, W, O, & E elements provide a convenient framework for investigating the systemic reality of the infrastructure project proposal. For instance, the role of the government as the ultimate owner (O) of the project must be recognised. The owner has the power to initiate, alter, suspend, or terminate the project, and the authority to award billion dollars worth of contracts. The purchasing and bargaining power of the commuters who are the customers (C) and end-users must be accurately assessed to determine the level of demands and fares to be charged. It is important to appreciate that there will be a political cost to the government if the charges are perceived to be excessive by the community.

In dealing with a complex project situation, a major human factor consideration is the value or worldviews (W) of the stakeholders. The understanding and goodwill of the community are needed specially in consideration of routing and locations for stations; land and fixed assets acquisition to make way for the project; and for the noise, dusty and other inconveniences caused during the heavy construction periods. The effectiveness and efficiency of the administering bureaucracy, and availability of experienced contractors and major equipment suppliers, must also be carefully considered as these are likely to influence the cost, quality and length of negotiation and implementation.

The project team must have good knowledge over the transformation (T) processes over the entire project life cycle. This includes initial project conceptualisation, pre-project definition, negotiation, contracts award, to finally owning and operating the completed system. The whole process can take a great many years and it represents a long-term commitment. The long-term investment and commitment represent both opportunities and uncertainties. The latter may come from the external macro-environment (E) in terms of political and economic stability. Changes in government and development priority and policies; and long term economic growth prospects are potential sources of risks. Finally in the CATWOE analysis, it is the competence and capabilities of the human actors (A) which are the leadership and the team that will ensure the ultimate project success .

The CATWOE analysis gives special considerations to the impacts of human factors in terms of human value, perception, problem learning and solving capabilities, and in effective stakeholders management.

5. Human Factors in Enlightened Project Management

Jack Lemley[5], the former Chief Executive, Transmanche-Link stated that the management of human factors was central to the completion of the Channel Tunnel project. He considered human factors as equally if not more critical than the technological, materials and engineering factors. Lemley defined human factors in the context of a large-scale transfrontier project management, as those factors which are concerned about the complex of managerial and organisational skills operating through an empowered and highly educated, trained, motivated, safety and quality conscious work force. Such managerial and organisational skills are necessarily influenced by the project management's problem-solving capability and his emotional intelligence in dealing with and motivating a new work force.

An enlightened management and an empowered work force are complementary; and are both needed to ensure the mastery over and harnessing of the power of new technologies and materials for construction. The challenges facing the project management are many when dealing with large complex projects. Some of these, giving emphasis to education, training and learning, include the following:

1. Managing both radical and incremental changes in the work place
2. Need to continually train and empower the work force to master new skills
3. Managing new and advanced technologies and practices
4. Continual management education to acquire new knowledge and experience.

The core of the challenge is for the management to create a truly learning organisation made up of a network of dynamic, continually learning, and complementary teams. Such learning capability will be required to deal with the growing complexity in managing large complex and increasingly high-tech projects. The ideas of learning organisation hinge on institutionalised learning which requires the organisation to tap its people's commitment and capacity to learn at all levels in the organisation, so as to produce extraordinary results.

The enlightened project management should encompass the key disciplines in the learning organisation as defined by Senge [6]. These key disciplines are systems thinking, personal mastery, mental models, building shared vision, and team learning. Most of these disciplines have already been illustrated in the earlier sections.

5.1 TQM AND ENLIGHTENED MANAGEMENT

TQC (Total Quality Control) is core to the Japanese management approach. Many Japanese construction firms have used TQC to reform and renew their operations and the industry as a whole, since the 1980s [7]. The Western equivalence is TQM (Total Quality Management). Shiba [8] in his American TQM identified four pillars in TQM namely, customer focus, continuous improvement, total participation, and societal networking.

TQM is a mass movement focusing on serving the customers well. The TQM approach when applied to project management is to assure baseline performance in quality and delivery to the satisfaction of clients and ultimate end-users. Continuous improvement or 'Kaizen' in the process of project management is required which pay great attention to details to achieve such excellent performance. Kaizen is also a systematic process of problem identification, learning and problem solving.

Total participation is about promoting team working and team spirit. Societal network gives emphasis to societal learning which is systematic and purposeful learning from the network of companies, lead customers, suppliers and others who are trying to improve quality practices.

The ideas of learning organisation and TQM are closely linked, and based on the same basic need for institutionalised learning and improving problem solving capability especially when dealing with complexity. Enlightened project management can be achieved through TQM and learning organisation concepts and practices. In essence, the enlightened management is about understanding how human beings behave and how they can be motivated.

5.2 LOCAL PROJECT MANAGEMENT CHALLENGES

With the on-going and future large-scale projects in Singapore and the region, developers and institutional clients are now looking for project managers who are professionally qualified and experienced in managing complex projects to take charge of their projects.

To date, most of the major local construction projects are undertaken by international construction firms, mainly the Japanese and Korean. These companies have the track records and experience in terms of management and systems, in undertaking most of the major complex projects. However, they would invariably use their own nationals to occupy key project management and engineering roles. Local construction management personnel are often relegated to subordinate roles. The Japanese firms, by policy, are not willing to train the locals for senior management positions for fear that they may quit after becoming experienced [9].

The local project managers would have limited opportunities to build up the necessary and relevant mental models in handling world class projects without having the necessary hands-on experience. This represents a major weak link in the local construction industry and professional training. This is a dilemma facing the local construction industry and the government. Human factors at the political and strategic planning levels as discussed earlier, are not equally matched by human factors at the project delivery levels. For instance, the two project directors for the Singapore MRT system project were indeed, senior expatriate professional construction managers who were experienced in the construction of similar mass transit systems in Hong Kong and Atlanta respectively.

The local project managers are still yet to be enlightened, and to build up independent capability to take on major complex projects. This is a chicken-and-eggs situation. Without the track records, they will not be given the tasks to manage such projects. The limited domestic market will continue to deprive the local professional project managers from becoming world class. But, hopefully they can do well in the region, as the regional markets are growing rapidly and offering new opportunities. But the local professionals must be willing to learn and have the courage and risk-taking attitude when venturing overseas, learning from their counterparts in Japan and the West.

One of the strategies is for the local companies to pursue more aggressively in international collaboration by forming joint venture partnership with international construction firms to venture into third countries in the region. Such partnership must have the understanding that there will be genuine transfer of technology and management know-hows to the local partners, who must of course, have something to offer in return. These may take the forms of financial resources, local and regional connections and market knowledge.

This has been the wish for the local construction industry and the government that in future, major infrastructure projects including the future phases of the MRT and LRT systems can be undertaken by local project management. For local contractors, they can not expect to be good in every aspect of a large complex projects, but to concentrate on what they know best and what can be developed into a repeatable business. For instance, it does make sense for local companies to compete in tunnel projects, but they can do very well in providing electronics control systems for the MRT stations. Indeed, local companies have succeeded in securing such projects in Taipei mass rapid system, with the experience and track records they have from Singapore's own MRT system.

6. Conclusion

This paper has discussed human factors at three levels: first, at the highest political leadership level; second, at the strategic planning and learning level; and third, at the project construction management level. Aspects of Singapore's experience and aspiration to build a world class land transport system have been discussed and used as illustrative examples to highlight the relevance and significance of human factors in dealing with the planning and managing of complex infrastructure projects. Concepts and ideas of systemic thinking, learning organisation, and TQM have also been explored as potentially powerful sources of ideas and mental frames of reference when facing the problems of complexity in large-scale system planning and implementation.

References

1. Land Transport Authority, Singapore (1996), A world class land transport system - a White Paper, Jan., 1996.
2. Yeo, K T (1995) " Planning and learning in major infrastructure development: systems perspectives", International Journal of Project Management, Vol 13, No 5, pp. 287-293
3. The Economist (1993), 6-12 Nov 1993, pp 86.
4. Yeo, K T (1993) "Systems thinking and project management - time to reunite", International Journal of Project Management, Vol 11, No 2, May, pp 111-118.
5. Lemley, J K (1995) "Human Factors in the management of the Channel Tunnel project", Proceeding of the 1st International Conference on Construction Project Management, Jan., 1995, Singapore.
6. Senge, P M (1990) The fifth discipline- the art & practice of the learning organisation, Doubleday Currency.
7. Takeuchi, R (1995) "Case study on Ngee Ann City construction project management," Proceeding of the 1st International Conference on Construction Project Management, Jan., 1995, Singapore. pp. 317-325.
8. Shiba, S (1993) A new American TQM- four revolutions in management, Productivity Press, Portland.
9. Paulson, B C (1995) "International construction- cooperation and collaboration in a world of contrasts, constraints, competition and conflicts", Proceeding of the 1st International Conference on Construction Project Management, Jan., 1995, Singapore. pp. 1-10.

ADDENDUM to "HUMAN FACTORS IN PLANNING AND MANAGING COMPLEX INFRASTRUCTURE PROJECTS-A Singapore Experience"

Noted by the editor

Prof Yeo completed his presentation by noting some of his current research. This concerns the study of the Leadership Process as a function of four aspects:

- the leader: education of enlightened Project Managers, who are not just system engineers, but understand the softer aspects of human perceptions and values and the meaning of work; and are capable of dealing with many stakeholders; this is a new aspect of PM education not currently dealt with much in the current Bodies of Knowledge;
- the team: empowered, motivated and highly skilled teams;
- the task: this must be categorised and the task structure specified;
- situational (external and internal) variables.

Prof. Yeo mentioned his particular interest for future research in the Dialectics of Change, an area in which Eastern-bloc countries are perhaps expert.

OPEN DISCUSSION: HUMAN FACTORS

Discussion chaired by:
MARTIN BARNES
Coopers & Lybrand,
London WC2N 6NN, UK

This summary by:
T.M. WILLIAMS
University of Strathclyde,
Glasgow, UK

Dr. Barnes introduced the discussion by emphasising the importance of the "soft" human factors in the practice of PM. They are if anything *more* tangible than the "harder" aspects: the mathematics of PM are "soft" in the sense that they are difficult to apply, whereas the human aspects are "hard" in the sense that they are easy to understand and to apply. This is partly because there are not very many human factors: they are to do with looking at how people involved in PM react, and looking to see what works well and what does not. People have been managing projects for a long time, and there are many more case-studies about the success of the human techniques or their failures than there are written-up case-studies about techniques such as network analysis or risk assessment. He pointed to perhaps the first written-up PM case-study: the biblical account of the Exodus; this includes some powerful commentary about the Project Manager (Moses) and his ability as a Project Manager - and the account includes points about how to delegate around a project team and how to maintain motivation within the team; in fact, it is an excellent case-study to read about the timeless project management of the human factors. This very ancient case-study illustrates that if we become masters/mistresses of the human factors, we are likely to have real impact on our projects, in a way that the relatively short-term "modern" techniques will have less impact. Dr. Barnes also discussed the more local project to construct the city of St. Petersburg in the early 18th Century - then, a very new-technology project, combining defence, accommodation and transport aspects, and a management of change in administration and government, all carried through in less than 20 years, with no modern techniques such as bar-charts. (A Russian participant later stated that the building of St. Petersburg was not a project. A project has financial and profitability constraints, and a life-cycle. Many so-called projects in Russia have no financial calculations, such as the Belomor-Baltic canal which was carried out by prisoners. Besides, some of Peter the Great's work fell apart after his death, which suggets an unsuccessful project.)

Dr. Barnes asserted that modern management of the human factors in PM is a matter of technique in decision-making: decisions affecting the planning and execution of the project are the actual expression of all the things discussed in this Workshop. If those decisions lead to successful completion of the project, this implies that the PM was good. So everything discussed here can be tested by its impact on the quantity and quality of the decisions that the people make: whether we have the right people can be judged by whether they take the decisions properly, and the information flows we set up can be judged by how well they help the decisions that affect the remaining work. Education of Project Managers could be discussed in relation to delegation of decision-making down to small parts of projects, so that people get training in the right sort of decision-making before they become responsible for large parts of projects or for single projects themselves.

To show how the "soft" aspects can be treated in the same way as the "hard" aspects are treated, the same sort of analysis can be used to decide what sort of leadership works in PM. Dr. Barnes had concluded, after observation and experimentation, that the collaborative style of leadership worked whereas the authoritarian style did not, expect in the case of an "emergency" project where decisions have to be made in a crisis. Similarly in the case of contracts, the delegates as project managers should decide what is important to go into those contracts, making sure that they stimulate good communication, good decision-making, and unification of motivation around the team, even if some lawyers find it difficult to put such things into contracts.

Other participants agreed on the fundamental importance of the human factors, since humans are at the basis of project failure and (perhaps with a bit of luck and good timing!) at the basis of project success. Competancy is a function of knowledge, skill (the ability to apply that knowledge) and attitude; and this latter might be the most difficult to change.

Motivation is important. An example was quoted of using PM in Siberia on reconstruction work at the gold-mines. The problem was that there was insufficient repair capability. In order to prioritise, latest start-times were collected and the activity with the earliest latest-start-time was the priority. Not surprisingly, all the sub-managers from different project areas claimed very early latest-start-times.

Prof. Voropayev felt that mathematics and human factors were not comparable. Of course projects are made for people, are finished by people and are managed by people. But mathematics is the means by which to take decisions in cases where those decisions are very complicated and when we can't solve them without mathematics. Human factors are extremely important, and can dominate. But in complex projects we have very complex relationships among the participants, hence very complex systems for making decisions arise. But while the projects are becoming more complex, the demands to make decisions are increasing; for that we need adequate information, and this information should be given fluently. We have to react quickly to changes in this complex project which happen quite often in all spheres of the project; very often we have to take precise and sensible decisions and in this case we need mathematics. Of course we need some methods which can help deal with the quick stream of information and make sensible quick decisions, and we can do that only if corresponding models are present and we have the necessary technological instruments. So without diminishing the role of people in this system, we have to take into account technology, means and methods and to enhance them because they increase the importance of the human factor in fulfilling the project successfully.

However, Dr. Barnes claimed that if it could be ensured that people within a heirarchy made decisions in a way that was forward-looking, that considered risk and the motivations of others within the project, is focussed on the ultimate objectives of the project (and one or two other points), perhaps no complex mathematics would be needed to run a project other than the ability to do addition! - or at least, mathematicians should recognise the relevance of the human models in the management of projects. Where the more sophisticated sources of information are available to the managers, they are seldom used, and where they are used, they often do not produce better decision-making than decision-making having these human-oriented characteristics. He did agree that the hierarchy is important - but this is not a problem, rather it is the only way to manage complex projects, otherwise only the person at the top could make the decisions, there would be too many to make, and he would not be able to get all the necessary information. It is the hierarchy that divides the project into manageable packages of decisions; it does create the problem of managing the interfaces between those packages, but how to do that is known.

Some, in particular Prof. Elmaghraby (see Section 3), disagreed with Dr. Barnes' initial down-grading of mathematics in favour of the "soft" factors: the importance of the latter does not imply the lack of importance of the former.

So the relative importance of the human factors and the mathematical "hard" factors was not fully agreed upon, with the emphasis to some extent being different between the NATO-country and CP-country delegates, the latter placing generally more emphasis on the "hard", strictly quantitative, factors.

SECTION 5
CORPORATE STRUCTURES

Complex projects may be best managed within new types of corporate structures. This section aimed at looking at some new ideas for such structures. There are two papers. One is from the West looking at a joint venturing scheme, and the second is from a CP country considering extensively the differences between PM in a market and a transition economy. This, and a similar arrangement in the following Section 6, attempts to help compare the attitudes and methods in the West with those in the CP countries.

JOINT PROJECT MANAGEMENT COMPANIES

KLAUS PANNENBÄCKER
Managing Director of GABO mbH
Michael-Vogel-Str. 1 e, D-91052 Erlangen, Germany

The more different interested parties are involved in a project, the more we speak of large projects. "Large" means in this sense e.g.

- complexity of definitions of e.g. aims, qualities, benefits;
- diverse financing institutes and bank offices with their own regulations ;
- multi-national participants;
- specific controlling procedures.

Somebody has to sit in the middle and to moderate all the engaged parties: the project manager. He has to manage the project under joint venture conditions, that means, all partners are in the same rank.

1. Justification of Requirements ensures the Aims

Each project manager deals with a limited budget and time to reach the more or less well defined project quality. Quality is the measurable and measured aim, which is set up by the client based on his requirements and demands.

Each project manager always wishes to run his project without any influences from outside. On the other hand each client likes to control his project manager, especially in cases where he cannot define the project aim at the early beginning.

Therefore a competent project manager should be able to simulate the project aims through project implementation with the newest knowledge of "today". The client should justify the requirements of "today" as the same as the first project start-up.

T. M. Williams (ed.), Managing and Modelling Complex Projects, 167–172.

If there are changes in the requirements the project manager must follow by changing officially the aims too and creating a new way of reaching the revised aim. These procedures are the roots of claim management.

Figure 1 shows the interrelations between the parameters.

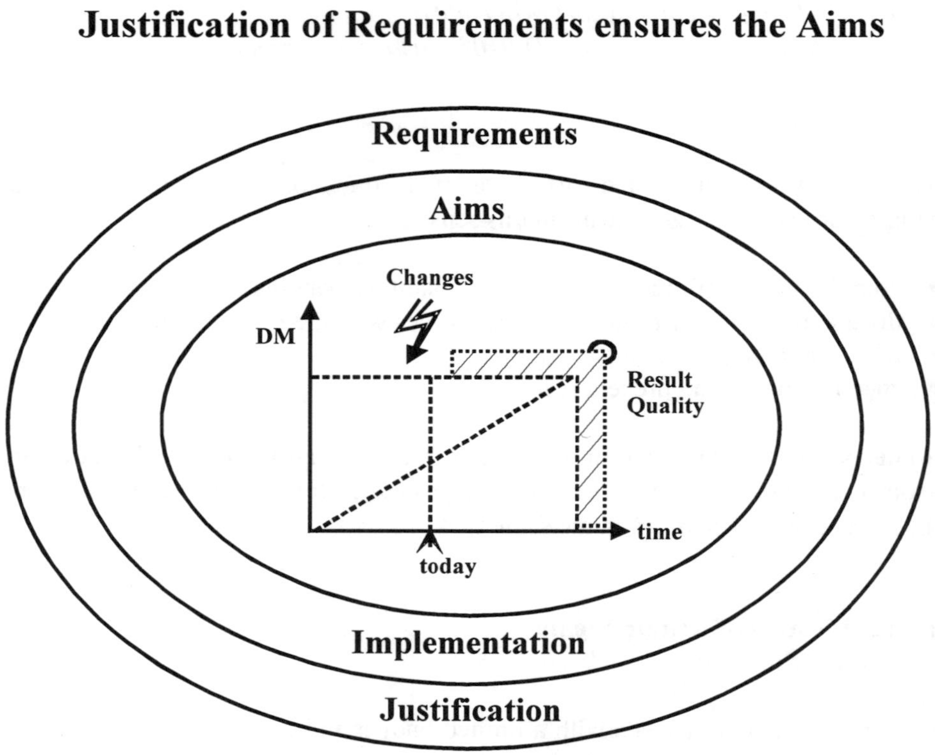

Figure 1

2. Project Joint Venture (Example)

In 1991 the American Army left with about 600 soldiers an area of about 1.2 sq.km. This area was more and more surrounded by the nearly located town in the last 45 years, so it was nearly in the middle of a new town district.

About 100 different bodies like companies, institutes, universities, housing and health organisations, public schools and kindergartens started applications to take over part of these buildings and installations. The town parliament was unable to approve the various proposals and applications because there was no plan for the future and any guidelines etc.

Of course the Owner was defined, but the German Ministry of Finance took over in this time more than 60 areas in East and West Germany from French, English, American and Russian Army, Navy and Air Force. Therefore the Minister was far away from my home town, where this project of revitalisation had to be started.

A Project Joint Venture was created with four major partners, the

- Owner with the representative of the Minister (hard to find him)
- Operator(s) who starts their business after finishing the reconstruction
- Sponsor(s) who finances the revitalisation projects
- Contractor(s) who does the reconstruction work.

But these bodies are partners, that means, independent with different interests and in consortial connection to each another.

Traditionally the Sponsor takes over the leadership, because he usually has the best lawyers. But it was decided to set up a "project table" in the middle on which all contracts had to be discussed and finally signed. The project manager is the moderator as well as the steerer and controller of the project managing parameters.

Figure 2 gives an overview of partners, interrelations, contracts and key tasks of the Project Joint Venture.

3. Project Joint Venture Management

The project manager defined in the starts up sessions the different project management phases as tasks for everybody in the project. He finally defined the "project language", so that everybody could communicate with everybody and without misunderstanding.

Figure 3 show the main project management tasks, which are the traditional ones in a set theory manner, readable from left to right.

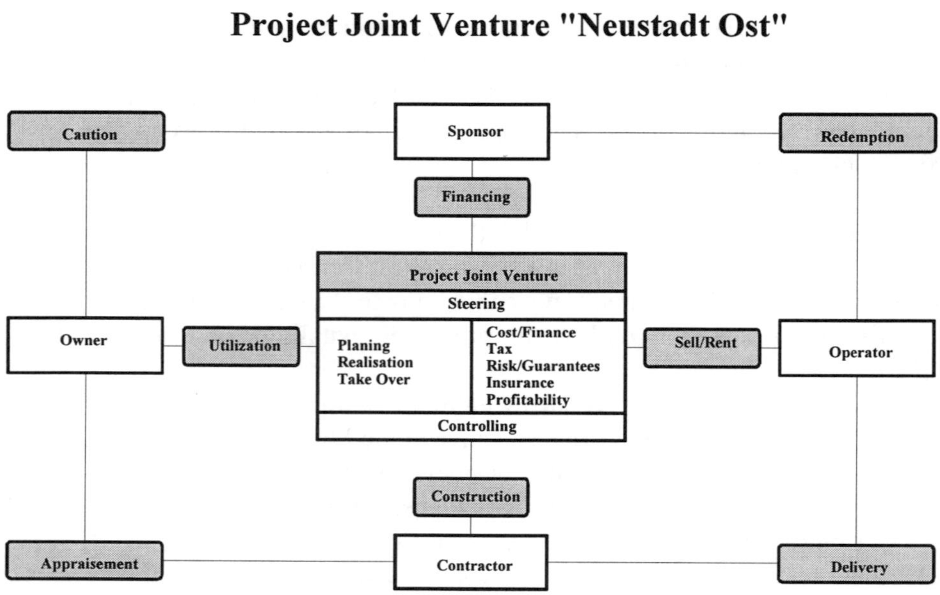

Figure 2

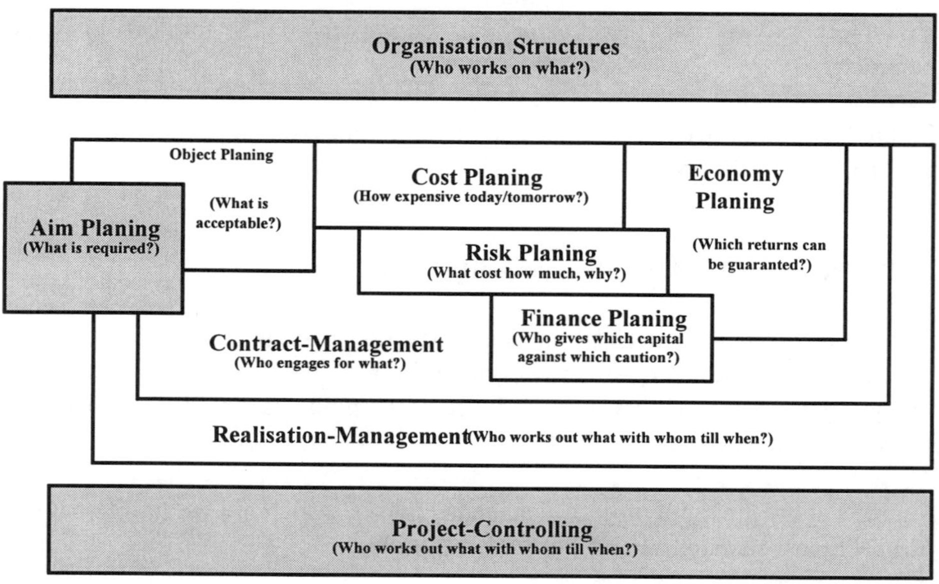

Figure 3

4. Organisation Structure (Who works on what)

The project manager of the Project Joint Venture reports directly to the Steering Committee. The Steering Committee with the Mayor as chairman is, with only two Division Managers, responsible to the Council. The Advisory Committee with representatives of the future operators works as "Consultant" to the Steering Committee.

In the Project Joint Venture as a legal body, permanent staff work as managers of subprojects and temporary staff of the Council Administration as well as of external experts for specific single tasks.

Figure 4 shows the basic organigram.

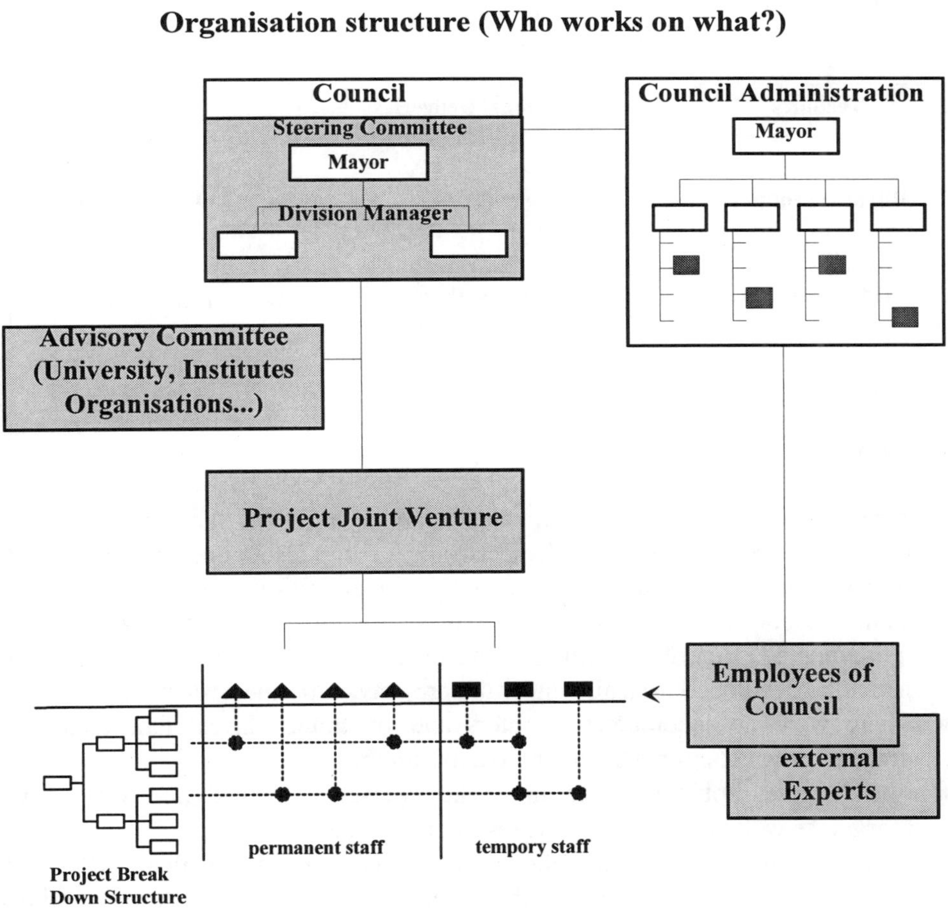

Figure 4

5. Development of Finance Schedule

Wherever a part of the complete revitalisation programme was started, the financial conditions and possibilities arose in the first discussions. After it had been agreed that the "finance schedule" is the last item in a (long) row of prior items, the project partner started to work efficiently. The work flow in figure 5 can also be used as a progress controlling scheme.

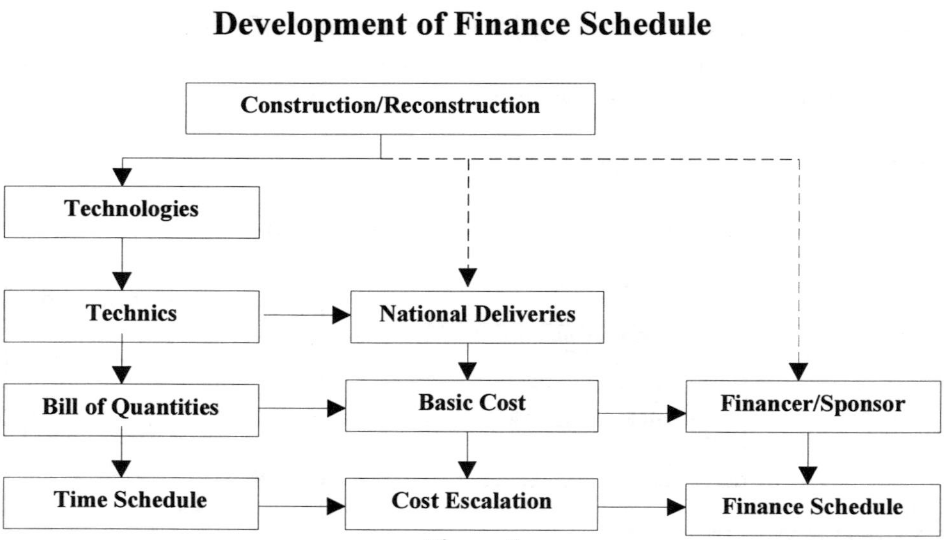

Figure 5

6. Summary

- After about one year of sometimes stressful discussions the partners spoke their own "project language". Project language implies an understandable and therefore accepted break down structure, but in a multi-dimensional way.
- Time, resources, cost, finance - all these traditional parameters could be implemented. But risk calculation, documentation and change management had to be developed basically under the project and partners' demands.
- There were no incompetent masters and frustrated slaves. Fair contracts dropped down claim management to a minimum.
- The biggest problems came up if key persons changed and wanted to reorganise well accepted and successful procedures.
- Turnkey contracts were only allowed if the risk was smaller than 10%. This was always hard to agree on by the sponsors, because they preferred turnkey contracts: the higher the risk the more finance repayment.

PROJECT MANAGEMENT GLOBALIZATION AND UNIFICATION FOR TRANSITION AND DEVELOPING ECONOMIES*

VLADIMIR VOROPAJEV
Vice president of IPMA
President of Russian Project Management Association
Moscow, Russia

1. Introduction

Modern Project Management (PM) understanding was formed in the 70s - 90s based on the experience of technically developed countries with relatively stable economic conditions. Among such countries we should mention: Great Britain, Germany, Holland, USA, Canada, Australia and Japan. The most complete description of PM understanding is given in collective works, connected with forming the PMBOK and some other similar research.: PMI [1,2]; INTERNET [3,4]; AIPM; ISO and so on.

In the present time project managers from developing and transition economy countries are actively joining the "World of Project Management". The movement for PM professional development bears now a global character. Two stable tendencies can be marked here:

- A growing number of firms and companies from developing countries involved in the international economic activity, including co-operation with developed countries.
- A growing number of investors and other project participants from developed countries engaged in projects and programs implemented in developing countries.

These processes show intensive integration of both the world economy and the whole world community. It is obvious, that this process requires development of a common language for business communication and creation of a professional business culture and joint creative activity. There is no doubt that modern PM can and, in our opinion, must be such a universal tool, which is confirmed by a long practice of its application.

---

* This paper is based on a presentation made on October 15, 1995 at the "PM Global Forum" in New Orleans, USA during the 26-th Annual PMI Conference

T. M. Williams (ed.), Managing and Modelling Complex Projects, 173–201.

Developing countries can choose between several ways of PM development and application:

- to take as a basis the experience of the North America - PMI [1,2];
- to take as a basis the experience of Europe - the former INTERNET (now IPMA) [3, 4];
- to choose orientation to neighbouring and other countries - Australia, Japan and so on;
- to develop its own approach to PM development and application basing on home experience, achievements, traditions and culture;
- to take the most advanced experience and knowledge of the world and develop the national PM basing on international co-operation on PM unification and globalization, taking into consideration and saving national traditions and culture.

The last way is much more preferable, efficient and practical for developing countries, as it allows them to overcome dragging behind the world level and avoid possible problems and mistakes.

For this purpose the world community should unite their efforts to solve the problems of PM globalization, unification and necessary international co-operation in this field. However, the problem stated gives rise to a number of questions:

1. What is the essence of developing countries' economies?
2. What are the principal differences between the social and economic systems of developing and developed countries from the PM point of view?
3. What is the influence of such differences on PM and its components?
4. Is it possible to form global PM understanding and recognition and if possible, then, to what extent?
5. Which PM spheres are subject to unification and which require taking into consideration special local cultural, economic, social, historical and other features?
6. By whom, when and for what money should PM global unification be realised?

It is obvious that solution to these problems requires a considerable period of time and concentration of efforts of all the world community of project managers. Here we only touch upon some of these problems and try to state our point of view and invite everybody concerned to further discussion.

2. Characteristics of the social and economic environment in developing countries

To our mind special features of the project environment in developing countries are mainly caused by changes connected with the dynamics of development. The possible influence of dynamic factors on PM depends on how deep and intensive the changes in the country and its economy are. The deepest large-scale and rapid changes have occurred since the early 90's in Russia, former USSR republics and countries of Eastern Europe. Therefore we make our statements mainly based on Russian data and experience, because they reflect general tendencies of modern world development and can be applied, to a certain extent, to all developing countries.

2.1. SOME PRELIMINARY COMMENTS AND STATEMENTS

I would agree with Dr. Martin Barnes [5] that nowadays all countries are developing and their division into developed and developing is just conditional and depends on the achieved level of development, intensity of development and level of social and economic environment stability. Proceeding from the above it is useful to emphasise 3 levels of changes in Earth's countries:

1. *Developed countries* - slow evolutionary development in stable social and economic environment (Great Britain, USA, Japan and so on).
2. *Developing countries* - rapid evolutionary development in relatively unstable social and economic environment (China, India, Indonesia and so on).
3. *Transition economy countries* - very rapid development with changes of revolutionary type in unstable social and economic environment (former USSR, Eastern Europe countries and so on).

Dynamic factors of development in all countries allow us to formulate a set of statements related directly to Project Management and project managers:

- In a sense, we all live in developing and changing countries.
- As far as it depends on our efforts we should assist in the management of such changes trying to use them for the good of our countries and the whole of mankind.
- Existing changes themselves form the "nutrient medium" for development of PM and the world community of project managers and serve as a basis for PM global unification.

2.2 A BRIEF COMPARATIVE CHARACTERISTIC OF DIFFERENT SOCIAL AND ECONOMIC SYSTEMS

It is certainly impossible to give a thorough analysis of different systems within the framework of the present paper. However, we would like to mention some of the most substantial characteristics and outline the dynamics of environmental change. It will give an opportunity to consider the changes and realise the following:

- where did we come from?
- where are we now?
- where are we carried to by the stream of events?

So, let us try to review 3 types of social and economic environment reflecting dynamics and mile-stones of Russia's development:

- centralised or planning economy (before 1991);
- transition economy with its crisis conditions in 1991 - 2000 (data as of the middle of 1995);
- market economy after 2000 (forecast).

Here we consider the following characteristics as the most important for PM development:

- general characteristics of social and economic environment;
- structure and characteristics of project environment and projects in different social and economic environment;
- main characteristics of managing projects in different social and economic environment.

The data given below are based on official statistics [6-12,17], but, nevertheless, they are approximate and serve to show a general picture of changes.

2.2.1. General characteristic of social and economic environment

The most substantial indices characterising the dynamics of changes occurred or to be occurred (forecast) in Russia in 1980 - 2010 are given in Table 1. All data are given in % or ranging from 0 to 1.

Fig. 1 shows the qualitative picture of tendencies of changing the most important characteristics of social and economic environment in Russia for the period of 1980 - 2010.

Table 1.

N	Characteristics, factors	Social and economic environment		
		Centralised (before 1990)	**Transition (1991-95)**	**Market (after 2000)**
1	2	3	4	5
1.	Property relations (%) - state - collective - private - mixed	 90 - 95 7 - 8 0 0 100	 31 - 35 30 - 33 20 - 24 15 - 18 100	 14 - 16 8 - 12 70 - 75 6 - 8 100
2.	Land property	State and collective with the right of usage	Mixed without the right to sell and partly private property	Mainly private with the right to sell and inherit
3.	Political stability (from 0 to 1)	0.95 - 0.99	0.4 - 0.5	0.75 - 0.85
4.	Economic stability (from 0 to 1)	0.9	0.2 - 0.3	0.7 - 0.8
5.	Level of monopoly in economy and production	0.85 - 0.95	0.6 - 0.8	0.2 - 0.25
6.	Rate of inflation (% per year)	8 - 10	150 - 170	2 - 5
7.	Dynamics of production (p.a. = per year)	Decrease by 4-6% p.a.	Decrease by 15 - 20 % p.a.	Increase by 3.5-5 % p.a.
8.	Annual increase of level of living	+ (2- 3 %)	- (8 - 10 %)	+ (5 - 10 %)
9.	Influence of shadow and criminal structures (% of state budget)	7 - 10	30 - 50	10 - 15
10.	Provision with normative and legal regulations	up to 0.95	0.4 - 0.5	up to 0.95
11.	Social maintenance	0.85 - 0.9	0.35 - 0.45	0.75 - 0.8
12.	Share of foreign investments (% of the total volume)	2 - 4	3.3. - 4.1	21 - 25
13.	Share of Russian investments abroad (% of the total volume)	10 - 12	3 - 5	15 - 20
14.	Annual increase of gross national income (% of prior period)	+5	- (10 - 15) Total decrease for 1991-94 is 35,2 % (*)	+ (6 - 7)

* According to corrected data of the State Statistic Committee and the World Bank.

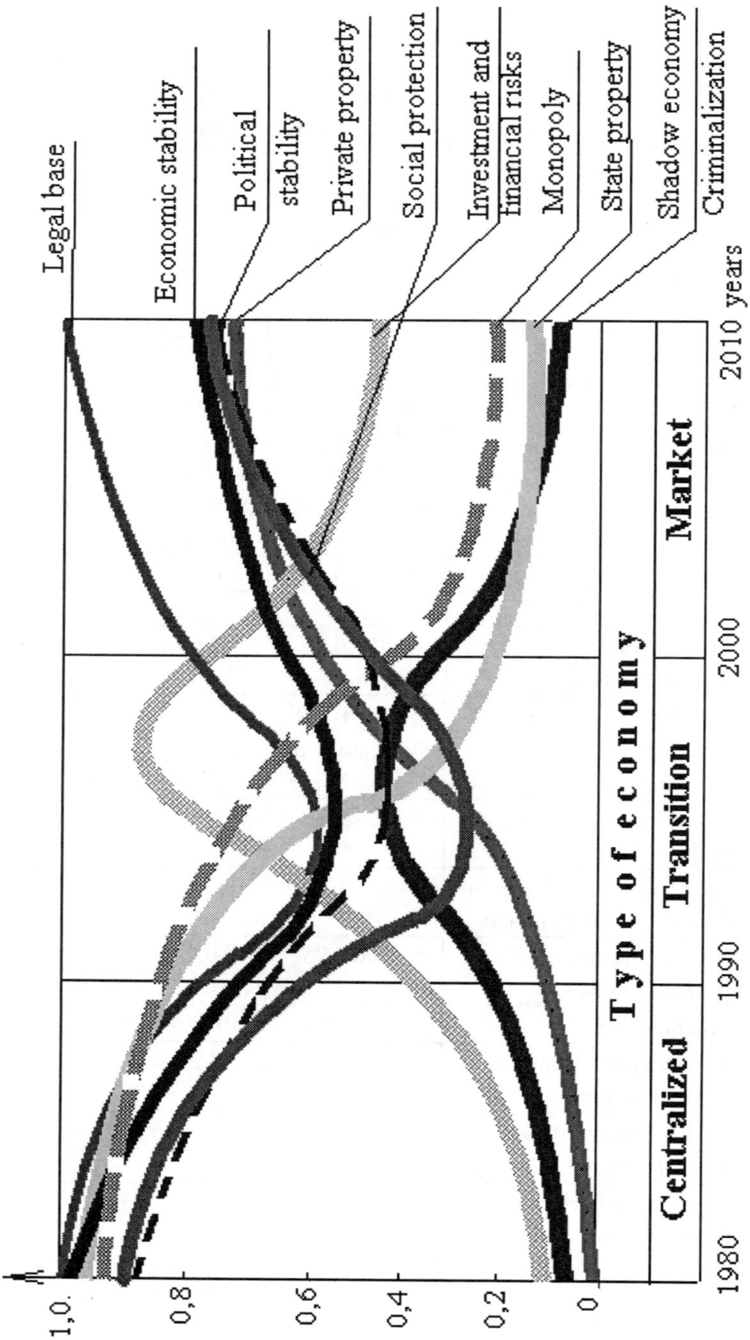

Fig. 1. Major tendencies of social and economic systems changing in Russia.

From the analysis of the data given above and other sources it is possible to draw some conclusions concerning the general status of social and economic environment of Russia. They are the following [9,10,13,17]:

- sharply changing property relations;
- political and economic non-stability;
- sharp differentiation and decrease of living conditions with simultaneous decrease of social and legal guarantees for the population;
- high tax rates;
- crisis of economy and production;
- lack of legal regulation, especially in the sphere of investment;
- non-co-ordinated activity of Federal power structures and subjects of Federation;
- underdeveloped market infrastructure.
- high degree of criminal structures influence.

From the PM point of view all these processes show a high level of close and distant project environmental changes, and financial and other risks at great demand of investments and the relatively cheap market of labour and services.

2.2.2. Structure and general characteristic of projects and project environment

Together with the above mentioned changes of the social and economic environment in Russia the restructuring of priorities, spheres of investments, programs and projects to be implemented is in progress. Table 2 shows statistic and expert data on dynamics of changes in the sphere of project-oriented activity.

The analysis of the data from Table 2 allows us to estimate the project-oriented activity in Russia in the transition period. This kind of activity is characterised by the following:

- rapid change in the structure of projects and multiple increase of the number of projects in non-traditional spheres: organisational, social, economic and mixed;
- decrease in the number of long-term programs and mega-projects in connection with risk increase and reducing share of centralised investments;
- change in traditional spheres of project activity application and their re-orientation to branches with short payback periods and to owners (customers) and users;

- change in the investment policy, structure and sources of investments including reducing investments to the sphere of production from 70% to 50% and increase of investments to non-production spheres (up to 50%).

2.2.3. Characteristic of management systems and practice in different social and economic environment

The changes that have occurred lately in social, economic spheres and project-oriented activity and given in Tables 1, 2 are closely connected with Project Management. Table 3 shows generalised data characterising the dynamics of changes in managing projects at different stages of the development of economic reform in Russia.

Analysis of Table 3 allows to realise the status of managing projects in the transition period and trace the dynamics of its changing on the way from a centralised to a market economy.

In this sphere we have unstable conditions and heightened risk, too. It is connected with such negative factors of the centralised economy as the monopoly, lack of competition, underdevelopment of economic methods of economy regulation, lack of legal and normative base, underdevelopment of market relations, infrastructure - which still affects the national economy.

N	Characteristics, factors	Type of social and economic system		
		Centralised (before 1990)	Transition (1991 - 1995)	Market (after 2000)
1	2	3	4	5
1	Types of projects - technical - organisational - economic - social - mixed (% of total volume)	 60 - 65 7 2 25 - 30 3 100	 30 20 - 23 15 - 20 22 - 25 10 - 15 100	 56 - 65 12 - 15 6 - 10 7 - 10 6 - 10 100

2	Classes of projects - mono-projects - multi-projects - mega-projects (% of total volume)	 17 25 58 ______ 100	 50 37 13 ______ 100	 60 - 65 25 - 30 10 - 15 ______ 100
3	Duration of projects - long-term (> 5 years) - medium-term (3-5 yrs) - short-term (1-2 years) (%total volume) Average	 75 20 5 - 7 - 12 years	 3 - 5 10 - 15 80 - 87 - 1 - 2 years	 5 - 7 25 - 30 63 - 70 - 3- 4 years
4	Major spheres of investments by branches of industry, % (priority order)	Military industry. Aero-space projects. Heavy industry, metallurgy. Fuel energetic. Chemistry and oil chemistry. Agriculture and light industry. Construction. Transport and communications. Social and cultural sphere. Other.	Fuel energetic (up to 40%). Transport/communi cations (14-17%). Agriculture, light industry (5 - 8%). Metallurgy(4-7%) Construction(4-6%) Machine building (4 - 5%). Chemistry and oil chemistry (2-3%). Forest and wood processing (1%) Others (23-27%). Total 100%	Competitive products, goods and services. New construction and reconstruction of residential areas. Social development programs. Aero-space, defence and state security.
5	Project orientation	State interests, manufacturers and contractors	Investors and consumers with consideration of interests of manufacturers and contractors	Owner, customers, users and consumers
6	Structure of capital investments - state - municipal - collective - private - mixed (%) by forms of property	 75 - 80 15 - 17 5 - 7 - - ______ 100	 31 7 15 17 30 ______ 100	 10 - 13 3 - 5 70 - 75 5 - 7 7 - 10 ______ 100

Table 2

N	Characteristics, factors	Type of social and economic system		
		Centralised (before 1990)	**Transition (1991 - 1995)**	**Market (after 2000)**
1	2	3	4	5
1	General style of management	Administrative, co-ordinated by party structures representatives, party headquarters	Elements of admin-istrative style. Forming professional project management	Professional management (project manager and team)
2	State influence on implementation of programs and projects	Directive planning, finance and resource allocation. Direct management of programs and projects	Planning and control of federal programs and projects. Forecast of development, co-ordination of investment process.	Forecast of national economy develop-ment. Regulation of tendencies of dev-elopment. Planning and control of federal programs and projects.
3	Investors	State structures and branches	Mixed - state - collective - private - foreign	Predominance of private investments (up to 80%)
4	Customers	Representatives of state structures	Representatives of owners	Juridical and natural persons representing owners
5	Procedure of work assignment, concluding contracts and agreements	State plan, decisions of party and state governing bodies	State orders, tradit-ional economy relations, initiating tenders and market competition	Market competition - tenders - competitions
6	Market competition	Monopoly of manufacturers and contractors	Reducing monopoly, starting competition	Market compet-ition. Monopoly is limited by the law
7	General contractor	Does not exist. Partly the functions of general contractor are fulfilled by general executors assigned by the state	Forming the institute of contractors, mixed methods of their assigning	Selected by the owner on the basis of competition

8	General executor	Assigned by the state in accordance with the plan	Assigned or selected on the basis of competition	Selected by the customer or general contractor on basis of competition
9	Principal suppliers	Assigned according to the order or state plan	Mixed way of selecting - old relations - state order - competition	Selected basing on market competition
10	Distribution of functions between project participants	Fixed in state normative documents	Predominance of fixed functions, forming contract relations	Regulated by the system of contract relations within the frames of the law
11	System of sanctions and motivation	Party administrative regulation with weak economic mechanisms	Administrative regulation, developing economic mechanisms	Economic motivation and regulation. Professional rating.
12	Contract discipline and liabilities observance	Party and state control and arbitrage	Transition to legal regulation. Partly using power methods	Legal regulation, economy responsibilities Arbitrage and court
13	Legal and normative base of project activity	State and branch standards, norms and rules, fixed rights and duties	Mixed - old norms & rules - forming new laws and regulations - contract relations	Regulation by the law, systems of contract relations
14	Prices and norms of resource expenditure	State, branch. Obligatory	Recommended norms. Contract prices. Regulated prices for a row of res-ources and services	Reference indices. Market prices.
15	Information, statistics, reporting	State informational system, statistics and reporting according to fixed standards	Mixed informational service. Simplified statistics. Reporting within the company.	Commercial informational systems. Limited statistic reporting. Reporting within the company or project

Table 3

The changes in project sphere influence the PM BOK structure and other PM components as follows:

- necessity of taking into consideration in PM intensive and often non-realised changes and heightened risk factors;
- necessity of introducing transition organisational structures of matrix and mixed type, with changing competence, responsibilities, economic methods of stimulation and sanctions;
- necessity of developing special methods and recommendations on PM application in conditions of transition economy, specific for each country

2.3. SOME CHARACTERISTICS OF THE INFLUENCE OF DIFFERENT ECONOMIES ON PROJECT MANAGEMENT AND ITS COMPONENTS

The data given in Tables 1,2,3 show substantial differences in project activity in different social and economic systems. These differences should influence PM itself and its components. Political and economic changes that have occurred in Russia caused changes both in project-oriented activity and in the style and methods of management. In the first place, it is seen in changes of the project context including:

- changing the structure, type, scale and duration of projects and programs as well as the spheres of their application;
- changing the structure of project participants, their mutual relations, property relation within the project, functions, motivation system, sanctions, etc;
- changing organisational forms and structures of project-oriented activity;
- changing close and distant project environment;
- changing the structure of expenses during the project life-cycle.

These changes, in their turn, caused considerable difference between such PM characteristics as:

- changing project purposes, scope, conditions, limits and requirements;
- changing systems of values and ethical norms, criteria of project analysis efficiency and project results estimation;
- changing integrative processes in PM: concept and project development, planning, control, system of documentation and reporting;
- changing the structure, role and methods of realisation of PM functions.

First of all, it relates to change, cost, risk, contract, procurement, resource, personnel and configuration management. Below we give an illustration of project context changing as for the life-cycle of construction project and some PM functions.

2.3.1. Specific features of project life-cycle in different economic systems

Defining project life-cycle and phases given in [1], [2], [3] is connected with the market economy. In conditions of the centralised and transition economies of Russia the life-cycle of project and its structure have specific features related to the context of project and its environment (see Tables 1, 2, 3). Differences in the structure of project life-cycle and in efforts distribution by its phases is illustrated as for construction projects in [15], [16]. The data of analysis of construction projects in conditions of centralised, transition and market economies have been analysed: the curves in Fig. 2 show the distribution of effort by phases of project life-cycle in different economies.

Fig. 3 shows typical forms of project life-cycles implemented in different social and economic environment (according to Tables 1 - 3). Differences in life-cycles are caused by different work and efforts structures connected with specific features of social and economic systems, styles and methods of project management and realisation in these systems. Simultaneously, the similarity of life-cycles shows that PM complex is invariant to the spheres of project implementation.

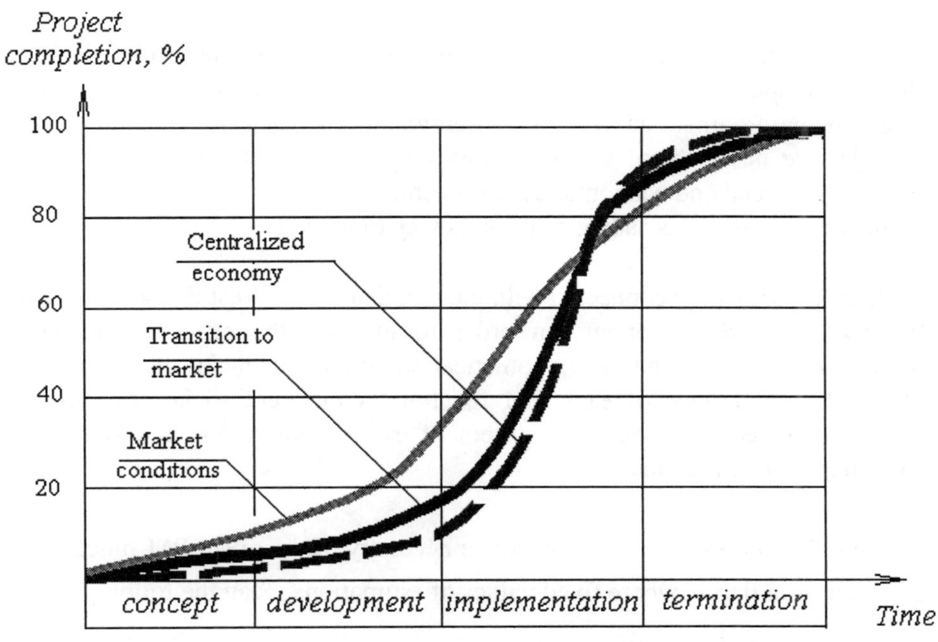

Fig. 2. Structure of effort by project phases in different economies.

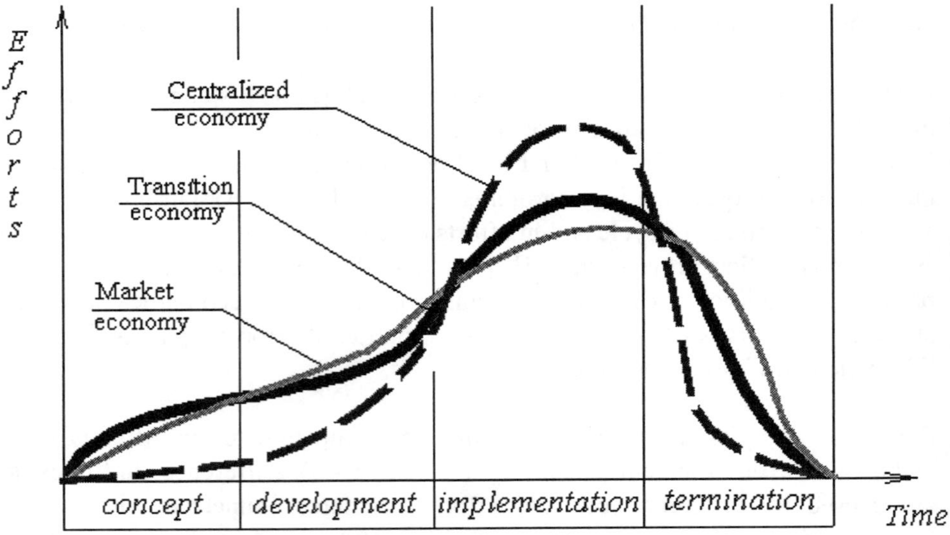

Fig. 3. Typical life-cycles of projects implemented in different social and economic environment.

2.3.2. Specific features of some PM functions in different economies

The results of research [14], [16] and experience of PM application in Russia show that specific features of certain PM functions in different social and economic environment depend, to our opinion, on two factors:

- objective necessity of the given PM function in conditions of the existing stage of social and economic development;
- how the economy is "advanced" on its way to market.

Both these factors are connected with the fact that in the West PM was brought to life by the market economy in order to increase the efficiency of project realisation. In order to build a hypothetical picture of tendencies of changing role, place and importance of each PM function in different social and economic environment let us introduce a system of expert estimations and some initial statements for further analysis.

Assume that the estimation of objective necessity of the given PM function F_i is changed from 0 to 1, where fixed values of estimations mean the following:

- $F_i=0$ - PM function is not required.
- $0.5 > F_i > 0$ -the function is required, but in combination with the other, more important functions. It is not used as a separate function.
- $0.8 > F_i > 0.5$ - the function is used separately.
- $1 > F_i > 0.8$ -the function requires close attention and development of special methods and tools for solving tasks in the given conditions.

The above statements allowed us to make expert estimates for principal PM functions in the centralised, transition and market economies based on Tables 1 - 3, Figures 2 - 3 and [14], [16]. The results are given in Table 4.

N	Characteristics, factors	Social and economic environment		
		Centralised (before 1990)	Transition (1991-95)	Market (after 2000)
1	2	3	4	5
1	Scope	0,6	0,8	0,7
2	Quality	0,4	0,65	0,8
3	Time	0,65	0,75	0,85
4	Cost	0,35	0,7	0,9
5	Risk	0,1	0,9	0,7
6	Labour resources	0,55	0,65	0,75
7	Procurement, contracts	0,5	0,8	0,7
8	Communications, information	0,4	0,7	0,6
9	Changes	0,15	0,95	0,55
10	Configuration	0,2	0,7	0,5
	Average	0,39	0,76	0,71

Table 4

The analysis of Table 4 allows us to make some interesting conclusions, as follows:

- The necessity of modern PM in centralised economy doesn't allow it to become a sphere of professional activity. Before the reforms in Russia started (in 1990) PM was not applied in practice in full volume.

- In transition period, when the risk and level of changes in project environment are very high, the degree of objective PM necessity is the highest, too.
- In market conditions the level of PM necessity remains very high, though it is lower than in transition period. This is connected with stable social and economic environment and reduced risk.
- The level of all PM functions necessity on the way to market increases. The functions most sensitive to changing are changes, risk, procurement, contracts, scope, configuration, communications, information reach the maximum in transition period. The functions less exposed to these factors - quality, time, cost, human resources - increase monotonously as they approach the market economy.

Figs. 4, 5 show curves built according to the data of Table 4. The curves show the dynamics of changing the role of PM functions in transition from centralised to market economy.

2.3.3. General conclusions

Based on the data of Tables 1 - 4 and Figures 1 - 5 it is possible to make some general conclusions concerning PM in different social and economic conditions.

1. Social and economic environments of all types are exposed to changes and, therefore, they can be considered as developing.
2. Transition economy by frequency and intensity of changes can be considered, from PM point of view, as the most common case of social and economic project environment.
3. All types of economies need PM application. However, in transition from a centralised economy to a market one, the necessity of all PM functions and, hence, PM methods and tools application, considerably grows.
4. The PM scientific basis, the PM BOK structure and core, including the context of PM functions, has evidently the common character, in spite of the difference between economies themselves.
5. The difference in PM for different social and economic environments is not in principle and does not affect the PM basis. As a rule, the difference is connected with differences in types and kinds of projects implemented, with specific features of spheres of their application, the natural environment, as well as methods of PM practical use. All these differences can be considered within the frames of local supplements to the PM BOK core and in special guides and recommendations on PM application.
6. In transition to market the probability of changes in project and its environment, as well as risks, increase sharply. The tendency is also peculiar to the market economy. In this case, common for all economies, we can define

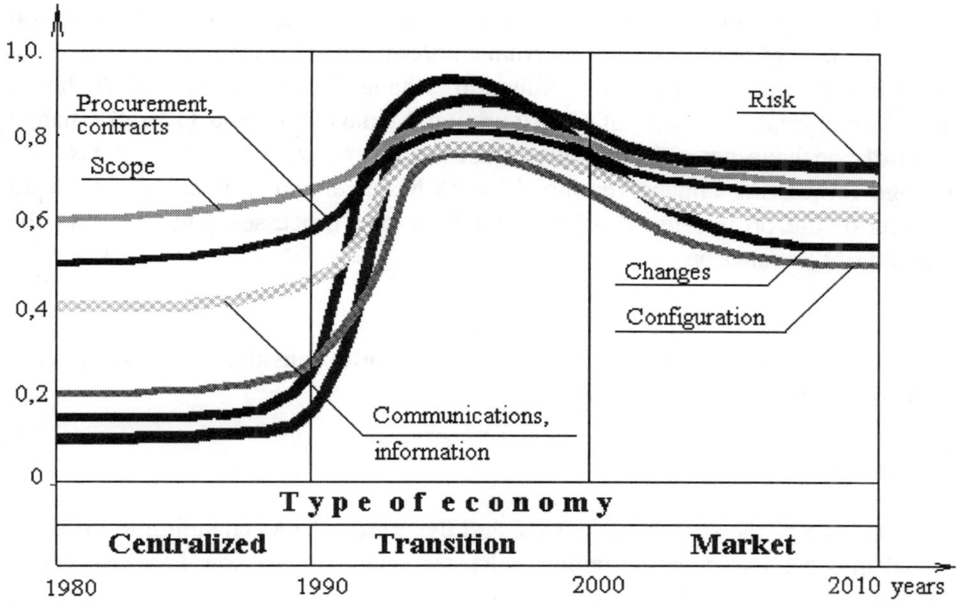

Fig. 4. Scheme of tendencies of changing necessity of PM functions, more sensitive to risk factors, in different economies as for Russia.

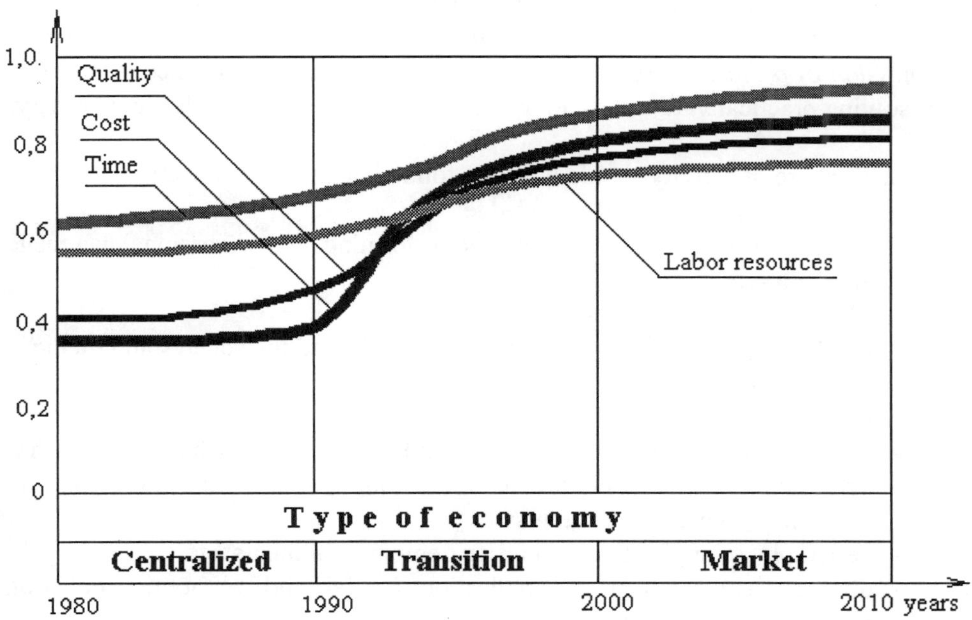

Fig. 5. Scheme of changing necessity of PM functions, less sensitive to risk factors, in different economies.

Project Management as managing changes in changing "parent" organisation under the influence of changing external environment. This circumstance requires introducing a new PM integrative function "Change Management" which should cover traditional functions of Configuration Management and Overall Change Control and have not only operative, reactive character, but also solve tasks of change forecasting and preventing. Change Management is connected with all aspects of the project itself, all PM functions and processes. Fig. 6 shows the context of this function.

3. Rational ways of PM development taking into consideration the specific features of developing countries

3.1. GENERAL

The analysis of the status, tendencies, and dynamics of PM changing in different social and economic environment in Russia, given in Section 1, allows us to formulate substantiated proposals on the trends of efficient PM development. The proposals are based on the fact that modern PM reached such a level of development that it made it possible and necessary to put the question of creating acknowledgement all over the world of *a specific sphere of PM activity, including a new complex applied scientific discipline "Project Management" and the world profession of project manager.* It is known that each applied science is intended for solving practical vital tasks and based on professional knowledge and activity.

PM as a sphere of activity should have the following basic components:

- Scientific basis and applied knowledge:
  - main notions, terms, their definitions forming the professional language and culture;
  - PM basic theory and methodology
  - PM BOK as the core of the discipline, invariant to the spheres and methods of PM application.
- Spheres of application:
  - various spheres of human purposeful activity (branches of industry, agriculture, economy, culture, administrating, public affairs and so on); all possible projects and programs;
  - geography of application (countries, regions, continents, etc.);
  - different environment of application (social and economic, political, national, cultural, technological, etc.);
  - project-oriented companies, organisations and their associations.

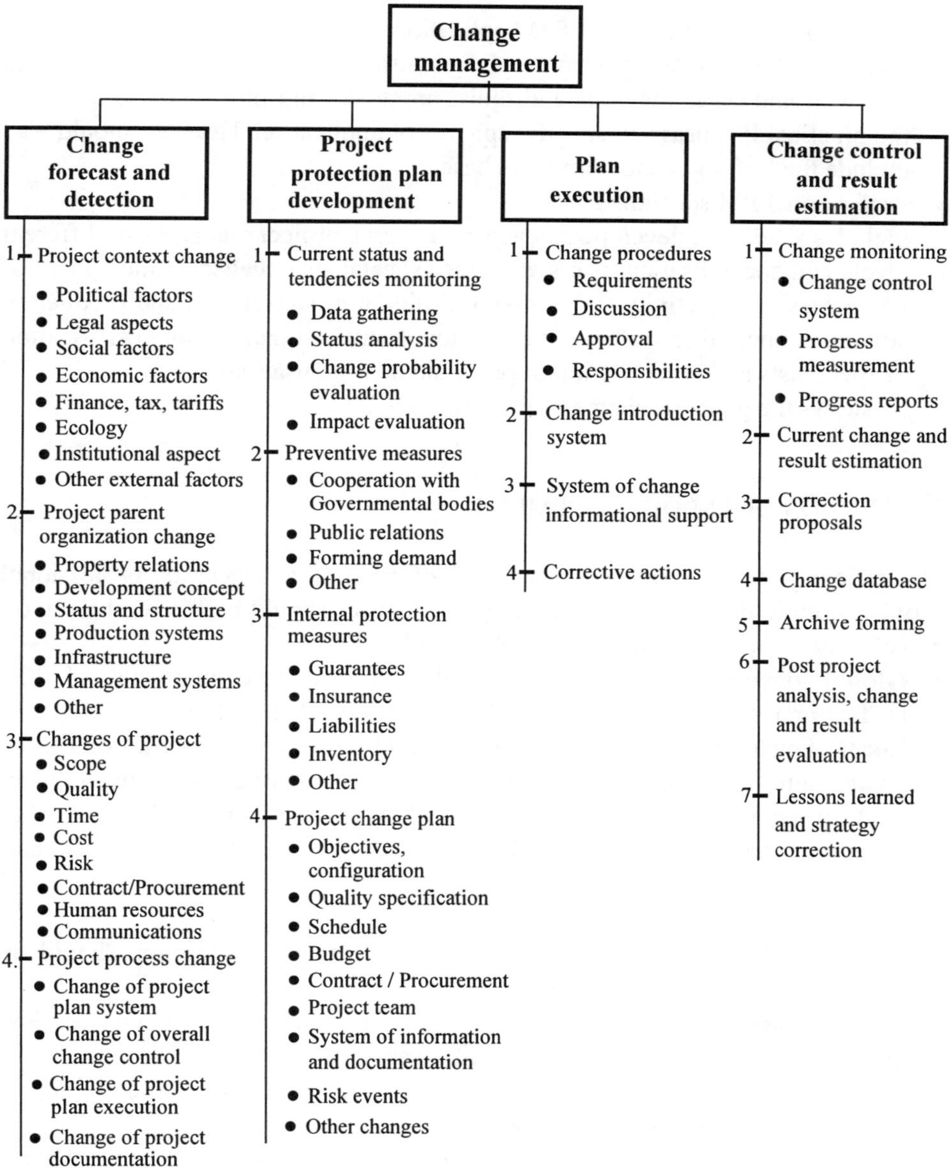

Fig. 6. Overview of Change Management.

- Methodology and practice of PM application
  includes the whole spectrum of scientific, methodical, organisational, technical and other methods and tools, processes and procedures oriented to the practical PM usage in specific spheres of application. Here we should also include the PM legal and normative base.
- Professional PM specialists
  includes scientists, developers, teachers, experts, project managers of different levels and specialisation, users and other specialists engaged in the sphere of PM professional activity and united in different organisational and production structures, including systems of education, training, qualification, certification of specialists and accreditation of professional organisations.

Fig. 7 shows the general scheme of PM professional activity.

3.2. TRENDS OF PM DEVELOPMENT

The efficient trends further development of PM, as a sphere of professional activity, both in developing and developed countries should be formed basing on two closely connected approaches: global and local.

- **Global development** is international and covers components of PM professional activity invariant to PM practice and spheres of application.
- **Local development** is national or oriented to another specific sphere of application (for example, country, region, trans-national company, branch, etc.). In local PM development we should take into account political, social, economic, technological and other specific features of PM practice and spheres of application.

According to this division further PM development, to our opinion, should be performed by means of the following:

- **PM globalization** is forming the global, common for all countries, PM understanding and world-wide recognition as:
  - special sphere of professional activity spread world-wide and covering all possible spheres of PM application;
  - separate, complex applied scientific discipline with its theory, methodology, spheres and practice of application;
  - profession of project manager which requires special knowledge, skills and competence for successful project management.
- **PM unification** is forming and developing the unified PM components, elements, processes and procedures for all countries.

- **PM localisation** is forming and working out various PM components, elements reflecting specific features of different spheres of PM application and practical experience.

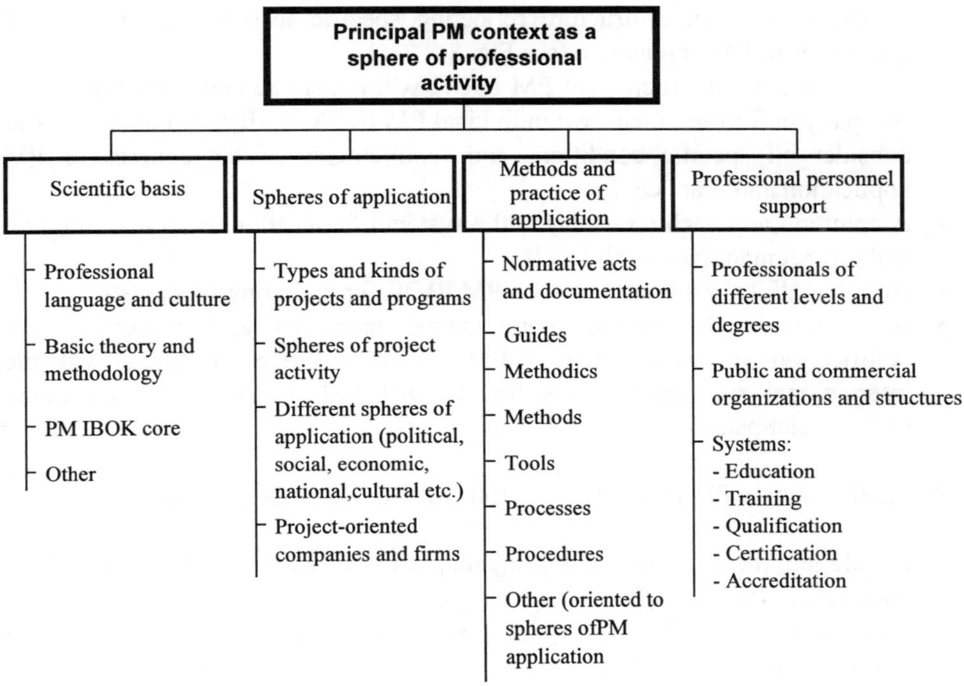

Fig. 7. The general scheme of PM professional activity.

3.2.1. PM globalization

It is seen from the above that PM globalization should solve the following problems:

- Forming and providing global PM understanding and recognition as a sphere of special activity, scientific discipline and profession.
- Forming scientific PM basis including:
  - international PM language (terms, definitions, notions, glossary, thesaurus, dictionaries, etc.);
  - developing basis of PM theory and methodology;
  - developing PM international core and PM BOK invariant to the spheres of PM application and practice.
- Organisation, support and co-ordination of global co-operation of PM organisations and specialists.

3.2.2. PM unification

PM unification should include the efforts to developing the following.

- The unified structure of PM BOK including the invariant core and distant sectors with flexible structure reflecting specific features of spheres and conditions of PM application (see Fig. 8).
  Note: the unified structure of PM IBOK will assist each country, branch, or company in forming their own individual PM IBOK configurations in order to consider all specific conditions and requirements of the spheres of PM application and practice.
- A common approach (concept) to the system of education, personnel training and project managers of all levels.
- An international program based on PM IBOK for preparing specialists.
- An international concept, requirements, programs and procedures for certification and accreditation of PM specialists and organisations with the international recommendations for national PM organisations and trans-national companies.

As the first step in PM unification the following could be performed:

1. Developing the international program and procedures of PM specialists' knowledge *qualification,* basing it on PM BOK and certification programs PMP worked out in PMI and updated with materials of IPMA, AIPM, ISO and so on (1-st level of certification).
2. Developing the international program and procedures of *certification* of project managers for confirmation of their competence basing on PM IBOK and IPMA certification programs updated with PMI and other international organisations materials (2-nd level of certification).
3. The same on the program of accreditation (see Sections 1, 2).
4. Appropriate national certification programs can be developed based on 1, 2, 3.
5. The system of international standards and normative bases on unified PM components and elements and putting them into practice through the international organisations.

The plan of creating the international system of qualification, certification and accreditation is given in Fig. 9.

3.2.3. PM LOCALISATION

Our understanding of PM localisation includes development of local supplements, adaptation of global and unified PM components, and elements of specific conditions of the spheres of PM application and practice. Such supplements include the following.

- Adaptation of PM IBOK to specific conditions.
- Creating the national PM language compatible with the international one.
- Creating the national system of education, certification, qualification of specialists and accreditation of PM organisations.
- Development of special PM methods and tools, methodological recommendations and guides on PM application taking into consideration local conditions and requirements.
- Development of guiding and methodical PM materials within the frames of a firm.

So, the specific trends and practical course of PM application in Russia should be defined in the process of localisation taking into consideration specific features of transition economy and results of PM unification and globalization.

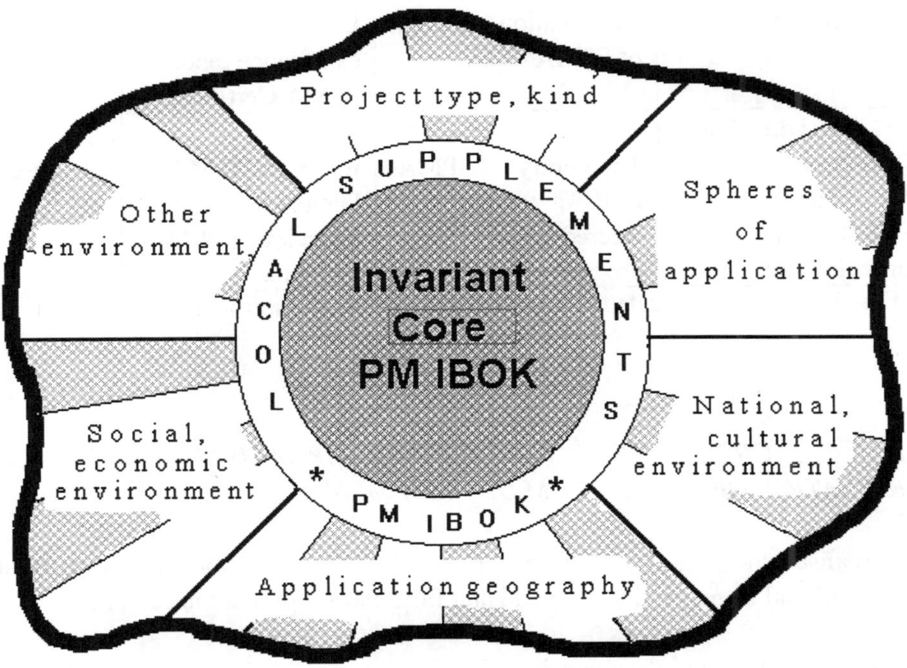

Local supplements to the PM IBOK core for specific spheres of P application (for example, a country)

Fig. 8. The scheme of the unified PM IBOK structure.

Stages	Scientific and technical base	Procedures and result	Period of realisation
PM Qualification			
a) Temporary variant	Temporary program based on PM BOK PMI, IPMA, ISO, AIPM	Examination, qualification PMP - PMI and IPMA	1 year
b) Permanent variant	PM IBOK by FPMA	Examination, qualification FPMA	2-3 years
PM Certification of project managers by qualifications			
a) Temporary variant	IPMA programs of certification with PMI supplements	Interview + experience report. IPMA Certificate	1 year
b) Permanent variant	Certification program of FPMA	Interview + experience report. FPMA Certificate	2 years
PM Accreditation			
a) Auditors (examiners)	Temporary IPMA, PMI Agreement	Interview, IPMA Diploma	2-3 years
b) Professional organisations	Accreditation programs FPMA	Interview, FPMA Diploma	3 years

Fig. 9. The plan to create the international system of qualification, certification and accreditation.

3.2.4. PRINCIPLES OF INTERNATIONAL ORGANISATION AND CO-ORDINATION OF WORK ON PM DEVELOPMENT

To realise the international program of PM development, appropriate organisational forms, and intellectual and financial support are necessary.

Organisational forms of supporting global complex PM development can differ much. The most acceptable form is creating the International Federation of PM Associations (FPMA), which can unite all existing international, regional and national PM associations and organisations. The main purposes of FPMA could be:

- global PM profession development and recognition as a professional sphere of activity and its promotion to new countries, regions and new spheres of application;
- PM unification and standardisation;
- development and realisation of international programs and projects on education,
- scientific research, qualification, certification of project managers and accreditation of PM organisations;
- organisation and holding PM world forums;
- information and publishing activity and services for FPMA members;
- organisation and co-ordination of FPMA members in PM development.

FPMA should be managed for 2-3 years by an elected co-ordination group and should not have rigid vertical hierarchic structure of bureaucratic type. To create FPMA it is necessary:

- to create the initiative group;
- to develop the concept and draft Charter of FPMA;
- to obtain consent of international and national organisation to join FPMA and come to agreement on constituent documents;
- to hold the Constituent Conference.

<u>Intellectual support</u> of global PM development and unification should be performed by and under the aegis of FPMA members. For this purpose the following should be done:

- forming temporary working commissions and groups from authorised experts for realisation of FPMA projects. The list of experts should be confirmed by FPMA members;
- periodical discussions on work results, projects and programs during FPMA working meetings and forums;
- review and approving the final PM documents on FPMA projects;
- presenting approved PM documents to competent international organisations.

Financial support of projects and programs on PM globalization, unification and localisation should be performed by international foundations, international financial and crediting institutions, governments of countries concerned, national organisations and trans-national companies.

The proposed approach to further PM development in the world and each separate country allows obtaining the following benefits:

- acceleration of the process of integration of the world community and its social and economic development;
- providing further efficient development of PM all over the world;
- acceleration of PM knowledge and experience transfer to developing countries and their social and economic, scientific and technical progress;
- expanding the experience, achievements, national and cultural values of all countries in the field of PM;
- opening new spheres of PM application in our developing world.

REFERENCES

1. Project Management Body of Knowledge (PM BOK), PMI, Drexel Hill, Pennsylvania, 1987.
2. A Guide to the Project Management Body of Knowledge (PM BOK), Exposure Draft, PMI, Pennsylvania, August, 1994.
3. Body of Knowledge. Association of Project Management, INTERNET UK, Buckinghamshire, 1995 (Version 2).
4. Turner J. Rodney. The handbook of project-based management: improving the process for achieving strategic objectives, McGraw-Hill Book Company, London , 1993.
5. Barnes M., Wearne S., (1993) The Future for Major Project Management, International Journal of PM. 11,3,135-143.
6. USSR National Economy in 1990. M.: Finance and Statistics, 1991.
7. Developing economic reforms in R.F. M.: Gosstatizdat of Russia, 1993.
8. Economy and life. 1994. NN 14-18.
9. Level of our life in 1991-1993. Analytical reference-book. (by V. Uralov). Moscow. 1995.
10. Rekitar Y. (1995) Actual problems of the investment process in Russia. Proceedings of the International Symposium, St. Petersburg, September 14-16, 1995. M.: SOVNET, "Alans", 1995, p.p. 42-49.
11. Fedorova E. (1995) Estimations of GDP of Russia became more reliable. Financial news N70 (199). September, 29, 1995.
12. Investment processes in the economy of Russia in the first half of 1995 (by D. Belyaev). Minstroy RF, 1995.
13. Balakin V., Voropajev V. (1995) Managing the investment and construction activity of Russia: status, problems, perspectives. Proceedings of the International Symposium, St. Petersburg, September 14-16, 1995. M.: SOVNET, "Alans", 1995, p.p. 19-25.

14. Voropajev V. Project Management in Russia (Russian language version). M.: SOVNET/"Alans", 1995, p. 226.
15. Morris, P.W.Q. (1994) The Management of Projects. Thomas Telford. London p. 358.
16. Poznyakov V. Project Management: context, actuality and special features of application in Russia. Book: "The World of Project Management". Translated from English. - Moscow: "Alans", 1993. p.p. 252-270.
17. Russia's reforming: Myths and reality (1989 - 1994). Collective work under the leadership of G.V. Osipov. M.: Academia, 1994, p. 384.

OPEN DISCUSSION: CORPORATE STRUCTURES

Discussion chaired by:
V. RUCH
Deputy Rector, Lugansk University, Ukraine
Director E.Ukrainian Centre of PM

This summary by:
T.M. WILLIAMS
Strathclyde University
Glasgow, UK

Prof. Ruch introduced a very brief discussion by giving three thoughts:

- Complexity. From the Systems approach, any project is a system of inter-dependent elements working towards a common objective with restricted resources. Perhaps a definition of the complexity of the project can be based on the most complicated element - clearly we must divide complex projects into smaller elements.
- Models. We need to model the customer requirements before accepting a project, because a full understanding of what the customer wants is essential.
- PM methods. There is only a small amount of knowledge that we can formulate, and it is only this that we can reproduce and pass on. If we describe the manager's activity as art, realised within certain laws, then the formulated part of knowledge is the laws, and the art is something personal; the only way to pass this on is to work together.

Ukraine has an economy being reformed, so a most important question is how to change the structure of corporate enterprises to reinforce the ideas of PM?

Belarussian work was discussed covering a strategic approach to projects, and a holistic approach to the organisation of anarchic structures (Kirpich), including formal mathematical methods, which have been tried out and applied in practice.

There is an old-fashioned view of projects in which the client tells the contractor what to do. Clearly for complex project this is not optimal, and the two parties should co-operate, but this can be difficult for different corporate structures. How can we have a commercial project where contractor and client work together?

Prof. Elmagrhaby brought to the Workshop's attention a book from 25 years ago entitled "Interpretative Structural Modelling" dealing with "Societal Systems" in which there are many owners interested (others would say many "stakeholders"); for example, building a hospital, school, roads, or airport.

SECTION 6
MANAGEMENT TECHNIQUES

The Workshop is concerned with both "Modelling" and "Managing" complex projects. Most of the material in Sections 4 and 5 (and indeed, the risk management part of Prof. Chapman's paper in Section 3) has dealt with the latter of these two aspects. This Section completes the discussion of managing single complex projects (the next Section concerns multi-project management) with two papers on specific management methods. Again, one of these papers is from a NATO-country perspective, and one from a CP-country perspective. Following the discussion at the end of Section 4, it is perhaps not surprising that these papers are very different. The first paper, by Prof. Cleland of the University of Pittsburgh, covers the use of alternative teams. The second paper, by Prof. Voropajev, looks at the analysis and management of projects where a deep hierarchical management structure is used.

ALTERNATIVE TEAMS IN STRATEGIC MANAGEMENT

DAVID I. CLELAND
University of Pittsburgh
Pittsburgh, Pennsylvania, U.S.A.

1. Introduction

All societies and all institutions are in a continuous process of transformation - change has become the only permanent force in contemporary times. Consider some of the changes taking place on industrial and other institutions: Global competitiveness, fluctuations in market demand, the advent of new technologies, changing social and political forces, the fall and rise of governments, and new paradigms for the organisation and management of the enterprise, are just a few of the more relevant changes that challenge contemporary managers.

For most managers the key challenge is to invent and develop new strategies, new products, new services, and improved organisational processes to deal with the inevitable changes buffeting today's enterprises. Managers today must invent and develop enterprise strategies that are able to assess the inevitable changes facing contemporary and future managers, and develop proactive plans to implement the use of resources to accommodate these changes.

2. Strategic Management

Strategic management is seen as a process whereby organisations can work through alternative teams to learn to cope with unpredictable and rapidly changing environments. Strategic management is a relatively new discipline for which there is limited precise language or process. How does the manager who is dedicated to "strategically managing" the enterprise plan for, monitor, evaluate, and control the use of resources dedicated to alternative strategic options for the enterprise's future? Simply put, strategic management means to manage the enterprise as if its future mattered - working synergistically within the concept of strategic planning to prepare for the enterprise's future.

T. M. Williams (ed.), Managing and Modelling Complex Projects, 205–219.

This paper examines how alternative teams can be used in the strategic management of an enterprise. The field of strategic planning has undergone a major change since its tremendous growth and development in the 1960s and 1970s. One of the key changes has been a growing realisation that strategic planning and strategic management are closely linked. While the concept of strategic planning is generally understood, strategic management is in its early evolutionary stages. Every manager must practice strategic management today, since an interest and the active study of likely future opportunities and challenges facing all enterprises has become necessary because of the inevitable changes facing organisations today - particularly those that are competing in the global marketplace. A manager wishing to pursue the practice of strategic management faces ongoing assessment of the enterprise's key strategies in terms of:

- New and/or improved product and service opportunities.
- New and/or improved organisational process changes to support product and service initiatives.
- Selection and commitment of resources to position the enterprise to take advantage of future opportunities.
- Development of action plans to implement organisational resources towards accomplishing future enterprise purposes.
- Ongoing assessment of how well strategies are supporting operational competence.

Strategic management means to manage the enterprise's future - selecting and setting the future direction of the enterprise is increasingly becoming the key challenge facing managers today. Those enterprises that are facing difficulties, as well as those that are enjoying success - all have the challenge of preparing for an uncertain future. In preparing for that uncertain future contemporary managers deal with several key questions:

- What has been our "business" in the past?
- What is our current "business."
- What should our "business" be in the future?
- What "business" should we not be in for future periods?
- How should organisational resources be committed and dedicated to the attainment of future organisational purposes?
- What means can be used to assess the progress being made towards the accomplishment of organisational purposes?

Gaining insight and answers into the preceding questions poses key challenges for enterprise managers, as well as professionals seeking to prepare the enterprise for survival and growth in the unforgiving global marketplace. To gain adequate insight into the potential future opportunities for the enterprise requires an ongoing assessment of anticipated environmental changes.

3. Assessing Environmental Changes

The development of strategies to position the enterprise for its future requires that the following environments facing the enterprise be continuously assessed to determine likely and potential changes in the following:

- Political changes likely to impact the enterprise's future competitiveness, such as political changes in Eastern Europe.
- Social changes likely to influence the enterprise's future products, services, and organisational processes. For example, affirmative action policies have had significant influence here.
- Economic changes that will impact market pricing, resource cost, or exchange rates in the global marketplace.
- Legal influences that can restrict the use of resources and products in international markets. For example, the freedom to close plants in a foreign country may be limited by local governmental policy.
- Technological innovations likely to set new competitive dimensions in products, services, and organisational processes. Computer innovations, telecommunications capabilities, and advanced manufacturing strategies are a few of the technological advancements that have set new dimensions in competition.
- Competitive strategies that can impact the success or failure of the enterprise's products and services. Competitors are found everywhere in today's fiercely competitive marketplace. Even a distinct technological advantage in a product or manufacturing process is soon copied and exploited by aggressive competitors.

Contemporary managers have to assess the possibilities and probabilities of the future likely to impact their enterprises. Simply stated, such managers need to:

- Build a forecast of expected environmental changes likely to impact the enterprise.
- Identify alternative strategies to support selected product, service and process changes.
- Select those alternatives that best fit the enterprise's capabilities.

- Develop action plans to execute the use of resources to attain future organisational purposes.

Contemporary managers require a "mental model" - a roadmap - of how to strategically manage the enterprise [1].

4. A Roadmap for Strategic Management

A roadmap is a representation of a route of travel to a desired destination. It provides a summary of where one wants to go, preliminary destinations along the route of travel that serve as checkpoints to measure progress, and the journey's endpoint. Strategic managers require roadmaps to prepare the organisation for its future. A roadmap that outlines a general route for the future of the enterprise contains several key checkpoints on the road to the final destination:

- Foresight
- Competitive assessment
- Organisational strengths and weaknesses
- Benchmarking
- Strategic performance standards
- Functional expertise
- Vision quest
- Stakeholder interfaces
- Market needs
- Product and/or service and process development
- Feedback

A brief description of these checkpoints follows.

- <u>Foresight:</u> The orderly development and anticipation of the possible and probable changes in the markets in which the organisation would like to provide products and services.
- <u>Competitive Assessment</u>: A rigorous examination and evaluation of the strengths, weaknesses, and probable strategies of competitors in their products and/or services, organisational processes, and market performance.
- <u>Organisational strengths and weaknesses:</u> Assessment of the strengths and weaknesses of the enterprise, particularly regarding the ability to deliver competitive products and/or services through the use of effective organisational processes.

- Benchmarking: Review of the performance of competitors and other organisations that are considered the best in the industry, to determine what operational and strategic abilities enable them to perform so well.
- Strategic performance standards: Identification, development, and dissemination of the elements required to establish strategic performance standards and guide the use of resources in the enterprise. These standards include the organisational mission, objectives, goals, strategies, and design; individual and collective roles; leader and follower style; supporting systems; and the quality and quantity of organisational resources.
- Functional expertise: The ability to develop and maintain centres of excellence in the disciplines needed to support operational and strategic initiatives.
- Vision quest: The ability to envision what the future should be for the enterprise--that is, the general direction the enterprise should follow to become what organisational leaders want it to be.
- Stakeholder interfaces: Discovery, development, and maintenance of common boundaries with those people and organisations that have a vested interest in the products and/or services and processes provided by the enterprise. Customers and suppliers are the principal stakeholders with whom close working relationships need to be maintained.
- Market needs: Assessment of the probabilities and possibilities of the marketplace, including the potential for products and/or services for markets that do not currently exist.
- Product and/or service and process development: The use of project teams to focus cross-functional and cross-organisational resources to conceptualise, design, develop, and initiate continuous improvements in existing products and/or services and organisational processes, as well as the development of new initiatives in such products and/or services and processes.
- Feedback: Ongoing insight into how well the strategic roadmap elements are being developed and implemented in positioning the organisation for its future purposes.

Figure 1 [2] shows a model of this strategic roadmap. The effective use of this model requires alternative types of teams to focus resources for operational and strategic purposes.

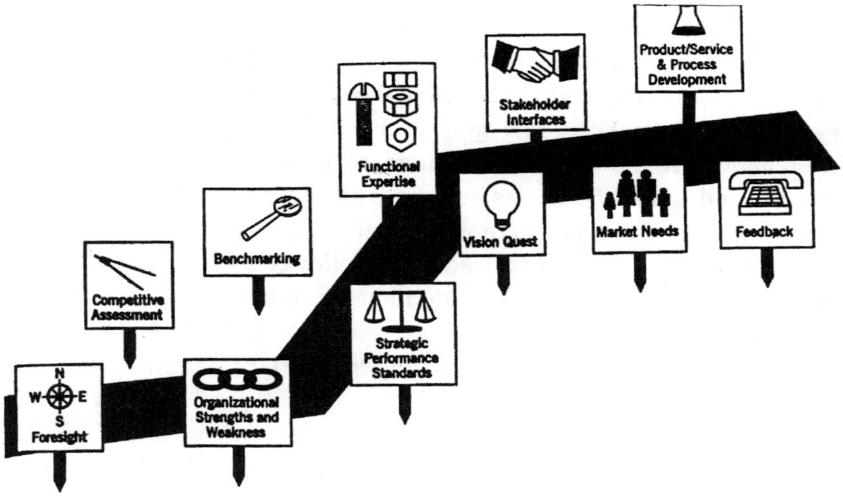

Figure 1. Roadmap for strategic manager

5. Alternative Teams

Strategic management provides for changes reflected in the development of improved or new products, services, and organisational processes. Other changes brought about by effective strategic planning and execution include new markets, expansion of existing markets, acquisitions, mergers, divestments and new customers, suppliers, and changes in the composition or application of the human and non-human resources used in the enterprise. Usually these changes come about following a modified life cycle starting with an idea or concept on how things can be done differently. As these ideas come forth an organisational focal point is required to manage organisational effort to bring about the needed change - as through the use of alternative teams such as the following.

Concurrent Engineering Teams to develop in a simultaneous fashion new or improved products, services, and required supporting organisational processes to facilitate earlier commercialisation.

At the Boeing company the design, development and manufacture of the newest member of the Boeing jetliner, the 777, project teams were used to work with customers, suppliers, and people from diverse disciplines within the company to share information, technological, marketing, and management expertise to create a more efficient process for producing the new aircraft.

Reengineering teams to bring about a fundamental rethinking and radical redesign of business processes to achieve extraordinary improvements in enterprise performance.
At the Chevron company, a major oil refinery, a "best-practices" discovery team was formed in 1994 to facilitate a reengineering process underway. It consisted of ten quality-improvement managers and computer experts from different functions of the enterprise to include oil production, chemicals, and refining. The team uncovered numerous examples of people sharing best practices. After a year of operation the company published a "best-practice resource map" to facilitate the sharing of knowledge across the company. The map contains brief descriptions of the various official and grass-root teams along with direction on how to contact them. The map and its information helps to connect people working on diverse things in the company [3].

Crisis Management teams that deal with critical situations and emergencies that impede effective and efficient organisational performance.
Some examples include major product liability legal actions, loss of key personnel or facilities, or major movements in the exchange rate in overseas markets.

Self-Managed Production teams that manage themselves typically resulting in the improvement of quality and productivity in manufacturing (production) operations. Such teams have made paramount improvements in enterprise competitiveness.
During the period 1989-1994, the General Electric Company, one of the most profitable companies in the world, experienced a 20% increase in sales while its head count has fallen more than 25%. By cutting inventories and production times, GE's working capital turnover has almost doubled since 1986. Executives at this premier company are strong believers in human potential, convinced that motivated and properly empowered people can work near-miracles. It means breaking down hierarchies and ensuring that information flows up as well as down. Rapidly disappearing from GE's plants are the old-line foremen. At the company's remarkably productive Grove City, Pennsylvania locomotive diesel engine plant, 600 non-union hourly workers manage their own jobs by teams - facilitated by five nonsupervisor salaried workers. Since the new team organisational design has been inaugurated assembly time on new locomotives has been cut in half since 1992. At a GE plant in Louisville, Kentucky the use of self-managed production teams for a new washing machine was the major contributor in going from a $47 million loss in 1992, to a projected $40 million profit in 1993, then on to another projected increase to $80 million in 1995. CEO Jack Welch believes that many of the problems that have cropped up at GE and

elsewhere trace to a common cause: People who know about problems won't or can't get the problem to the boss because of fear [4].

Benchmarking teams used to carry out an ongoing measurement of enterprise products, services, and processes against the most formidable competitors and industry leaders - resulting in the design of improved strategies and performance standards that will lead to enhanced capabilities.
A large shipyard had for decades dedicated its resources to competing in the military marketplace. As a consequence of being a one-customer company the enterprise was totally unprepared for the end of the Cold War. By launching a Benchmarking project the senior managers were able to obtain directional information by reviewing the strategies of best-in-class companies such as Germany's Thyssen for pipe bending, Walt Disney World for preventive maintenance, and L.L. Bean for order processing. Included in the benchmarking activities were ten shipyards in Halland. Through insight gained from the benchmarking process the company established a new mission in the design, manufacture, integration, and testing of complex structures. Other teams within the enterprise look at where core competencies have applicability, leading to the development of strategic priorities and a new direction for the enterprise [5].

Facilities Construction teams that integrate organisational resources to design, engineer, and build facilities and equipment to support enterprise operational and strategic needs. A major food processing company shut down its traditional major factory concurrently with the building of a new factory that had approximately 553,000 square feet of space. By using many teams, workers were retrained, plant design was carried out, unions, local communities, customers, and suppliers were integrated into a massive investment in making the transition from the old factory to the new one without losing a single day of production. Membership on the teams included professionals, suppliers, customers, union representatives, local community groups, and workers. The factory transformation was characterised by an unprecedented degree of employee involvement in team activities, leading to a strategic renewal for the company [6].

Quality teams to provide a focus for total quality management accomplished through working with organisational units to improve the quality of products, services, and enterprise processes.

Audit teams that evaluate the competency of organisations, programs, projects and functions to delivery quality products, services, and organisational processes.

Plural Executive teams that work together in a strategic management context in the design and execution of major strategies to prepare the enterprise for its long-term future. These teams also maintain surveillance over how well the enterprise is meeting its short-term objectives and goals, thus contributing to the accomplishment of the overall mission of the enterprise.
At the General Motors Corporation the top 100 executives are organised into a plural executive - a leadership council that meets four or five times a year. These meetings are interactive - the council takes operational and strategic issues apart, and puts them back together again. The use of such a plural executive has been significant in moving the company towards its vision of being a world leader in transportation products and services.

6. The Limits of Downsizing

In the U.S. after a decade of downsizing, corporation executives face a growing realisation that efficiency approaches directed to cutting costs has limits in gaining competitive advantages. Strategic planning is again a growing major focus in the quest for higher revenues and profits. Strategic planning has become now the single most important management issue facing contemporary enterprises, and will likely remain so for the next several years. Companies are being encouraged to take the process of strategic planning out of the senior offices of the enterprise and handing it over to teams of line and staff members to include junior professionals who have the ability to think creatively - along with competent senior "old timers" who are willing to tell it like it is -to keep realism into the design of future strategies. Strategic planning today should be seen as an opportunity to transform the enterprise and change the rules of the game in products, services, organizational processes, and competitive dimensions. Key suppliers and customers are also being brought into the process - even rivals to gain a competitive advantage, and are taking on important roles in the teams dedicated to improvement of future strategies. Some examples:

- At the General Electric Company responsibility for strategic planning has been delegated to each of the company's twelve major operating units. Each summer CEO Jack Welch meets with his top management team for all-long planning sessions, with a focus on both near-term and four-year look at the future to lay out their strategic direction, what noteworthy new products and services are needed, and what the competition is doing.
- At the H.J. Heinz Company, strategic planning related project teams were used to research and identify global trends that could have impact on their business. The team identified four major trends that boded well for Heinz and

consumer products companies that could take advantage of them. These trends were:

- The movement toward free market economies, particularly in the world's developing nations.
- The spread of democracy in Eastern Europe, Asia, and South Africa.
- The growth of the consumer class and the rise in disposable income, especially in the Pacific Rim and Southwest Asia.
- The inexorable growth of global media and satellite/cable television, which are transforming the world into a global village and reinforcing a powerful consumer culture [7].

- At Nature Conservancy, a very successful environmental protection organisation, acceptance of change was slow because the organisation was doing so well. The Conservancy has always had a clear mission: to preserve plants and animals and special habitats that represent the diversity of life. Nevertheless, even in the face of such success a project team was appointed to work on a strategic plan for the organisation. Through extensive research in the literature, many interviews with people in the Conservancy as well as outsiders such as scientists and people in other organizations, a strategy evolved under the oversight of the project team for an integration of economic growth with environmental protection. One of the major obstacles the team had to overcome was the popular perception that economic growth and environmental protection were at odds with each other. An important lesson came out of their successful strategic planning--how to measure and manage success [8].
- The new CEO at the Kodak Company launched a major strategic renewal of the enterprise which had several major threads of team management to include:
  - Revitalisation of the company "matrix" organisation;
  - Emphasised shorter product development time through the use of concurrent engineering;
  - Instituted a formal project review system; and
  - Tightened up the co-ordination of new product development [9].

Other examples of the use of teams in the strategic context of the enterprise include the following.

- At the Ford Motor Company global product teams design cars to be sold around the world. Other improvements in the use of project teams have been instituted - for example just getting approval for a new-car project used to take up to 22 meetings over two months. Now approval comes in less than a month [10].

- At Asea Brown Boveri (ABB), teams are used in a global matrix encompassing 1,300 separate operating companies, with ABB's matrix requiring the managers of each frontline unit to report to both a regional manager and a worldwide business head. The CEO and his senior management team has taken the lead in redefining key organizational relationships and basic management behaviours [11].
- Additional insight into the effectiveness of teams is found in the experiences of *Industry Week* magazine which profiles the winners of its "Best Plants" awards and has drawn conclusions about the commonalities among these plants. In its most recent survey, *Industry Week* reported that all 25 winners and finalists in 1995 relied on empowerment practices and that 88% of those plants used self-managed production teams to some extent [12].

7. How Are We Doing?

A key challenge facing managers is how to maintain an overview of how well the enterprise is preparing for its future. Oversight of operational or short-term activities is much easier to carry out. Cash flow, productivity and quality assessments, financial ratio analysis, and the evaluation of short-term financial, marketing, and manufacturing goals can be done in an ongoing basis - on a daily and weekly basis as needed to see how things are going. Assessment of the effectiveness of emerging future long-term strategies is more difficult. The major considerations to be evaluated here include:

- Conceptualisation and development of new products, services, and supporting organizational processes
- Development of new customer bases, markets, and promising new businesses
- Development and updating of environmental forecasts and likely future scenarios regarding potential new ventures for the enterprise.
- Acquisition of new equipment, facilities, and human resources for anticipated long-term usage.
- Periodic audit of the effectiveness of current operations as well as a comprehensive review of the strategic direction of the enterprise.
- Interfunctional and co-ordinated review of the strategic direction of the enterprise through an analysis of the likely efficacy of organizational strategies to support strategic purposes.

Teams can be used as organizational focal points to develop the data bases and intelligence needed to evaluate future initiatives. As the managers review each team's progress, an informed judgement can be reached concerning the adequacy of preparation and progress for strategic purposes. By using teams as the cutting edge of evolving strategic initiatives, an ongoing assessment of the likely "strategic fit" of the team's anticipated results can be measured. An overall prescription of the key questions to be asked during the review of each team's work includes:

- Where are the team's results with respect to anticipated cost, schedule, and technical performance capabilities?
- What is the likely strategic fit of the team's results into the overall strategic purposes of the enterprise?
- Are there any anticipated problems developing in the team's activities that could have a negative impact on the timely delivery of the team's results?
- During the team's deliberations are there any opportunities coming forth that could suggest future opportunities for the enterprise?
- Finally, given the other teams' work that is being carried out in the enterprise, can the probability of being adequately prepared for the enterprise's future be determined?

8. Summary

Strategic planning is seen as a process through which individuals and teams within an enterprise can learn to deal with unpredictable and rapidly changing environments. If a forecast turns out to be wrong, it is not an indication that management is incompetent, but rather a confirmation that we are living in an uncertain world where environments are changing and strategic management needs to assess and reassess continually the enterprise's current and potential future situation. Strategic planning is one element in a wider program of organizational change. It is the role of strategic management to integrate strategic planning into an overall perspective of the enterprise's resources as if its future mattered.

Typically the implementation of strategic management is usually carried out, at least in part, by people that were not involved in the original formulation. Ownership of the plans can become a problem - as well as full interpretation and understanding of the strategies. All too often managers make the simple assumption that implementation is not a problem. However, once this is brought out into the open and its implications recognised, problems associated with implementation can be understood leading to removal of implementation difficulties. Having alternative teams involved in both the design implementation of strategies can help to better utilize the commitment and use of resources to meet team, and organizational strategic objectives.

References

1. Description of the strategic management roadmap is taken from Cleland, D.I. (1996) *The Strategic Management of Teams*, John Wiley & Sons, Inc., New York, NY.
2. *Ibid.*
3. Martin, J. (30 Sep 1996) Are You As Good As You Think You Are? *Fortune*, 150-152.
4. Norman, J.R. (10 Oct 1994) A Very Nimble Elephant, *Fortune*, 88-92.
5. (29 Sep 1993) Strategic Benchmarking, *Financial World*, 30-31.
6. (1993) Working Smart, *The H.J. Heinz Company 1993 Annual Report*, 16-17.
7. O'Reilly, A.J., (12 Sep 1995) *H. J. Heinz Company Annual Meeting*.
8. Howard, A. and Magretta J., (Sep-Oct 1995) Surviving Success: An Interview with the Nature Conservancy's John Sawhill, *Harvard Business Review*, 108-118.
9. Maremont, M. (30 Jan 1995) Kodak's New Focus, *Business Week*, 62-68.
10. Kerwin, K. (3 Apr 1995) Ford, Cover Story, *Business Week*, 94-101.
11. Ghoshal, S. and Bartlett, C.A. (Nov Dec 1994) Changing the Role of Top Management Beyond Purpose, *Harvard Business Review*, 79-88.
12. Sheridan, J. H. (19 Feb 1996) Lessons From the Best, *Industry Week*, 13-20.

QUESTIONS on "ALTERNATIVE TEAMS IN STRATEGIC MANAGEMENT"

Prof. Cleland answered a few questions about his work.

- Concerning how to select members of a team, Prof. Cleland asked who knows best how to do a job? - the people actually doing it. Therefore, he has found the best results come from allowing the team themselves to select the members, and even to some extent assess the team-members' merit. The managers still have the overall responsibility, but they should listen to the views of their team. The senior management provide the vision, the foresight, and the guidance as to where might the organisation go and where it can go.
- Concerning the borderline between areas such as strategic management, marketing and project management, Prof. Cleland saw not borderlines but interfaces, representing coming together at common boundaries of interest.
- Concerning the borderline between the management of change and the management of on-going operations, again it is important not to have a line of demarcation, since the two reinforce each other and new ideas from each must feed into the other.
- Prof. Cleland emphasised the need for training programs, particularly for senior management on the benefits that PM can bring.

MANAGING COMPLEX PROJECTS BY ACTIVE HIERARCHICAL SYSTEMS

V. I. VOROPAJEV
S. M. LJUBKIN
Russian Project Management Association
(SOVNET), Moscow

1. Introduction

In recent years the development of Project Management (PM) has been characterised, firstly, by an essential increase of the projects' complexity and by an expansion of the areas of their application. Secondly, both methods and techniques used in PM are not developed sufficiently and lag far behind practical demands. This results in decreasing the efficiency of the projects' realisation. In this paper we suggest a new concept to create a PM system which is adequate to today's conditions for realising modern and complicated projects. The basic conceptions developed comprise both methods and techniques, which enable raising the activities of all project participants into proper on-line decision-making.

We suggest implementing an active hierarchical system in project management (AHSPM) which enables:

- increasing both activity and efficiency of the project managers.
- supplying each managing level and all project participants with proper information, methods and techniques to undertake on-line decision-making, in accordance with the participants' responsibility and competence.
- reduction of useless links and stages in processing information and decision-making.
- minimisation of PM costs.

The suggested concepts of creating the AHSPM are based on the results obtained by the authors and their colleagues in the last two decades [1-7]. The results obtained deal both with the projects of high complexity and with developing for these projects efficient managing systems. Certain results deal with the human role in a project and with attempts to increase that role. This refers to the

T. M. Williams (ed.), Managing and Modelling Complex Projects, 221–236.

suggested active hierarchical system, which enables undertaking decision-making by all participants in a project.

The history of our results is as follows. The basic results in the paper have been developed by large scientific institutions, which have been supervised by one of the authors over more than twenty years. Thus some achievements are 25 years old, while others seem to be younger. It is well known that scientific institutions of the former USSR, which have been financed from the state budget, have carried out many developments. Some of them are of scientific value nowadays too. Unfortunately, these results have not been implemented in practice, but some of them have been in advance of their time and have been developed for the future. We are sure that some of the former developed fundamental results will be useful for creating the PM systems of the future. Those results have been obtained by a group of mathematicians from Moscow State University. Among those participants have been such well known mathematicians as Adelson-Velskii G.M., Kalinovskaia S.S., Lebed B.Ia., Sheinberg M.V., Orel T.Ia., etc., together with specialists on information technology and management systems. Those results have been published in numerous books and journals, but, unfortunately, all of them in Russian. Thus, most of those results are unknown to the broad western readership.

The results include also a number of normative and methodical documents in the area of the construction industry. In Russia those results have been successfully tested on several construction and investment projects. In such a systematised form this research is presented for the first time. Four topical questions are discussed in the paper:

- Why active hierarchical systems in PM?
- How can they be created?
- What is their application area?
- What is the status of their research and development?

2. The Problem

As mentioned above the need to create a new system is caused by a delay in developing PM methods and techniques from the growth of projects' complexity and other practical demands.

2.1 HOW A COMPLEX PROJECT IS MANAGED NOWADAYS

According to traditional managing systems the burden is laid upon the project' team. The existent management system is shown on Fig. 1.

Hierarchical Management system

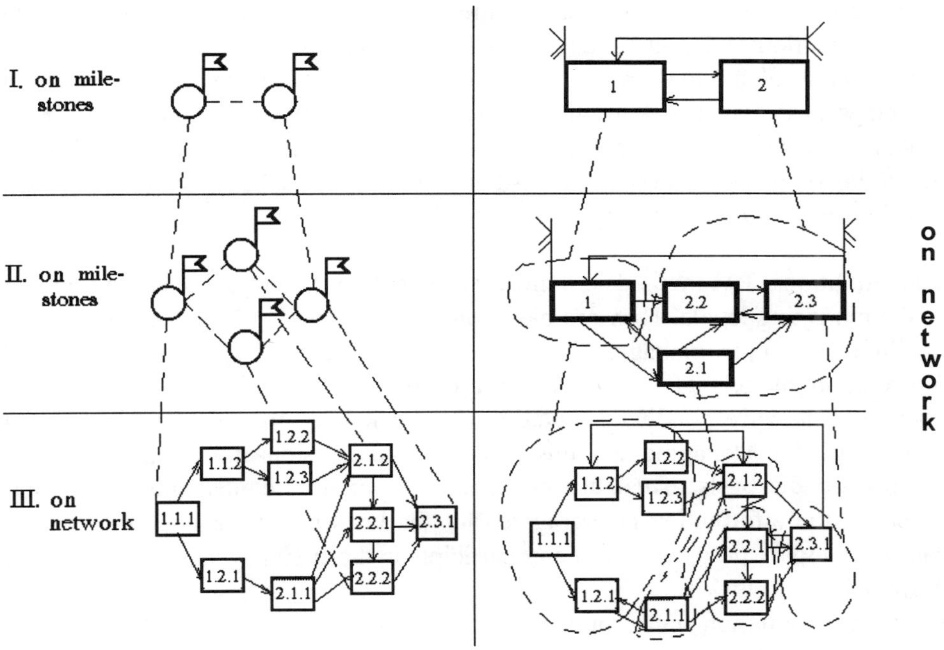

Fig. 1. existent system   Fig. 2. suggested system

For such a system the managing processes are carried out as follows. A basic detailed model of the project's realisation is developed. Let it be a network model. Assume that there are three managing levels. The information for the higher levels is received after calculating the main parameters of the basic detailed model. The information is aggregated and enlarged relative to certain key events, which are of interest to the managers of the upper level. The aggregation may be cost-oriented, target-oriented, etc. The information may be given in the form of the Gantt diagram to present the timetable of the activities. Such a modified information is not a network model since it does not include the logical links between the activities. This is nothing but a reflection of the basic model via information. For example, for decision-making at the upper level one has to:

- descend to the lower level and calculate the parameters of the basic model;
- reflect the information on the second level;
- enlarge and reflect that information for the first level;
- at that level all the managers undertake decision-making according to the received information.

Unfortunately they are unable to undertake an efficient decision-making on the basis of that information. In order to analyse the latter to calculate certain parameters together with forecasting, they have to introduce certain changes in that information. Afterwards they are obliged to descend again to the lower level to undertake calculations and later on to return "upstairs" again. This results both in delays in decision-making and in losses of the system's activity. The basic model is mostly an acyclic precedence network model. The latter usually are deterministic time-oriented models. Alternative or stochastic network models are mostly used in unique projects.

Thus traditional PM methods are mostly oriented on the project team and on one of the managing levels. They are based on a detailed network project model. For medium and upper managing levels we generally use a set of non-connected information models given in the form of milestones and an enlarged Gantt diagram. The corresponding calculations are carried out on the basis of the initial network model. Managing all kinds of reports and decision-making at all levels is realised on the basis of calculating parameters of the detailed initial network. Later on, the information is aggregated for the level under consideration. In the case of complex projects with many participants those PM systems have evident drawbacks as follows:

- many information and computation procedures;
- delays in decision-making due to long information routes and numerous information aggregations and filtrations;
- losses in accuracy and quality of the information;
- low reactivity of the PM system.

This results both in decreasing the participants' reaction to current changes at all levels and in increasing the cost of the project's management.

To develop PM methods and techniques one has to create new systems without the drawbacks outlined above (Fig. 2). Unsuccessful attempts have been made to create PM systems (for all levels) on the basis of traditional networks. That is because it is impossible to undertake an accurate aggregation of the project's activities and to undertake co-ordinate decision-making on all levels via simulating acyclic networks.

2.2 WHAT DO WE INTEND TO UNDERTAKE? WHAT'S THE PROBLEM?

We intend to create a system in which each participant and each managing level could operate with an individual model. The latter would provide the necessary adequate information to undertake decision-making.

In the case of managing the project, the OBS has not only a vertical management hierarchy, but includes horizontal structures since we deal with matrix structures. Certain participants, being responsible only for a part of the project, are interested to work only with a certain fragment of the model, which corresponds to their local interests. In order to provide them with such a possibility one has to know how to aggregate models or to realise their equivalent transformation. Moreover, it is a quasi-equivalent transformation since it is impossible to undertake an equivalent transformation for networks with precedence. It is possible to enlarge only chains of activities or certain fragments, which have common start and finish events. In the course of aggregation the links are lost together with the network information. Thus we obtain entirely different models.

If we could succeed in developing for network models their proper aggregation and equivalent transformation, then we would succeed in delivering to each managing level its individual model. Such a model could be analysed and decision-making could be carried out on the basis of the model. The gain from solving that problem could be tremendous. For example, in creating a power-station with up to 100,000 activities at the lower level: at the second level the number of activities would be less by a factor of 10^{-1} to 10^{-2} (up to 1,000 activities); at the next level 100 activities are enough; while at the upper managing level 5-10 enlarged stages to be controlled are enough. There is no need to analyse and to calculate at all levels, i.e. to descend to lower levels every time and to undertake calculations by using the detailed model.

To create an AHSPM it is necessary:

- to develop the theoretical grounds of simulating the project activities, which enable equivalent transformation of the project model;
- to develop both formalised methods of the project's decomposition and the project's information models;
- to develop a hierarchical model with required properties according to the requirements of the project's participants (investors, customer-users, contractors, subcontractors, project managers, etc.) on different managing levels;
- to develop methods of network models' equivalent transformation together with methods of their analysis and calculations;

- to develop both methods of hierarchical time-oriented scheduling and models to control the project's realisation;
- to organise an effective PM system's functioning.

In the opinion of the authors, an effective way to develop an AHSPM is to use the theoretical grounds of simulating generalised network models and various methods of decomposition together with developing appropriate information technology and its application.

3. What is the basis of the suggested approach? How can it be realised?

The authors suggest using the four main components.

A. The method of the project's structural decomposition together with methods of developing different structures with given properties. This is carried out nowadays by well-known PM programs such as Primavera, Artemis, etc. However our approach is based on strictly formalised decomposition methods applied to various information objects. Later on, various other required structures may be composed in the course of realising the project.

B. Simulation techniques are based on the so-called generalised network model (GNM). The latter possess mathematical properties, which enable an equivalent network aggregation.

C. Algorithms to aggregate GNM.

D. Scheduling and monitoring methods applied to hierarchical GNM.

These components are considered in greater detail below.

3.1. THE PROJECTS BREAKDOWN STRUCTURE

To create an AHSPM, various decomposition procedures (developed by the authors [4]) to present and to analyse the project's information are used. Those procedures enable:

- multiple project breakdown structures oriented on various decompositions attributes (structural spatial, cost, resource, organisation, etc.);
- creation of various breakdown structures with given properties;
- solution of optimisation problems to develop hierarchical networks with given properties.

In order to realise all this one has to:

- choose a complete section of the structural model answering the managing levels' requirements;
- choose non-complete sections (network fragments) with given properties. The corresponding attributes here are either the executors for each type of activity, the types and the moments of payment, the required resources, etc.;

- develop a hierarchical GNM in accordance with the chosen graph sections on breakdown structure.

The project's decomposition structure is presented on Fig. 3.

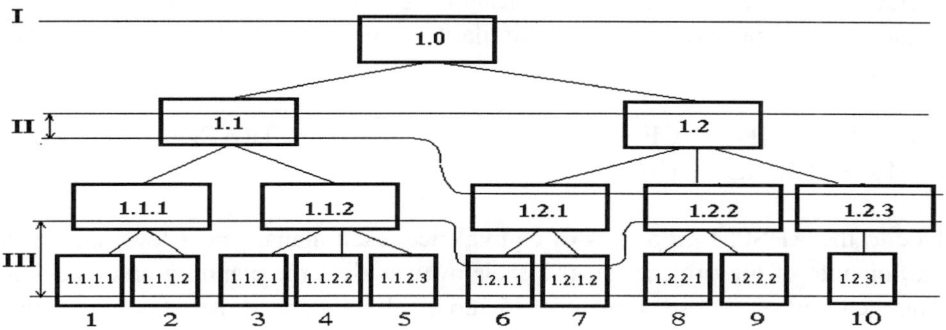

Fig. 3. Work breakdown structure

3.2. GENERALISED NETWORK MODELS

To simulate hierarchical PM systems we suggest the use of GNM [1-4]. The main difference of GNM from other well-known types of network models (CPM, PERT, GERT, VERT, etc.) is that in GNM inclusion of non-positive cycles is allowed. This enables implementing practically all kinds of time-oriented links as well as other types of linkage, which can be transformed to time-oriented links. Besides this the links between arbitrary project events enable aggregation of network information, i.e. enable the models' equivalent transformation.

The GNM properties outlined above enable carrying out the networks' equivalent aggregation. The latter is needed for creating an AHSPM. Moreover, a GNM enables an adequate descriptions of various conditions and links, which cannot be described by other models. GNM models have a strict mathematical basis and a widespread spectrum of methods and algorithms to analyse and to calculate those models.

3.3. THE GENERALISED NETWORK MODEL'S AGGREGATION

Using aggregation methods developed by the authors [6,7] provides the equivalence, i.e., the similarity in AHSPM. The essence of the method is as follows: on the basis of the detailed initial GNM at the lower level we develop an enlarged model of the adjacent/hierarchical level. Later on the procedure is repeated. Each activity entering the enlarged model is a unification of activity fragments of the detailed model, including start and final links' parameters, various restrictions. Such an unification is obtained by using aggregation algorithms to provide the nets' similarity according to the outlined above conceptions.

3.4. METHODS OF SCHEDULING AND TRACKING USING HIERARCHICAL NETWORKS

Due to the AHSPM properties of equivalence, each management level and each participants of the project can use their own models for planning and tracking. They can also choose from the hierarchical network those models, which correspond to their current tasks. During project realisation some changes and solutions demand correction of models and recalculation. In this case all management levels agree upon proper decisions and then modifications are made simultaneously in all AHSPM models.

The transition to GNM, i.e. to cyclic models, led to essential changes in the mathematical conceptions. We have used as yet only deterministic time-oriented models. Those models do not include probabilistic parameters or alternative events. In our opinion the corresponding research can be undertaken and we are planning to do it.

Using GNM led both to the development of various new algorithms to analyse the models as well as to the development of optimal schedules for these models. A conclusion can be drawn on the basis of the results obtained that we possess a set of required algorithms to undertake a time-oriented analysis of GNM as well as various scheduling algorithms for the case of restricted resources. Those algorithms include [3,5]:

- the algorithm of minimising the utilized resource level with given due dates ("smoothing" algorithm);
- the algorithm of minimising the project's duration with given restricted resources (the "calibration" algorithm);
- the algorithm to minimise the project's cost;
- various algorithms to undertake time-resource analysis;
- the multicriteria optimisation algorithm (of Pareto type);
- the method of penalty functions, etc.

4. The Hierarchical Scheduling Problem

Consider one of the most important PM problems: the hierarchical scheduling problem within the framework of the AHSPM (see Fig. 4). Each contractor and subcontractor with his own goals, criteria and restrictions is seeking a solution by use of his individual network model. All the requirements criteria and restrictions are passed to the computer. On the basis of the detailed initial network by using multicriteria optimisation algorithms proper feasible co-ordinated solutions (optimal in a certain sense) are obtained. The multicriteria optimisation problem can be solved by using:

- the method of penalty functions and discrepancies on the basis of Scheinberg and Kositskii method [2]. The method has been realised on a PDP-11;
- modified decision-making techniques, which are developed by Burkov [8];
- expert systems and methods of artificial intellect [9].

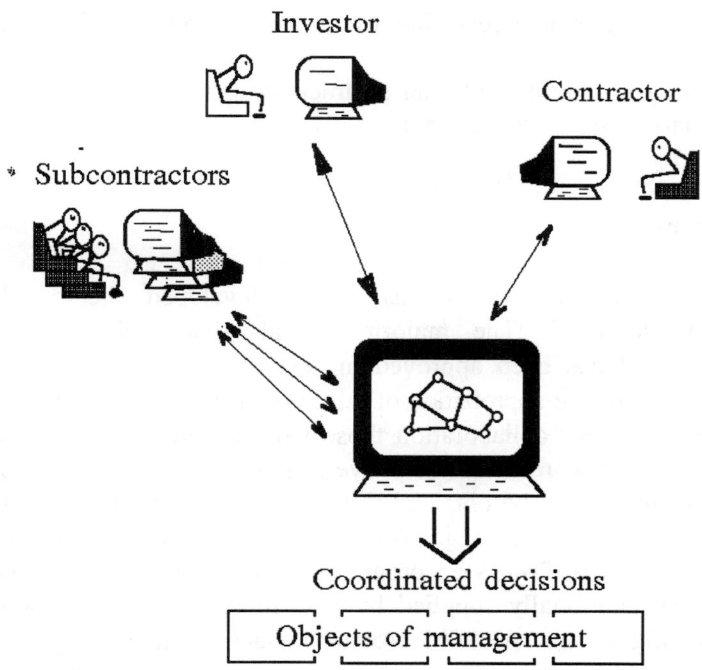

Fig 4: Scheme of hierarchical scheduling and monitoring

5. Creating an AHSPM

Developing an AHSPM comprises the following steps.

A. Creating the project WBS with pre-given parameters of activities' structural elements.
B. Creating the project structure models and their compositions according to the requirements of the PM system.
C. Creating the project OBS in accordance with project composition models (step B).
D. Developing the requirements to the detailed activities' (steps B, C).
E. Creating the GNM at the lower level in accordance with steps B, C and D.
F. Creating the hierarchical GNM on the basis of step E and using aggregation algorithms.
G. Analysing the hierarchical GNM and obtaining schedules to be tested for their equivalence. Developing the schedule for the basic network activities.
H. Planning on-line controlling and reporting using the hierarchical GNM in accordance with step C.
I. Recalculating the project's multilevel plan by tracking and decision-making on the basis of deviations from the basic plan.

6. State of Development

The proposed hierarchical PM system has a well-developed theoretical background published in [1-7]. The majority of algorithms have been implemented on a PC and has been approved in several projects. However, complete program realization and approbation of this system in actual practice is not done yet. Authors invite for collaboration those who are interested in this problem. In our opinion the research outlined above is a very actual one and we are sure that financial support can be obtained to proceed and to complete that research in the form of a program package. We suggest using for that purpose existing information stores, e.g. Primavera program packages or other similar. Those packages can be additionally supplied by new modular blocks, which could be used for solving various AHSPM problems. Such complex program packages can also be used for developing new packages with given properties to solve problems in hierarchical managing. The one which would first develop such packages will obtain essential advantage on the world markets.

7. Application Areas and Efficiency

The results of this paper have been used for managing large irrigation and melioration projects, construction of nuclear electrical power stations, etc. In our opinion, those results could be of a certain use for managing large investment and innovation projects and programs both in technical and social-economic areas. In modern Russia these projects may be related to:

- conversion of defence industry;
- development of the oil and gas industry and systems;
- aerospace and telecommunication systems;
- privatisation and reforming of state enterprises;
- social security and health projects;
- environment systems.

Using AHSPM enables:

- realisation of current trends in PM systems;
- improvement of PM efficiency;
- improvement in the usage of top management levels and main project participants;
- significant reduction in the cost of PM.

The problems of PM systems are common in many countries. The solution of those problems will result in future progress of PM methods and techniques in raising the efficiency of using AHSPM systems. Such a research will certainly be an object of international collaboration of specialists in the PM area.

References

1. Adelson-Velsky, G.M., Voropajev, V.I., and Kalinovskaia, S.E. (1971) Generalised networks in the construction industry. Methods of designing for generalised network, J. Russian construction 4, 5, 28-32, 23-26
2. Voropajev, V.I., and Scheinberg, M.V., et al (1974) Generalised network, Methodical instructions, CNIIPIAS, Moscow.
3. Voropajev, V.I. (1974) Models and methods of scheduling in information management system, Moscow.
4. Voropajev, V.I.et al (1988) Construction project`s decomposition into technological modules (Methodological grounds), VNIIGIM, Moscow
5. Voropajev, V.I., and Ljubkin, S.M. (1989) Hierarchical system of generalised networks for melioration objects, Improving of management and organisation systems for melioration constructions/USSR, SIBNIGIM, Krasnojarsk, 111-123.

6. Voropajev, V.I., and Ljubkin, S.M. (1994) Usage of Hierarchical Networks for Managingt of Complex Projects, PMI Annual Symposium, Proc., Vancouver.
7. Ljubkin, S.M. (1991) Aggregation of generalized networks in construction, Management and information problems in information area, CNIIEUS, Moscow, 77-82.
8. Burkov, V.V., and Irikov, V.A. (1994) Models and methods for controlling organization systems, Nauka, Moscow.
9. Bolotova, L.S., and Ljubkin, S.M. (1995) Expert Systems and Project Management, Proceedings of the International Symposium "Improving Project Management", September 14-16, 1995, ALANS/SOVNET, Moscow.

OPEN DISCUSSION: MANAGEMENT TECHNIQUES

Discussion chaired by:
V. BURKOV
Vice President, Russian Project, Management Association

This summary by:
T.M. WILLIAMS
Strathclyde University
Glasgow, UK

Introducing a very brief discussion of Management Techniques, Prof. Burkov began by discussing the human nature of project workers. He said that any human has an element of greed and laziness in his nature, so we need a PM system to motivate the workers. He contrasted his experience with that of Prof. Yeo, who in discussing his paper had described a fully committed project team, all working together towards the overall success of the project. He questioned the applicability of Prof. Cleland's concept of "alternative teams" in his domain of experience, as he felt disputes would arise within the team. Other CP-country participants agreed that the idea of teamwork was important for CP countries, as workers currently resist working in teams. So, Prof. Burkov asked, what are management techniques - just an art?

Prof. Bushuyev said that there was a gap between PM as a science and the techniques of PM. Some complex mathematics do not work in managing complex projects: for example, he was dubious that Prof. Voropajev's hierarchical methods could be implemented.

In a rare cross-cultural agreement, Dr. Reiss, who was to take part in the next Section on Programme Management, said that the models he used were very similar indeed to Prof. Voropajev's. It seemed to Dr. Reiss that PM software has been an external model, used only by Project Managers, and ignoring the interests of many people in the PM process - external in the sense that it is difficult to prove that the model has any effect at all. The model his company has developed, very similar to the CP-country hierarchical models, was designed for the multi-project environment - which is very similar to the "Complex Project". Tools are needed by the Project Manager, the Resource Manager, the senior management, and the individual resource: so, his reasoning had gone, if we could bring those people together in a model and provide them with tools, we would bring the human element back into PM, and if everyone involved can play a role in the planning and monitoring of the project, this produces a better team-spirit and a sense of ownership within the team

Prof. Golenko-Ginzburg had previously expressed concern that networks could not support planning, scheduling and controlling simultaneously. Prof. Voropayev's methods, by carrying out these functions at different levels of the hierarchy, possibly provided a solution to this concern.

It was suggested that the culture in CP countries was actively to discourage teamwork, apart from in some sports. Structures are hierarchical, and this makes projects less efficient.

Prof. Chapman pointed back to the discussion at the end of Section 6, which shows a bridge between the formal hierarchy and the team. Where there are a number of organisations involved, and there is necessarily a hierarchy (because different organisations do different things), it is important in the *planning* stage to use hierarchical approaches to understand how the project is to be carried out; in the *implementation* stage, even deeper levels of hierarchical structures are needed; but for *risk management,* this needs to be done at a fairly high level team to begin with, using a small team that cuts across the hierarchy. This also provides an benefit in improved teamworking between the organisations. You must have these cross-hierarchical teams, otherwise each part will plan their activities based on one predicted time for all the other activities, which is very likely bound to be wrong. It is the dialogue about the variations that might happen that get people ready to meet those variations, and in particular to meet those variations together as a team with a common objective.

So once again there was a difference in emphasis between NATO countries and CP countries, with team-work much more widely accepted in the former, both in actual practice and in the attitudes necessary for implementation.

ADDENDUM TO "MANAGEMENT TECHNIQUES"

A. del CANO
Escuela Tecnica Superior de Ingenieros Industriales,
Universidad Politecnica de Madrid, Spain

Reflections written after the Workshop (concerned with management of complex projects), edited by the editor of the Volume.

In almost all type of projects, functionality, security, safety or secrecy may be extremely important objectives, impacting the project results in similar ways to the standard objectives (scope, time, cost, quality). This is more important in complex projects, so there is a corresponding need for correct management of these objectives; it can even constitute a separate PM area in certain projects.

In general, integration, communication, organisational and human factors and risk areas are especially important in complex projects. Additionally, complex mega-projects normally need to be very long, so when time is critical, concurrent engineering can be a strategic alternative to adopt. This is a source of changes all along the project life cycle, so in this case quality, organisational and human aspects, communication and (especially) risk areas become even more important.

Large and complex projects need an organisational structure with a concept of decentralised operations and centralised control. Normal decision making must be very near to, or even at, working levels, always conditioned to achieving general project objectives and respecting general project guide-lines or meta-rules. Fluor Daniel and Bechtel (USA) or Spie-Batignolles (Europe) are some examples of companies including this concept in complex projects. High flexibility in project procedures is needed. The project manual is a relatively short document including only project objectives and a set of general guidelines to be detailed by the several sub-teams in the project management team. In international projects the organisational aspects can even be more complex than in other "normal" complex projects. In these cases, for instance, project team family care management by the company can be a separate highly important subproject, because family non-adaptation (whether the family is staying or not in the foreign country) can make excellent workers ineffective; Dragados (Spain) is an example of continuous concern in this field. International projects many times also involve circumstances resulting in a special need for management of cultural differences. When this is important, organisational management can be improved through:

- careful project management team selection towards finding persons with a special ability to perceive and manage cultural differences, i.e. able to understand others and respect and take into account their habits and values;
- if there are several locations, organising visits to the different places where the project is being developed, in order to achieve a better understanding on the specific environment surrounding the other project participants (including both work and cultural / recreational activities).

In general, it's important to "sow and reap" for innovation and creativity and for partner-feeling, enthusiasm, trust and commitment, and to make preparation when sources of complexity include company and/or general cultural differences.

Communication and information systems are often a key success factor. Using advanced communication systems (especially when distance or conflicts (terrorism, war) are a problem) and the broad set of available (qualitative and quantitative) decision-modelling tools (as described in this Volume) in an environment of easy-to-use Computer Aided Project Management results in more time for reflection, innovation and creativity to solve problems and to generate the optimum environment for the project. Some years ago Dragados (a leading Spanish construction contractor) began bidding for a very large public works construction project in a N. African country with an initially stable situation. The contracting process was very long and Dragados finally won just after a radical Muslim terrorism problem exploded. Then the project manager refused to continue at the same conditions: for several reasons he would be a strategic target for terrorists in the foreign country; he also refused to negotiate on the basis of improving his economic conditions. It was impossible to find another project manager with the adequate profile and wishing maximum responsibility on site for the project. The obvious solutions were to refuse to develop the contract in a risky environment, or to contract a project manager from outside the company, with corresponding cultural differences, and behavioural uncertainty and risk. In fact, it was possible to negotiate and obtain the commitment of all the other team members to work and live in the foreign country, so Dragados then conceived a solution where the project manager will work in Madrid, only visiting the country and the site when absolutely necessary (with special safety conditions), designing a complete real-time satellite video communications system with cameras in any relevant site and with video-conference for daily contact with the project team. This is an example of a complex environment where an advanced communications system can make feasible an apparently infeasible project.

SECTION 7
PROGRAMME MANAGEMENT

The problem of managing several large- or medium-sized projects simultaneously has some similarities with that of managing the "complex project", and some important differences. The following section contains one paper on the subject by Prof. Sergey Bushuyev, Workshop Co-Director. Before this paper, Dr. Reiss, who was to lead a brief Open Discussion afterwards, presented a short introduction to the topic.

INTRODUCTION to PROGRAMME MANAGEMENT

Introduction by:
G.REISS
Hydra development Corp,
37 Kirkgate, Otley, W.Yorkshire, UK

This summary by:
T.M. WILLIAMS
University of Strathclyde,
Glasgow, UK

The term "Programme Management" is used in many situations, which can be organised into four groups:

- an organisation with many projects, using common resources but with no common objective
- a mega-project (such as a space project)
- an organisation carrying out many projects for one client
- an organisation with many projects specifically designed to change the organisation itself (this is the UK government's definition).

So Dr. Reiss' definition is: "The planning and monitoring of tasks and resources across a portfolio of projects (and perhaps leading to a common objective)."

The problems of planning many projects are different from those of planning a single project, and the methods used are different. In a single project, there is one project at a time; the concentration is on time and method - the origin of these projects is in large complex projects where a single mind could not understand the whole process; each project is dissimilar, each one being a new challenge; the plans are often complex, and CPM can often help people understand what the project contains; there is a need to minimise the demand for resources, and those resources will generally be classifications of resources (e.g. specific trades); there is a finish and a start; there is usually a single objective; and the resources generally work full-time on one project at a time. In the multi-project environment there are many resources; the concentration is on resources; the projects themselves are usually simpler and are often similar to each other; the plans are often simpler (bar-charts often being used) and the need is to maximise the utilisation of resources (and the resources will often be individuals); there is no start and no end; there are many objectives and the resources will work on many projects.

What has been found being used is a model that takes the existing tools in the market and adapts them to this new problem. The model most used is as follows:

- each project manager plans his/her own project;
- these plans are transmitted to a central location;
- these plans are consolidated into one very large plan;
- decisions are made as if this plan was one big project (evaluating, planning, trying options, making decisions);
- decisions are disseminated to the individual projects;
- achievement is modelled.

But there are very significant problems with this model, as will be seen in this Section.

MANAGEMENT OF MEDIUM-SIZE PROJECTS PORTFOLIO

S.D. BUSHUYEV
Dr.Sc., Professor, Kiev State Technical University
of Construction and Architecture,
and President of Ukrainian Project Management Association
Kiev, Ukraine

The epoch of global changes and projects, conducted in the former socialistic states, stimulates not only the development of project management culture in these states, but the global process of knowledge, methods and models also. The creation of a uniform basis of knowledge for professionals in the branch of project management and the transition of Western culture in the joint projects also provide the motivation of Ukrainian project managers to active study of the newest methods and means of project management. The following functions of projects portfolio management systems are considered in this paper:

- creation and administering of projects portfolio;
- creation of dynamic links between projects and those structures (layouts, structure of activities, project groups and so on);
- preparation and monitoring of projects;
- analysis of current state and planned parameters of projects complex completion;
- use of wizards technique;
- preparation and making of solutions by projects portfolio and each portfolio management.

One of the projects portfolio administration methods is the method of dynamic priorities. The method is based on the creation of project dynamic priorities based on knowledge about the current state of complex, organisational- technological links, limited financial and material / technical resources, deviation of the project's implementation from target plans and so on. The rules of creating the dynamic priorities are put in a knowledge base. It provides a considerable flexibility for preparing solutions and making algorithms in a multi- project environment.

T. M. Williams (ed.), Managing and Modelling Complex Projects, 241–254.

1. Introduction

The development of a project management (PM) culture in Ukraine is conducted with the transformation of economics and society. Social changes stimulated the development of professional activities connected with investment projects, and preparation and maintenance became important. The main role in this process is played by technological transfer projects, which provide a fast exchange of knowledge and the newest technologies for the development of Ukrainian enterprises [1,2,3].

The implementation of large projects (for example, the close-out of the Chernobyl nuclear power station in Ukraine) requires the development of appropriate tools for their decomposition by groups and creating effective information systems for projects portfolio management.

2. Structure and functions of modern systems of projects portfolio management

Modern methods and means of complex projects' management are based on the methods of their decomposition by parts (subprojects, project groups and so on), with further creation and administering of projects portfolio. Now, the necessity of non-typical projects management methods of dimensional planning and distribution of limited resources has arisen.

The structure and functions of projects management systems depend on their place in the development of organisations. The industrial process, output production, creation and implementation of projects, their influence on the development of organisation are analysed and the tools of general and strategic management interact with the appropriate project management tools [4].

The general scheme of projects portfolio management system is defined based on the interaction of manufacturing, output production, and projects development cycles (fig. 1).

We define the following functions of projects portfolio management:

- creation and administration of the project portfolio;
- creation of the dynamic links between projects and their structures (layouts, structures of activities, project groups and so on);
- preparation and monitoring of projects;
- analysis of the current state and planned parameters of project completion;

- the preparation and making of decisions by the management of the project portfolio and the separate projects.

The decomposition of a complex project or projects by parts is connected with the necessity of simplifying decisions and defining tasks during the project's preparation and monitoring. The practice of project preparation, supported by modern information technologies, is based on the principle of inheritance and the creation of libraries of typical network model fragments in the frames of the current problem area (scope). The practice of project monitoring is created by the necessary variety of project representation forms and methods, which are used for distinct presentation of the situation of management solutions' preparation and making (fig. 2).

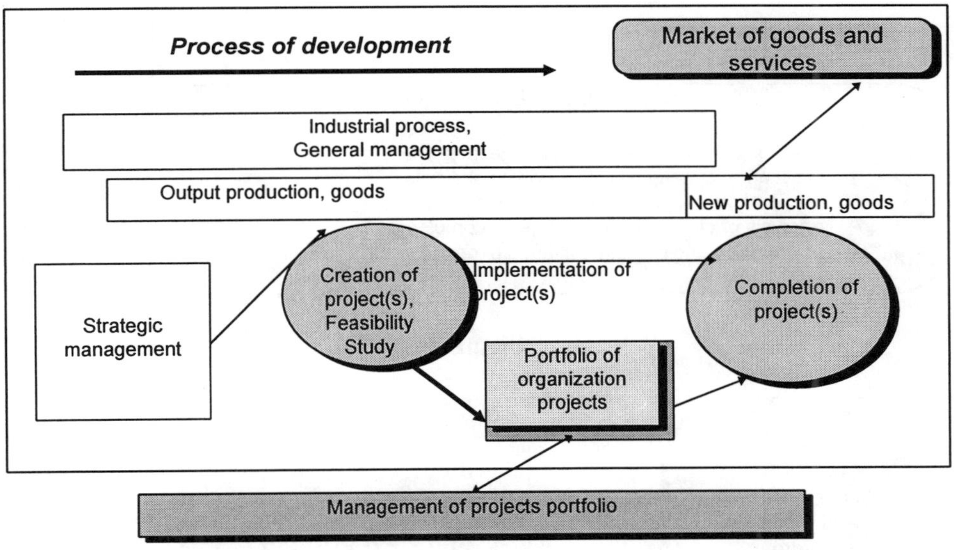

Figure 1

3. Administration of projects portfolio

The proposed structure of the projects administration system is shown in fig. 3.

Consider some of portfolio projects management methods, implemented based on previous functions. Define two different stages by contents - *preparation of models* of projects, creation of interproject connections, constraints of portfolio and *monitoring* of projects.

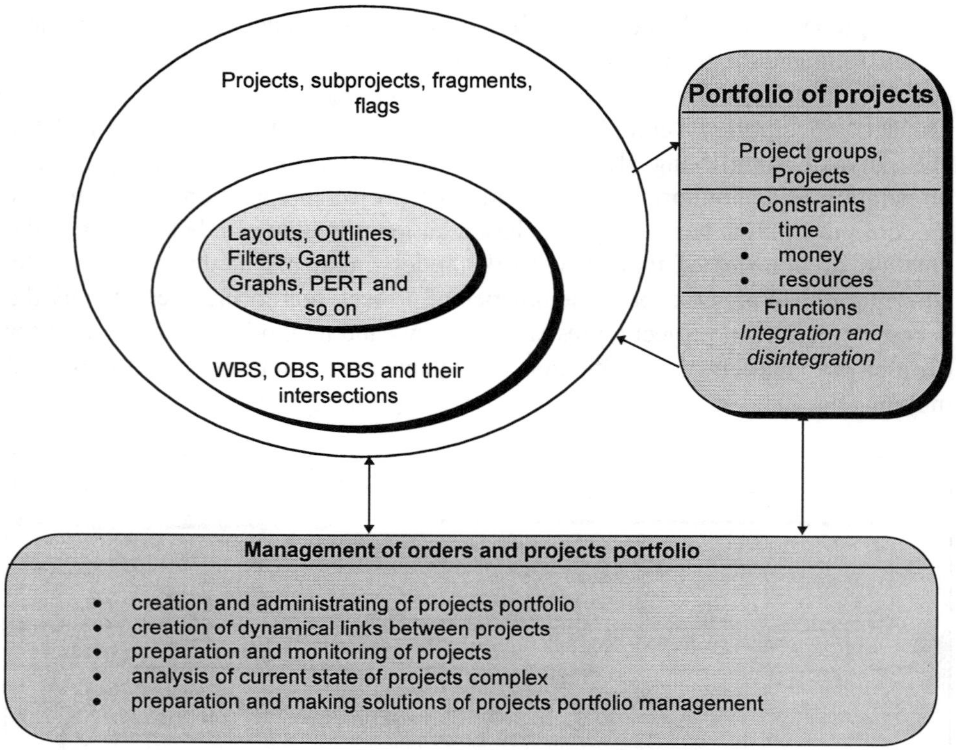

Figure 2

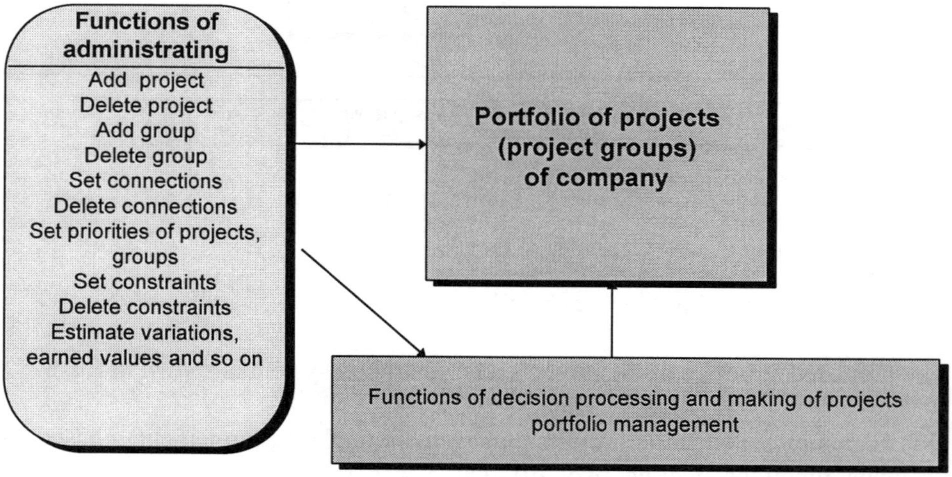

Figure 3

The preparation of projects models is conducted by the scheme: *activity ⇒ group of activities ⇒ fragment ⇒ WBS-structure ⇒ subproject ⇒ project ⇒ project group/interproject connections ⇒ portfolio of projects/ constraints*. The methods for projects in portfolio monitoring are connected with fixing of current states for all projects, variations of project states from target projects by time, cost and used resources, constraints by time, cost and/or resources, priorities of projects, interproject connections and connections inside project, possibilities of project(s) implementation suspension and so on. Well-tested procedures of levelling, smoothing, stretching and splitting of resources may be used. The general scheme of projects portfolio monitoring is shown in fig. 4.

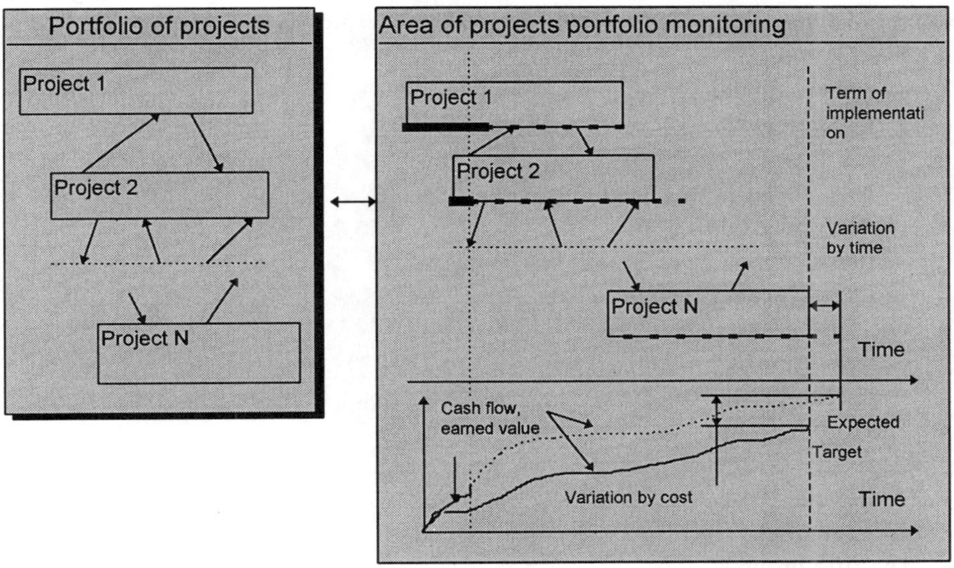

Figure 4

The simplest situation in projects portfolio implementation, given estimations of variations between expected time and costs, requires deep analysis with processing of an effective solution, as shown on fig. 4. Possible solutions include:

- intensification of critical and near- critical activities of projects by changing the calendars of activity and resources;
- reconstruction of the model of the remaining part of the project with the least priority by the way of changing of technology or activities organisation;
- search of possible projects intersection;
- undo of part of resource constraints;
- and so on.

An accurate account of constraints by resources is provided by multiple usage of such procedures at the levels of project groups, projects and activities. A group of projects, project, or group of activities may be represented as a complex activity with preliminary optimised graphs of resource using. A good example of such project management system can be seen in concentric project management [5].

4. Creation of dynamic links between projects and their structures

Project(s) as a vital body require flexible intelligent links, working in the environment of various situations, requiring managerial solution-finding. The main links may be divided by groups and kinds.

- The first group
  - organisational
  - technological
  - resource
  - cost
- The second group
  - linear
  - hierarchical
- The third group
  - deterministic
  - probabilistic
- The fourth group
  - fixed
  - flexible
  - intelligent
- The fifth group
  - linear
  - non- linear (synergetic).

A most important role in creating model flexibility, managed in the base of the projects portfolio, is played by the interproject links. Among a great number of such links we pay attention to the groups of organisational-technological links, cost and resource links, intelligent links of state, and links between managers of subprojects, project groups and project. The main tasks of such links are sampling the interaction between project managers, projects and resources (including monetary resources) with the account of time, cost and quality constraints and so on.

We consider each group of links as follows.

- The links of the first group - organisational, technological, resource and cost - are described and worked in the environment of traditional methods of network sampling. Their relation with time is defined by activity and resource calendars.
- The second group of links - linear and hierarchical - is defined by the composition and decomposition structures of projects, groups of projects and projects portfolio as a whole. These links are defined the integrity of projects portfolio and may be interpreted physically by links of the third and fourth group.
- The links of third and fourth group are one of the forms of physical representation of interproject interaction. Deterministic and probabilistic links are allowed to sample different methods of projects' interaction at the physical level. The modern practice of management is to use as a rule only deterministic links. Such deterministic links are described as fixed interactions between activities of the PDM - model in the following forms: *finish-start, start-start, start-finish and finish-finish.*

  Flexible links are oriented on *What-If* technology. Many conditions defining the links of project model with resources and constraints may be joined by logical *AND/OR* statements.

  The important feature of *intelligent* interproject links is their interpretation by the knowledge base - proof machine. They are described by fixed and flexible links, and the knowledge base has the possibilities of describing different algorithms of decision-making depending on the situation in the process of projects from organisation portfolio implementation. Examples of the possibilities of intelligent links sampling may be found in the systems REGULA [6] and MISS-EX [7].
- The links of fifth group sample the non-linear effects of projects interaction. Usually such effects are defined via synergetic interaction of portfolio projects.

The author has implemented the structure of typical situations and created the groups of output rules in the frames of a production knowledge base model.

5. Preparation and monitoring of projects

A variety of methods and possible technologies of projects models in modern computer systems is concerned with the complex tasks of processing the most effective technologies by the project manager. The key place in the development of such technologies is played by the technology of wizards, providing a high flexibility of information technology and the possibility of its tuning by the requirements of users. Three main methods of projects portfolio models preparation in the frames of such technology may be used: “up-to-down”, “down-to-up” and “splitting”. The general schemes of the preparation of models are shown on fig. 5.

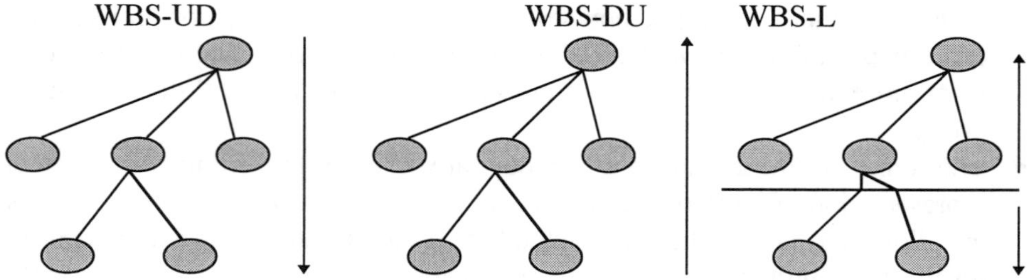

Note: arrows are shown the direction of projects models WBS structures processing.

Figure. 5.

At the stage of projects’ monitoring the key role is played by the questions of current situation identification, decision preparation and managing project portfolios from the point of view of pre-defined corporate aims. The structure of typical situations was described and groups of output rules were formed by the author in the frames of productive knowledge base. The main groups of such situations are described in table 1.

N	Situation	Main factors, Time/Cost/Links	Estimation of priority
1	Variation of project by time, delayed all complex of projects	Y/N/N	1
2	Overflow of project budget, caused the overflow of overall projects portfolio	N/Y/N	1
3	Variation of project by time with the overflow of portfolio cost	Y/Y/N	2
4	Change of one or several projects contents, and then change of time, cost and links between projects of portfolio	Y/Y/Y	3
5	Add/delete one or several projects, and then change of time, cost and links between projects of portfolio	Y/Y/Y	4

Table 1. Main groups of projects portfolio monitoring typical situations.

In the cells of each situation, groups during decision-preparation and -making are highlighted, and the appropriate groups of knowledge base output rules are working.

6. Decision-preparation and -making by the management of project portfolio

We consider the method of dynamic priorities proposed by the author for the management of organisation projects portfolio.

The methods of model preparation for each project in the organisation are used at the stage of projects portfolio models preparation. The interproject links are created during the conjunction of projects models. The portfolio consists of following set of projects

$$P=\{ p_1, p_2, ..p_i,.. p_n \},$$

where: n - the number of chosen methods. Each project is characterised by the indicator of efficiency in the frames of projects portfolio implementation

$$E=\{ e_1, e_2, ..e_i... e_n \}.$$

Where,

$$\forall p_i \exists e_i.$$

The projects of the portfolio are implemented by means of interconnected work complexes, which have synergetic effects during interaction between each other. This synergetic effect increases or reduces their general effect. We represent the synergetic links between pairs of set P as a square matrix S_{ij}. The efficiency of the pointed chain implementation is defined by following expression

$$C_{ij} = e_i + e_j + s_{ij}$$

This chain of projects forms the base of future complex model of projects portfolio. We define as the objective function of projects portfolio implementation

$$C_{opt} = \max_{i,j \in n} C_{ij}$$

This task of creating the projects portfolio model is related to the class of combinatorial problems and can solved by the branch-and-bound method.

As a result of the task solution, the most rational model of portfolio projects interaction is built. Then, the area of non-sensitivity is estimated near each interproject link, including the synergism. This area is estimated by expert opinion (fig.6).

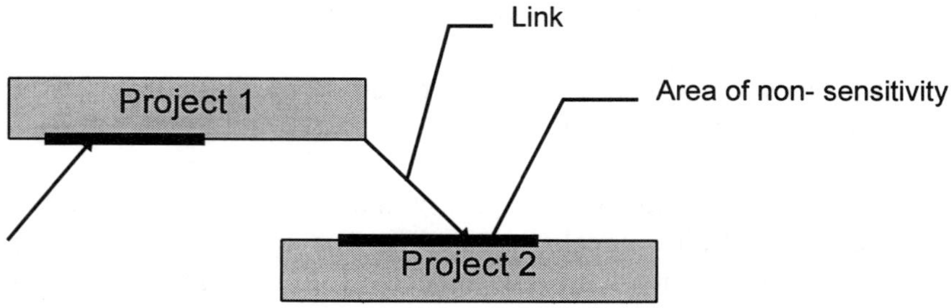

Figure 6

6.1 STAGE OF PROJECTS PORTFOLIO MONITORING

The stage of projects portfolio monitoring is implemented by following scheme:

fixing of current projects state (earned values of activities and used resources) ⇒ *estimation of results deviation of projects implementation from target parameters* (time, cost, resources) ⇒ *analysis of interproject links state* (timing floats of non-sensitivity area) ⇒ *identification of situation for decisions making* (knowledge base and output rules) ⇒ *creation of dynamic priorities* (prior value of priority + priority of current situation, caused by the conflict).

The solution of conflicts during the monitoring of projects portfolio is conducted on the base of sequential consideration of conflicts by priorities.

7. Practice of inculcation of corporation projects portfolio management culture

The main steps to the inculcation of such system in the practice of project management in Ukraine are:

- analysis of corporate needs
- creation of inculcation plan
- processing of methodology of corporation projects portfolio management
- creation of models structures
- processing of report system

- creation of situations and decisions prototypes
- training
- final estimation.

We are considering each step of projects portfolio management system in the examples of a number of pilot projects of the Ministry of Chernobyl organisations.

8. Conclusions

1. The development of project management culture in CIS states is connected with active technological transfer and the creation of a critical mass of projects managers with the knowledge of international methodologies of PM and skills of pilot projects management in international teams.
2. The methods of projects portfolio management plays an important role in project management and is characterised by high complexity of tasks and intelligent complexity of models.
3. The method of dynamic priorities of projects portfolio management for reaching defined aims is proposed.

9. References

1. Project Management Institute (1987). Project Management Body of Knowledge (PM BOOK), PMI, Drexel Hill, Pennsylvania,.
2. Project Management Institute (1996) A Guide to the Project Management Body of Knowledge, PMI, Standards Committee, Drexel Hill, Pennsylvania.
3. Association of Project Management (1995) Body of Knowledge. APM, UK, Buckinghamshire (Version 2).
4. Turner J. R. (1993). The handbook of project-based management: improving the process for achieving strategic objectives, McGraw-Hill Book Company, London .
5. Primavera Systems Inc. (1996) Steps to implementing concentric project management, Three-Day Course 701, Primavera Systems Inc., PA USA.
6. IPMA Bremen University (1991) REGULA Regeltechniken und -Anwendungen,
7. Bushuyev S.D. (1992) Manual for using of expert system MISS-EX, Kiev, Kiev State Technical University.

OPEN DISCUSSION: PROGRAMME MANAGEMENT

Discussion chaired by:
G.REISS
Hydra development Corp,
37 Kirkgate, Otley, W.Yorkshire, UK

This summary by:
T.M. WILLIAMS
University of Strathclyde,
Glasgow, UK

Dr. Reiss opened the discussion by saying that although he had begun (in the Introduction to this Section) by looking at the differences between Programme Management and the single-project case, clearly we should here be looking at the similarities, to help in the management of complex projects. In particular, perhaps we should be looking at the effects on human issues and on teams in this environment.

It was seen that it was important to link the objectives of each project with the strategic objectives of the whole portfolio, and this was perceived as a gap in current practice. A portfolio should be greater than the sum of its projects. Dr. Reiss agreed, saying that the Programme Manager should be "flying above" the projects: each Project Manager is aiming single-mindedly towards the finish of his project, but the Programme Manager needs to be able to understand the requirements for each project in the light of the overall objectives.

It was reported that projects within the UK Ministry of Defence are managed as a whole, in terms of affordability, resource planning and resource allocation, and also as groups of projects which have either inter-dependencies in their implementation, or which combine in some synergistic way to improve user capabilities. In this latter sense, the aim is to optimise the overall contribution of a group of projects, and this does require a level and a form of management which is separate from the management of each individual project (an exception is where there is a major technical dependency between projects, in which case one project is often designated as the "lead" project). This is particularly so for Information / Computer Systems, which achieve their greatest benefit when their user organisation adapts its business processes. But there has to be an initial stage of establishing the strategic Programme objectives, before the individual projects can be planned, as the strategic objectives form the framework within which the individual project managers plan.

There are some packages produced by CP countries covering multi-project PM. These have not been well-publicised because there has not been a competitive market, rather big centralised organisations carrying out programmes.

SECTION 8
CONCLUSIONS AND FURTHER RESEARCH

The main aim of the Workshop was to present a variety of approaches to complex projects, both their modelling and management, in specific domains and from both NATO and CP viewpoints. 54 selected participants from 15 countries attended, representing academics, consultants, contractors and clients, from a variety of domains and with a variety of different technical approaches. It is to be hoped that the different approaches and different cultural views will inform each other.

In terms of modelling, PERT-type methods are inadequate because they do not incorporate management actions, adequately consider uncertainty or consider effects that are systemic or "soft". The Workshop discussed papers extending PERT methods (using controlled networks and diffusion networks), risk analysis techniques and the System Dynamics technique.

The complexity of proposed methods was an important issue. It was recognised that complex methods may be necessary, but:

- they need to have simple implementations in order for the practising project manager to use them;
- modellers must be able to communicate their results to project managers:
- furthermore, the order of the words in the title of the Workshop is important: "managing and modelling complex projects"; the management of the projects is the key issue, and the modelling must be subservient to this.

In particular, managers must look at the potential for de-complexing projects rather than simply using complex methods to analyse them as they stand; this was only touched upon in the Workshop and was identified as an important area for further research.

The position of mathematical modelling within the practice of actual project management was the subject of some discussion; there could clearly be extremes of, on the one hand, academic modellers who have no impact on projects and, on the other hand, project managers who reject all mathematical techniques to the detriment of their planning; steering a middle course, development of technique must always be driven by the needs of project managers and the desire to improve performance. Differences were partly cross-cultural, with the position of

T. M. Williams (ed.), Managing and Modelling Complex Projects, 255–257.

mathematical modelling being much more secure in CP eyes; but there were distinct differences within the NATO contingent.

Modelling of the “hard” effects is of course only the start, and the human factors must also be considered. This is true firstly in the modelling. One paper described the technique of System Dynamics, which provides a simulation model of projects, capturing systemic effects and effects that are “soft”. One particular factor, mention of which recurred frequently during the Workshop, was the effect of schedule pressure on a project, and study of this effect and its modelling clearly warrants further research.

Secondly, the human factors must be considered in the management of projects, as discussed in the paper on human factors in Singapore infrastructure projects. There was clearly a strong cultural difference between NATO and CP approaches. The CP approach was generally to seek how to motivate workers seen as essentially selfish. This was in distinct contrast to, say, the Singaporean approach, in which management tried to understand how human beings behave, and how they can become committed to a project. This difference carried through into a marked difference in management structures. CP project management structures were highly hierarchical (indeed, it was suggested that the culture in CP countries was actively to discourage teamwork); NATO approaches were much more likely to be teamwork-centred, with teams being self-motivated and empowered, although again this was not entirely consistent across the NATO contingent.

As well as cultural differences, CP economies are of course in transition between being planned and becoming market economies, and the Workshop considered the characteristics of the social and economic environment in transition economies. At the most basic level, there is a need for educating managers in the CP countries in market-economy project management: one conclusion from the paper of the CP Co-Director is that “the development of project management culture in CIS states is connected with active technological transfer and the creation of a critical mass of projects managers with the knowledge of international methodologies of PM and skills of pilot projects management in international teams”. But further to this, there is a need to consider how the project management models and management techniques can be modified to take into account the particular characteristics of transition economies and further research is needed here. (And even at the most basic level, support is needed to help the market in Russian-language project management software.)

Some cultural differences and similarities came as a surprise to the Directors. For example, although much interest was shown by NATO participants in methods for managing large software projects, there appeared to be less interest from the CP participants. On the other hand, hierarchical methods of programme management turned out to be very similar between NATO and CP.

New technology projects were of interest to the Workshop, and there was considerable discussion about what constituted "high-" or new-technology projects, and their particular requirements. In particular, there was discussion about the need for flexible systems and suitable systems of information flow. These were problems faced by both NATO and CP project managers alike, and there is a need for further research.

Although progress was made, there is much need for further work. In particular, while ideas were shared within the heterogeneous group at the Workshop, there is a need for consideration of synthesising the various techniques and looking at the reasons for, and benefits of, the different cultural approaches. But the point was made frequently in the Workshop that while it might or might not be appropriate to look for better tools, it is essential that we seek to make the current, basic, tools, better understood and applied. Education and training in fundamental project management is important, for both CP and NATO-country project managers.

Methods and Protocols in Food Science

Series Editor
Anderson S. Sant'Ana
University of Campinas
Campinas, Brazil

For further volumes:
http://www.springer.com/series/16556

Methods and Protocols in Food Science series is devoted to the publication of research protocols and methodologies in all fields of food science.

Volumes and chapters will be organized by field and presented in such way that the readers will be able to reproduce the experiments in a step-by-step style. Each protocol will be characterized by a brief introductory section, followed by a short aims section, in which the precise purpose of the protocol will be clarified.

Leaves as Functional Foods

Edited by

Mitesh Kumar Dwivedi

C.G. Bhakta Institute of Biotechnology, Uka Tarsadia University, Mahuva, Surat District, Gujarat, India

A. Sankara Narayanan

School of Sciences, GSFC University, Vadodara, Gujarat, India

Humana Press

Editors
Mitesh Kumar Dwivedi
C.G. Bhakta Institute of Biotechnology
Uka Tarsadia University
Mahuva, Surat District, Gujarat, India

A. Sankara Narayanan
School of Sciences
GSFC University
Vadodara, Gujarat, India

ISSN 2662-950X ISSN 2662-9518 (electronic)
Methods and Protocols in Food Science
ISBN 978-1-0716-5225-1 ISBN 978-1-0716-5226-8 (eBook)
https://doi.org/10.1007/978-1-0716-5226-8

This Humana imprint is published by the registered company Springer Science+Business Media, LLC, part of Springer Nature.
The registered company address is: 1 New York Plaza, New York, NY 10004, U.S.A.

Preface to the Series

Methods and Protocols in Food Science series is devoted to the publication of research protocols and methodologies in all fields of food science. The series is unique as it includes protocols developed, validated, and used by food and related scientists as well as theoretical basis are provided for each protocol. Aspects related to improvements in the protocols, adaptations, and further developments in the protocols may also be approached.

Methods and Protocols in Food Science series aims to bring the most recent developments in research protocols in the field as well as very well-established methods. As such, the series targets undergraduate, graduate, and researchers in the field of food science and correlated areas. The protocols documented in the series will be highly useful for scientific inquiries in the field of food sciences, presented in such way that the readers will be able to reproduce the experiments in a step-by-step style.

Each protocol will be characterized by a brief introductory section, followed by a short aims section, in which the precise purpose of the protocol is clarified. Then, an in-depth list of materials and reagents required for employing the protocol is presented, followed by a comprehensive and step-by-step procedures on how to perform that experiment. The next section brings the do's and don'ts when carrying out the protocol, followed by the main pitfalls faced and how to troubleshoot them. Finally, template results will be presented and their meaning/conclusions addressed.

The *Methods and Protocols in Food Science* series will fill an important gap, addressing a common complain of food scientists, regarding the difficulties in repeating experiments detailed in scientific papers. With this, the series has a potential to become a reference material in food science laboratories of research centers and universities throughout the world.

Campinas, Brazil ***Anderson S. Sant'Ana***

Preface

Plants have played a foundational role in human nutrition, culture, and medicine for centuries, with leaves standing out as vital sources of energy, healing, and cultural meaning. Beyond their fundamental role in photosynthesis, leaves are increasingly recognized as rich repositories of bioactive compounds with considerable therapeutic promise. In recent years, scientific interest has surged around the potential of leaves as functional foods—agents capable of promoting health, preventing disease, and modulating immune functions. However, despite this growing recognition, comprehensive literature exploring the varied benefits of edible and medicinal leaves remains scarce.

This book, "Leaves as Functional Foods: Assessment of Bioactive Compounds, Immunomodulation and Human Health Benefits," endeavors to close this gap. Drawing together current research and emerging interdisciplinary perspectives, this volume offers an in-depth exploration of the therapeutic properties of leaf-derived bioactives in the form of laboratory protocols.

Organized into four sections, the book provides both a broad foundation and focused analysis:

- Part I: Extraction of Bioactive Compounds from leaves details modern methodologies and analytical advances for isolating phytochemicals from diverse leaf sources.
- Part II: Immunomodulatory Effect of Leaves examines how leaf extracts interact with innate and adaptive immunity, revealing experimental and clinical insights on immune regulation.
- Part III: Leaves and Human Health surveys the expansive health-promoting properties of leaves, including antioxidant, antimicrobial, antidiabetic, anti-inflammatory, and organ-protective effects.
- Part IV: Biosafety of Leaves addresses toxicity, safety standards, regulatory frameworks, and evidence-based guidance for health applications.

The thirty chapters of this collection reflect the contributions of leading experts and innovative researchers, blending foundational knowledge with recent advancements. Each chapter balances scientific rigor with practical methodology, making the book an essential resource for researchers, educators, nutritionists, healthcare professionals, and students alike.

By integrating scientific data and applied perspectives, this work aims to not only enrich academic discourse but also inspire further research and innovation in natural products, nutrition science, and integrative medicine.

This format highlights the book's scope, addresses the need for research in the field, and positions the volume within academic and practical contexts, all hallmarks of an effective scientific preface.

Mahuva, Surat District, Gujarat, India — *Mitesh Kumar Dwivedi*
Vadodara, Gujarat, India — *A. Sankara Narayanan*

Contents

Part II Immuno Modulatory Effect of Leaf

Part III Leaf and Human Health

Contributors

K. Ananthi • *PG and Research Department of Microbiology, Dhanalakshmi Srinivasan College of Arts and Science for Women (Autonomous), Perambalur, Tamil Nadu, India*

K. Arun • *Department of Microbiology, Srinivasan College of Arts and Science, Perambalur, Tamilnadu, India*

Devjani I. Banerjee • *Department of Life Sciences, GSFC University, Vadodara, Gujarat, India*

H. Vajiha Banu • *PG and Research Department of Microbiology, Jamal Mohamed College (Autonomous), Tiruchirappalli, Tamil Nadu, India*

Piyali Basak • *Guru Nanak Institute of Pharmaceutical Science and Technology, Kolkata, India*

Jeenatara Begum • *Guru Nanak Institute of Pharmaceutical Science and Technology, Kolkata, West Bengal, India*

Franciane Pereira Bernardes • *Secretaria de Estado de Educação do Espírito Santo, Vitória, Espírito Santo, Brazil*

Vaibhavi Bhatia • *Department of Chemical Sciences, School of Science, GSFC University, Fertilizer Nagar, Vadodara, Gujarat, India*

Sobhanjan Bhunia • *Guru Nanak Institute of Pharmaceutical Science and Technology, Kolkata, West Bengal, India*

Maria do Carmo Pimentel Batitucci • *Laboratório de Genética Vegetal e Toxicológica, Departamento de Ciências Biológicas, Universidade Federal do Espírito Santo, Vitória, Espírito Santo, Brazil*

Prapti Chakraborty • *Guru Nanak Institute of Pharmaceutical Science and Technology, Kolkata, West Bengal, India*

Tamalika Chakraborty • *Guru Nanak Institute of Pharmaceutical Science and Technology, Kolkata, West Bengal, India*

Nitin Chawaria • *Plant Microbe Interaction Laboratory, Department of Microbiology, Maharshi Dayanand University, Rohtak, Haryana, India*

Peng Chen • *School of Pharmacy, Lanzhou University, Lanzhou, People's Republic of China; State Key Laboratory of Applied Organic Chemistry, Lanzhou University, Lanzhou, People's Republic of China*

Priyanka Dahiya • *Plant Microbe Interaction Laboratory, Department of Microbiology, Maharshi Dayanand University, Rohtak, Haryana, India*

Amita Suneja Dang • *Centre for Medical Biotechnology, Maharshi Dayanand University, Rohtak, Haryana, India; Centre for Medical Biotechnology, Maharshi Dayanand University, Rohtak, Haryana, India*

Siddhi A. Darji • *Department of Life Sciences, GSFC University, Vadodara, Gujarat, India*

Rajarshi Das • *Guru Nanak Institute of Pharmaceutical Science and Technology, Kolkata, West Bengal, India*

Jean Carlos Vencioneck Dutra • *Laboratório de Genética Vegetal e Toxicológica, Departamento de Ciências Biológicas, Universidade Federal do Espírito Santo, Vitória,*

Espírito Santo, Brazil; Instituto Federal do Espírito Santo—Campus Ibatiba, Ibatiba, Espírito Santo, Brazil

MIROSLAV EFTIMOV • *Faculty of Medicine, Medical University "Prof. Dr. Paraskev Stoyanov", Varna, Bulgaria*

HONG FANG • *School of Pharmacy, Lanzhou University, Lanzhou, People's Republic of China; State Key Laboratory of Applied Organic Chemistry, Lanzhou University, Lanzhou, People's Republic of China*

JEAN MOISES FERREIRA • *Secretaria de Estado de Educação do Espírito Santo, Vitória, Espírito Santo, Brazil*

SILVIA GANCHEVA • *Faculty of Medicine, Medical University "Prof. Dr. Paraskev Stoyanov", Varna, Bulgaria*

MANJARIMA GANGULI • *Guru Nanak Institute of Pharmaceutical Science and Technology, Kolkata, West Bengal, India*

MANJARIMA GANGULY • *Guru Nanak Institute of Pharmaceutical Science and Technology, Kolkata, West Bengal, India*

SARAH DE JESUS CANTARINO • *Secretaria de Estado de Educação do Espírito Santo, Vitória, Espírito Santo, Brazil*

K. R. JEYA • *Department of Biotechnology, Government Arts and Science College (Women), Veppur, Perambalur, Tamilnadu, India*

EMMANUEL LETSYO • *Department of Food Science and Technology, Ho Technical University, Ho, Ghana*

FELIX KWASHIE MADILO • *Department of Food Science and Technology, Ho Technical University, Ho, Ghana*

MPHO MASHAU • *Department of Food Science and Technology, Faculty of Sciences, Engineering & Agriculture, University of Venda, Thohoyandou, South Africa*

NDIVHUWO J. MATODZI • *Department of Food Science and Technology, Faculty of Sciences, Engineering & Agriculture, University of Venda, Thohoyandou, South Africa*

FRANCESCA M. MITTON • *Instituto de Investigaciones Marinas y Costeras (IIMyC), FCEyN, Universidad Nacional de Mar del Plata-CONICET, Mar del Plata, Provincia de Buenos Aires, Argentina; Consejo Nacional de Investigaciones Científicas y Técnicas (CONICET), Mar del Plata, Argentina*

DEEPA NAUTIYAL • *Institute of Advanced Research, Koba Institutional Area, Gandhinagar, Gujarat, India*

JIGISHA ROY PANDA • *Guru Nanak Institute of Pharmaceutical Science and Technology, Kolkata, West Bengal, India*

DHARA PATEL • *Institute of Advanced Research, Gandhinagar, Gujarat, India*

ROSHANI PATEL • *Institute of Advanced Research, Koba Institutional Area, Gandhinagar, Gujarat, India*

ATHARVA PATIL • *Department of Life Sciences, GSFC University, Vadodara, Gujarat, India*

CAROLINA D. PEREZ • *Consejo Nacional de Investigaciones Científicas y Técnicas (CONICET), Mar del Plata, Argentina; Grupo de Investigación en Ingeniería en Alimentos (GIIA), Instituto de Ciencia y Tecnología de Alimentos y Ambiente (INCITAA), Facultad de Ingeniería, Universidad Nacional de Mar del Plata, Mar del Plata, Argentina*

MARIA T. PETROVA • *Faculty of Pharmacy, Medical University "Prof. Dr. Paraskev Stoyanov", Varna, Bulgaria*

T. POOVIZHI • *Department of Nutrition & Dietetics, Vellalar College for Women (Autonomous) Thindal, Erode, Tamil Nadu, India*

MAYA P. RADEVA-ILIEVA • *Department of Pharmacology, Toxicology and Pharmacotherapy, Faculty of Pharmacy, Medical University "Prof. Dr. Paraskev Stoyanov", Varna, Bulgaria*
SHONISANI E. RAMASHIA • *Department of Food Science and Technology, Faculty of Sciences, Engineering & Agriculture, University of Venda, Thohoyandou, South Africa*
KGAOGELO E. RAMATSETSE • *Department of Life and Consumer Sciences, College of Agriculture and Environmental Sciences, University of South Africa, Florida Campus, Johannesburg, South Africa*
SIMRAN RANI • *Plant Microbe Interaction Laboratory, Department of Microbiology, Maharshi Dayanand University, Rohtak, Haryana, India*
MMAPHUTI A. RATAU • *Department of Food Science and Technology, Faculty of Sciences, Engineering & Agriculture, University of Venda, Thohoyandou, South Africa*
CHETNA RATHI • *Plant Microbe Interaction Laboratory, Department of Microbiology, Maharshi Dayanand University, Rohtak, Haryana, India*
PRIYANKA RAY • *Guru Nanak Institute of Pharmaceutical Science and Technology, Kolkata, West Bengal, India*
SUMANA ROY • *Guru Nanak Institute of Pharmaceutical Science and Technology, Kolkata, West Bengal, India*
A. SANKARA NARAYANAN • *Department of Life Sciences, School of Science (SoS), GSFC University, Fertilizernagar, Vadodara, Gujarat, India*
T. SAVITHA • *Department of Microbiology Tiruppur Kumaran College for Women, Tiruppur, Tamil Nadu, India*
SRIPARNA KUNDU SEN • *Guru Nanak Institute of Pharmaceutical Science and Technology, Kolkata, West Bengal, India*
NAZNIN SHAIKH • *Department of Chemical Sciences, School of Science, GSFC University, Fertilizer Nagar, Vadodara, Gujarat, India; Department of Chemistry, SRICT—Institute of Science and Research, UPL University of Sustainable Technology, Vataria, India*
ABHISHEK SHARMA • *Institute of Advanced Research, Koba Institutional Area, Gandhinagar, Gujarat, India*
MAYANK SHARMA • *Department of Chemical Sciences, School of Science, GSFC University, Fertilizer Nagar, Vadodara, Gujarat, India*
ZHENZHEN SI • *School of Pharmacy, Lanzhou University, Lanzhou, People's Republic of China; State Key Laboratory of Applied Organic Chemistry, Lanzhou University, Lanzhou, People's Republic of China*
STANILA S. STOEVA-GRIGOROVA • *Department of Pharmacology, Toxicology and Pharmacotherapy, Faculty of Pharmacy, Medical University "Prof. Dr. Paraskev Stoyanov", Varna, Bulgaria*
POOJA SUNEJA • *Plant Microbe Interaction Laboratory, Department of Microbiology, Maharshi Dayanand University, Rohtak, Haryana, India*
RUCHIKA UPADHYAY • *Institute of Advanced Research, Koba Institutional Area, Gandhinagar, Gujarat, India*
M. VEERAPAGU • *Department of Microbiology, Srinivasan College of Arts and Science, Perambalur, Tamilnadu, India*
NING XU • *School of Pharmacy, Lanzhou University, Lanzhou, People's Republic of China; State Key Laboratory of Applied Organic Chemistry, Lanzhou University, Lanzhou, People's Republic of China*
AAKANSHA YADAV • *Department of Life Sciences, GSFC University, Vadodara, Gujarat, India*

BHUPENDER YADAV • *Plant Microbe Interaction Laboratory, Department of Microbiology, Maharshi Dayanand University, Rohtak, Haryana, India*
FANTING ZHAO • *School of Pharmacy, Lanzhou University, Lanzhou, People's Republic of China; State Key Laboratory of Applied Organic Chemistry, Lanzhou University, Lanzhou, People's Republic of China*
LICHEN ZHAO • *School of Pharmacy, Lanzhou University, Lanzhou, People's Republic of China; State Key Laboratory of Applied Organic Chemistry, Lanzhou University, Lanzhou, People's Republic of China*
RUNYU ZHU • *School of Pharmacy, Lanzhou University, Lanzhou, People's Republic of China; State Key Laboratory of Applied Organic Chemistry, Lanzhou University, Lanzhou, People's Republic of China*

Part I

Extraction of Bioactive Compound from Leaf

Chapter 1

Extraction of Bioactive Compound from Leaf: Saponins

Francesca M. Mitton and Carolina D. Perez

Abstract

Saponins, natural glycosides found in a variety of plants, are known for their unique ability to form stable foams due to their surfactant-like structure. These bioactive compounds hold immense potential in the food, pharmaceutical, and cosmetic industries. However, the extraction of saponins can be challenging due to their sensitivity to heat and the lengthy processes associated with conventional methods. In this chapter, we present a detailed protocol for extracting saponins from plant leaves using advanced techniques—ultrasound-assisted extraction (UAE) and microwave-assisted extraction (MAE). These non-conventional methods offer significant advantages, including reduced solvent use, shorter extraction times, and improved yields while preserving the integrity of heat-sensitive compounds. This protocol can be applied across various plant species, making it a versatile tool for researchers looking to efficiently extract saponins for diverse applications. By leveraging these innovative techniques, this method enhances both the practicality and sustainability of saponin extraction.

Key words Saponins, Bioactive compounds, Food purpose, Ultrasound-assisted extraction, Microwave-assisted extraction, Plant extract

1 Introduction

Saponins are a diverse group of natural plant and animal glycosides known for their ability to form stable foams. These compounds have a polycyclic aglycone backbone, or sapogenin, which is connected to one or more sugar side chains. The sapogenin can be either a steroid (C27) or a triterpene (C30) and the special foaming properties of saponins originate from the union of their hydrophobic sapogenin and hydrophilic sugar fractions [1]. Saponins are typically categorized by the number of sugar chains attached to their core structure, known as the aglycone. They can be classified as monodesmosidic, having a single sugar chain, or bidesmosidic, with two chains. There are also tridesmosidic saponins, which have three sugar chains, although these are quite rare [3]. The length and structure of these sugar chains greatly affect the solubility of saponins, and they function in food systems. This

Mitesh Kumar Dwivedi and A. Sankara Narayanan (eds.), *Leaves as Functional Foods*, Methods and Protocols in Food Science,
https://doi.org/10.1007/978-1-0716-5226-8_1,

variation makes some saponins better suited for food applications than others. In this sense, in recent years, saponins have emerged as commercially important compounds with expanding applications in foods, cosmetics, and pharmaceutical industries due to their naturally occurring surfactant properties (also called biosurfactants) and growing evidence of their biological activity (such as anticancer properties and hemolytic and hypocholesterolemia activity) [2, 3].

Even though saponins have several beneficial properties, their bitter taste can be a drawback when it comes to using them in food. This bitterness stems from their interaction with specific taste receptors. To address this, researchers have been exploring ways to mask or reduce bitterness, such as by using flavor-masking techniques or lowering their concentration in food products [4]. Additionally, saponins can sometimes decrease the bioavailability of food nutrients like proteins, fats, vitamins, and minerals in the human body [5]. So, while saponins offer advantages, their potential downsides mean they need to be carefully considered before being used in food formulations.

Saponins are found in a wide range of plants, including legumes, vegetables, and herbs. Commercially, they're often extracted from plants like *Quillaja saponaria* and *Yucca schidigera*, which are well known for having high saponin levels [1]. However, the quantitative amount and quality composition of saponins in plants depend on factors like local climate, seasonal changes, and external conditions such as light, temperature, humidity, soil fertility, and cultivation techniques [6].

The extraction of bioactive compounds, including saponins, can be divided into conventional and non-conventional techniques. Conventional extraction methods have been used for decades, but they come with certain limitations, such as longer extraction times and higher solvent consumption [7]. These traditional methods often lead to the degradation of heat-sensitive compounds due to the prolonged processing time and elevated temperatures involved. In contrast, ultrasound-assisted and microwave-assisted extractions are considered non-conventional approaches and offer promising alternatives [8].

In this chapter, we describe the procedural steps for extracting saponins from the leaves of plants. This protocol can be used in different kinds of plant species. The protocol described utilizes two non-conventional approaches, ultrasound- and microwave-assisted extraction. These are promising techniques because they require lesser solvent, shorter extraction time, and produce higher extraction yield, in addition to minimizing the degradation of thermosensitive compounds.

2 Materials

2.1 Reagents, Standards, and Materials

1. Distilled and/or ultrapure water.
2. Ethanol (GR Grade).
3. Methanol (GR Grade).
4. Alkaline detergent.
5. Liquid nitrogen.
6. Glass flasks.
7. Büchner funnel.
8. C18 column (6 cm × 10 cm, 55 μm, 50 g).
9. Glass flasks and bottles for freeze dryer (500, 1000 mL).
10. Mortar.
11. Whatman qualitative filter paper No. 1.
12. Filtration flasks (500 and 1000 mL).
13. Support, glass funnel, and clamps for 25 and 47 mm filters.
14. Graduated cylinder 10 mL, 100 Ml.
15. Glass beakers for disposals and washes (100, 250 and 300 mL).
16. Glass flasks and bottles for solution storage (500, 1000 mL).
17. Aluminum foil.
18. Polyethylene bags.

2.2 Equipment

1. Fume hood.
2. Vacuum pump.
3. Water bath.
4. Ultrasonic homogenizer.
5. Rotary evaporator.
6. Freeze dryer.
7. Microwave oven.
8. Furnace (>400 °C).
9. Oven (300 °C).

2.3 Cleaning Materials

1. Glass material: gently clean with an alkaline detergent, then rinse with distilled water. Cover the items with aluminum foil to prevent any contamination. Dry them in an oven at 250 °C overnight or in a furnace at 450 °C for 4 h. Keep them covered until needed. (Note: Do not heat volumetric flask—simply rinse them thoroughly with distilled water, dry with ethanol, and store them covered.)

2. Aluminum Foil: cut pieces of aluminum foil into various sizes and heat them in a furnace at 450 °C for 4 h. Keep them covered until ready to use for bottle tops or any other material.

2.4 Solutions

1. 70% Ethanol: Mix 70 mL of ethanol with 30 mL of distilled water in a clean glass cylinder. Keep sealed until ready to use.
2. 85% Methanol: Mix 85 mL of methanol with 15 mL of distilled water in a clean glass cylinder. Keep sealed until ready to use.

3 Methods

3.1 Ultrasonic Extraction of Saponins

1. *Weigh and Grind Leaves*: Start by weighing 200 g of fresh leaves and grinding them in a mortar with liquid nitrogen until a fine powder is obtained (*see* **Note 1**). Store the powdered leaves at −20 °C in vacuum-sealed polyethylene bags for future use in experiments.
2. *Prepare Extract Mixture*: Weigh 50 g of the fine leaf powder and transfer it to an Erlenmeyer flask. Add 1 L of 70% ethanol to the flask, maintaining a 1:20 (material to solvent) ratio.
3. *Extract with Ultrasound*: Place the Erlenmeyer flask into a water bath and set it up with an ultrasonic homogenizer at 50 °C. Allow the extraction process to continue for 60 min.
4. *Separate Solid Phase*: After the extraction is complete, separate the saponins from the solid phase by filtration using a Büchner funnel.
5. *Concentrate and Remove Solvent*: Transfer the liquid extract to a glass flask and use a rotary evaporator at 50 °C, with a pressure of 160 mbar and rotation sped of 100 rpm, to remove the solvent and concentrate the extract until the volume is reduced to 10% of the original (100 mL).
6. *Dilute and Purify*: Once concentrated, dilute the extract with methanol to a final concentration of 35% (v/v). For purification, load the solution onto a pre-conditioned C18 column (6 cm × 10 cm, 55 μm, 50 g) that has been conditioned with 36% methanol. First, wash the column with 200 mL of 60% methanol with a flow rate of 1 mL/min to remove sugars and phenolics, followed by another sequentially 200 mL of 70% and 85% methanol to elute the saponins.
7. *Remove Methanol and Dry*: To eliminate any remaining methanol from the saponin extract, use a rotary evaporator at 50 °C. Then, transfer the extract to a freeze dryer at −40 °C and 0.001 mbar for one day to yield a yellowish powder. Store the saponin extract at −20 °C until use (*see* **Note 2**).

3.2 Microwave-Assisted Extraction (MAE) Protocol

1. *Prepare the Leaf Powder*: Weigh 200 g of fresh leaves and grind them in a mortar with liquid nitrogen until a fine powder is obtained (*see* **Note 1**). Store the powdered leaves at −20 °C in vacuum-sealed polyethylene bags until further use.
2. *Extraction Process*: Weigh 20 g of the powdered leaves and transfer them to an Erlenmeyer flask. Add 1 L of 85% methanol, maintaining a 1:50 (w/v) ratio. Let the mixture soak at room temperature for 30 min. Then, perform microwave-assisted extraction (MAE) using a microwave oven set to 600 W power, applying radiation for 6 s/min, with a total extraction time of 40 min.
3. *Filtration and Concentration*: Quickly cool the extract in an ice-water bath, then filter it through Whatman qualitative filter paper No. 1. Concentrate the filtered extract using a rotary evaporator at 45 °C, with a pressure of 160 mbar and a rotation speed of 100 rpm, until the volume is reduced to 10% of the original (100 mL).
4. *Drying and Storage*: Dry the concentrated extract in a freeze dryer at −40 °C and 0.001 mbar for 24 h to obtain the powdered extract. Store the final saponin extract at −20 °C until use (*see* **Note 2**).

4 Notes

1. Make sure all the material is thoroughly homogenized. This step is crucial to ensure a good extraction process later on.
2. The material can be stored at room temperature, but storing it at −20 °C will extend its shelf life.

References

1. Kregiel D, Berlowska J, Witonska I, Antolak H, Proestos C, Babic M, Babic L, Zhang B (2017) Saponin-based, biological-active surfactants from plants. In: Najjar R (ed) Application and characterization of surfactants, vol 6. Intech Open, Rijeka, pp 184–205
2. Timilsena YP, Phosanam P, Stockmann R (2023) Perspectives on Saponins: food functionality and applications. Int J Mol Sci 24:13538
3. Xu M, Wan Z, Yang X (2021) Recent advances and applications of plant-based bioactive Saponins in colloidal multiphase food systems. Molecules 26:6075
4. Schreiner TB, Dias MM, Barreiro MF, Pinho SP (2022) Saponins as natural emulsifiers for nanoemulsions. J Agric Food Chem 70:6573–6590
5. Sharma K, Kaur R, Kumar S, Saini RK, Sharma S, Pawde SV, Kumar V (2023) Saponins: a concise review on food related aspects, applications and health implications. Food Chemistry Advances 2:100191
6. Szakiel A, Pączkowski C, Henry M (2011) Influence of environmental abiotic factors on the content of saponins in plants. Phytochem Rev 10:471–491
7. Agregan R, Munekata PES, Feng X, Astray G, Gullon B, Lorenzo JM (2021) Recent advances in the extraction of polyphenols from eggplant

and their application in foods. LWT Food Sci Technol 146:111381

8. Fu X, Wang D, Belwal T, Xie J, Xu Y, Zou L, Zhang L, Luo Z (2021) Natural deep eutectic solvent enhanced pulse-ultrasonication assisted extraction as a multi-stability protective and efficient green strategy to extract anthocyanin from blueberry pomace. LWT Food Sci Technol 144: 111220

Chapter 2

Extraction of Bioactive Compounds from Leaf: Dietary Fiber

Mmaphuti A. Ratau, Ndivhuwo J. Matodzi, Kgaogelo E. Ramatsetse, Mpho Mashau, and Shonisani E. Ramashia

Abstract

The growing global emphasis on health has boosted interest in leaf-derived bioactive compounds rich in dietary fiber, phenolics, vitamins, and antioxidants associated with protective effects against cancer, inflammation, and cardiovascular diseases. Soluble (such as pectin, inulin) and insoluble (such as cellulose, lignin) dietary fibers are responsible for the enhancement of gut microbiota, regulation of body weight, and reduction in mortality risk, and their leaf extraction is a prerequisite for food and pharmaceutical industries. This chapter presents an enzymatic-gravimetric method for isolating total dietary fiber from leaf material, addressing the need for green and efficient extraction techniques. It involves enzymatic hydrolysis in a sequential fashion by heat-stable α-amylase, protease, and amyloglucosidase on dried leaves, followed by precipitation and filtration processes to determine the fractions of insoluble, soluble, and total dietary fiber content. The method ensures maximum yield with integrity of compounds ensured and involves green chemistry principles of reducing the use of chemicals. The major steps include preparation of the sample at 60–70 °C, incubation at pH-controlled conditions, and analysis of protein and ash residue content, thus making the quantitation of the fiber more accurate. Overall, this process encourages sustainable harvesting, offering an avenue for the development of health-enhancing leaf-based nutraceuticals and augmenting nutrition security worldwide.

Key words Enzymatic-gravimetric extraction, Bioactive compounds, Dietary fiber, Insoluble dietary fiber, Leaf sources, Soluble dietary fiber, Sustainable extraction

1 Introduction

Recently, there has been an increased awareness of health, leading to a heightened interest in bioactive compounds from plants within food science, nutrition, and pharmaceutical research [1]. Among various sources, leaves are particularly rich in bioactive compounds such as dietary fiber, minerals, phenolic compounds, phytoestrogens, phytic acids, terpene derivatives, vitamins, sulfur compounds, and polyunsaturated fatty acids [2]. These compounds have diverse health-promoting properties, including antioxidant effects and activities against cancer, cholesterol, glycemia, inflammation, and

Mitesh Kumar Dwivedi and A. Sankara Narayanan (eds.), *Leaves as Functional Foods*, Methods and Protocols in Food Science, https://doi.org/10.1007/978-1-0716-5226-8_2,

microbes [3, 4]. Dietary fiber stands out for its association with improved gut microbiome composition, favorable body weight regulation, overall metabolic health, and reduced risk of cardiovascular disease and mortality [2, 5]. Consequently, the extraction of bioactive compounds from leaves is crucial for their utilization in food and pharmaceutical industries [6].

Dietary fibers are a key class of bioactive compounds found in leaves. They consist of non-digestible carbohydrates and lignin and are essential for promoting gut health by modulating gut microbiota and enhancing the production of short-chain fatty acids through fermentation in the large intestine [7, 8]. Based on water solubility, dietary fibers can be categorized into soluble fibers (e.g., inulin, resistant oligosaccharides, pectin) and insoluble fibers (e.g., cellulose, lignin, hemicellulose) [7, 9]. Additionally, dietary fibers may be classified by molecular weight into high-molecular-weight fibers (e.g., non-starch polysaccharides) and low-molecular-weight fibers (e.g., resistant oligosaccharides) [7].

The efficient extraction of dietary fiber from leaves is essential for food and nutraceutical applications. This chapter presents an analytical method for extracting dietary fiber from leaves, highlighting commonly used chemical, physical (mechanical), and enzymatic approaches to optimize yields and preserve compound quality. The enzymatic–gravimetric method involves sequential hydrolysis with α-amylase, protease, and amyloglucosidase, followed by precipitation and filtration to determine soluble, insoluble, and total dietary fiber [10–12]. Moreover, emphasis is placed on developing environmentally friendly extraction methods to ensure sustainability in industrial applications.

Enzymatic Gravimetric Extraction of Total Dietary Fiber
Principle: In summary, 1 g of dried leaves (in duplicate) (*see* **Note 1**) undergoes a series of enzymatic digestions using heat-stable α-amylase, protease, and amyloglucosidase (*see* **Note 2**).

2 Materials

1. Oven or drying chamber.
2. Analytical balance.
3. Ethanol, 95% v/v.
4. 78% Ethanol (*see* **Note 3**).
5. Acetone (*see* **Note 4**).
6. Enzyme assay: α-amylase, protease, and amyloglucosidase (*see* **Note 2**).
7. Deionized water.

8. Filtration apparatus (e.g., crucibles with celite, analytical grade).
9. Buffer solutions (e.g., phosphate or MES-TRIS buffer) (*see* **Note 5**).
10. 0.561 N Hydrochloric acid solution (*see* **Notes 6** and 7).

3 Methods

1. *Sample preparation*: Collect fresh leaves and dry them at 60–70 °C until constant weight is achieved. Prepare the sample by grinding it to pass through a 0.5 mm sieve.
2. *Weigh the sample*: Weigh 1.000 ± 0.005 g of defatted (*see* **Notes 1** and **8**), dried leaves into a 400 mL beaker, digestion flask, or suitable container. Add 40 mL of MES-TRIS (*see* **Note 5**) blend buffer solution at a pH of 8.2 into each beaker. Place a magnetic stirring bar in each beaker and mix on a magnetic stirrer at 170 rpm until the sample is completely incorporated into the solution, ensuring there are no clumps that might obstruct enzyme access to the sample (usually for up to 10 min).
3. *Sequential Enzyme Incubation.*
   - (a) *α-Amylase (pH 6.0, 100 °C, 30 min)*: Add 50 μL of heat-stable α-amylase solution while gently mixing to hydrolyze the starch. Then, cover each beaker with pieces of aluminum foil. Place the covered samples in a shaking water bath maintained at 98–100 °C and incubate for 30 min with ongoing agitation to gelatinize the starch, commencing the timer once all beakers are placed in the hot water bath.
   - (b) *Cooling*: Take all sample beakers out of the hot water bath and allow them to cool to 60 °C. Remove the foil covers. If needed, use a spatula to scrape any residue from the sides and bottom of the beakers. Rinse the walls of the beakers and the spatula with 10 mL of distilled water using a pipette. Finally, adjust the water bath temperature to 60 °C by draining some hot water and adding cold water.
   - (c) *Protease (pH 7.5, 60 °C, 30 min)*: Add 100 μL of protease solution to every sample to hydrolyze protein. Reseal them with aluminum foil. Place them in a shaking water bath set to 60 ± 1 °C, maintaining continuous agitation for 30 min. Start the timer as soon as the water bath reaches 60 °C to denature proteins.

(d) *pH Check*: Take the sample beakers out of the shaking water bath and remove the foils. While stirring, add 5 mL of 0.561 N HCl solution to the sample (*see* **Notes 6** and **7**). Measure the pH, which should range from 4.1 to 4.8. If needed, modify the pH by adding either 5% NaOH solution or 5% HCl solution.

(e) *Amyloglucosidase (pH 4.5, 60 °C, 30 min)*: While stirring on a magnetic stirrer, add 200 μL of amyloglucosidase solution. Then, cover the mixture with an aluminum lid. Incubate in a shaking water bath at 60 °C for 30 min, ensuring constant agitation. Begin timing once the water bath reaches 60 °C.

3.1 Determination of Insoluble, Soluble, and Total Dietary Fiber

3.1.1 Insoluble Dietary Fiber

1. *Filtration setup*: Tare two crucibles containing Celite to the nearest 0.1 mg for each sample. Wet and redistribute the Celite bed in the crucible using 15 mL of 78% (v/v) ethanol (see **Note 3**) or 3 mL distilled water from a wash bottle. Apply suction to the crucibles to evenly distribute the Celite onto the fritted glass. Discard the resulting filtrate.
2. *Filter enzyme*: Use a vacuum to filter the enzyme digestion solution through a crucible. Use a wash bottle to quantitatively move any remaining particles to the crucible using 21 mL or less of deionized water warmed to 60–70 °C and rinse any residue. *Retain the filtrate and water washings for soluble dietary fiber (SDF) analysis.*
3. *Wash residue:* Wash the residue twice with 15 mL portions of 78% ethanol, 95% ethanol (*see* **Note 6**), and acetone using a vacuum.
4. *Dry and cool crucible*: Place the crucible containing the residue in a 103 °C oven to dry overnight. After cooling the crucible in a desiccator for about 1 h, weigh it along with the dietary fiber residue and Celite to the nearest 0.1 mg.
5. *Protein and ash determination*: Analyze the residues from two duplicates. One residue is examined for protein using either the Kjeldahl or combustion methods, taking care to avoid clogging the combustion analyzer's transfer lines with volatilized Celite. To calculate the protein content, multiply the nitrogen content by a conversion factor of 6.25, which assumes that proteins are 16% nitrogen. The second residue undergoes ash analysis by incinerating it for 5 h at 525 °C. After cooling in a desiccator, weigh it to the nearest 0.1 mg and subtract the weights of the crucible and Celite tare to find the ash content.

3.1.2 Soluble Dietary Fiber

1. Weigh the filtrate and water washing from 3.1.1(1).
2. *Precipitate SDF*: Add four volumes of 95% ethanol that have been preheated to 60 °C. You can use some of the ethanol to

rinse the filtering flask using the insoluble dietary fiber procedure. Alternatively, you can adjust the weight of the combined filtrate and water washings to 80 g and then add 320 mL of the preheated 95% ethanol. Let the precipitate form at room temperature for 60 min.

3. *Filtration setup*: Tare the crucible with Celite to the nearest 0.1 mg. Moisten the Celite bed in the crucible with 15 mL of 78% EtOH (*see* **Note 3**) from the wash bottle and redistribute it. Vacuum the crucible to draw Celite onto the glass frit as a uniform mat.
4. *Filtration*: Filter the enzyme digest that has precipitated from 3.1.2(2) through a crucible. To ensure all residual particles are collected, use a wash bottle filled with 78% ethanol and a rubber spatula to transfer them completely into the crucible.
5. *Wash*: Use a vacuum to wash the residue with two 15 mL portions of each of the following solvents in order: a. 78% ethanol, b. 95% ethanol, and c. acetone (*see* **Note 4**). After washing, dry the crucible with the residue in an oven at 103 °C overnight.

3.1.3 Total Dietary Fiber

1. *Precipitation of fiber with ethanol*: To each sample, add 225 mL of 95% ethanol that has been warmed to 60 °C. After the heating process, check the volume; the proportion of ethanol to sample volume should be 4:1. If the 95% ethanol is mistakenly heated to 65 °C, use 228 mL to adjust for the increased volume of alcohol. Then, cover each sample with large sheets of aluminum foil and allow the precipitate to form at room temperature for 60 min.
2. Follow steps 3.1.2 *(3–5)*.
3. *Calculation*: Determine the amount of total dietary fiber using the equation below:

$$\text{Total dietary fibre } (\%) = \frac{\frac{R_1 - R_2}{2} - p - A - B}{\frac{m_1 - m_2}{2}} \times 100$$

Where, R_1 = residue weight 1 from m_1; R_2 = residue weight 2 from m_2; m_1 = sample weight 1; m_2 = sample weight 2; A = ash weight from R_1; p = protein weight from R_2

$$\text{Blank} = \frac{\text{BR}_1 - \text{BR}_2}{2} - \text{BP} - \text{BA}$$

Where, BR = blank residue; BA = blank ash from BR2; and BP = blank protein from BR1.

4 Notes

1. Blanks: For each assay, include two blanks alongside the samples to assess any potential residue contribution from the reagents.
2. Store α-amylase, protease, and amyloglucosidase enzymes at 0–5 °C.
3. For 78% Ethanol: To prepare 78% ethanol, measure 821 mL of 95% v/v ethanol and pour it into a 1 L volumetric flask. Then, fill the flask to the 1 L mark with deionized water and mix thoroughly. If needed, add more deionized water to reach the 1 L mark accurately.
4. Acetone, reagent grade.
5. For MES/TRIS buffer at a concentration of 0.05 M for each component, with a target pH of 8.2 at 24 °C, dissolve 19.52 g of 2(N-morpholino) ethane sulfonic acid (MES) and 12.2 g of tris(hydroxymethyl)aminomethane (TRIS) in 1.7 L of deionized water. Adjust the pH to 8.2 using 6.0 N NaOH, then dilute the solution to 2 l with additional water. Ensure the pH is approximately 8.3 at 20 °C or about 8.1 at 27–28 °C.
6. Wear a mask and gloves when working with concentrated ethanol and hydrochloric acid.
7. For hydrochloric acid solution, 0.561 N: Mix 93.5 mL of 6 N HCl with about 700 mL of water in a 1 L volumetric flask, then fill the flask with water to reach a total volume of 1 L. Prepare buffer solutions with pH values of 4.0, 7.0, and 10.0.
8. If the fat content exceeds 10%, defat the leave samples to prevent any interference.

References

1. Negi R, Jan T, Kaur T, Khan SS, Sharma B (2024) Bioactive compounds as plant-based functional foods for human health: current scenario and future challenges. J Appl Biol Biotechnol. Published online January 1, 2024:1–23. https://doi.org/10.7324/JABB.2024.180889
2. Pateiro M, Domínguez R, Munekata PES, Nieto G, Bangar SP, Dhama K, Lorenzo JM (2023) Bioactive compounds from leaf vegetables as preservatives. Foods 12(3):1–20. https://doi.org/10.3390/foods12030637
3. Du B, Xu B (2025) Natural bioactive compounds exerting health-promoting effects by ameliorating oxidative stress. Antioxidants 14(1):1–4. https://doi.org/10.3390/antiox14010085
4. Samtiya M, Aluko RE, Dhewa T, Moreno-Rojas JM (2021) Potential health benefits of plant food-derived bioactive components: an overview. Foods 10(4):1–25. https://doi.org/10.3390/foods10040839
5. Barber TM, Kabisch S, Pfeiffer AFH, Weickert MO (2020) The health benefits of dietary fiber. Nutrients 12(10):1–17. https://doi.org/10.3390/nu12103209
6. Bitwell C, Indra SS, Luke C, Kakoma MK (2023) A review of modern and conventional extraction techniques and their applications for extracting phytochemicals from plants. Sci Afr 19:1–19. https://doi.org/10.1016/j.sciaf.2023.e01585
7. Ye S, Shah BR, Li J, Liang H, Zhan F, Geng F, Li B (2022) A critical review on interplay

between dietary fibers and gut microbiota. Trends Food Sci Technol 124:237–249. https://doi.org/10.1016/j.tifs.2022.04.010

8. Hu YC, Hu JL, Li J, Wang J, Zhang XY, Wu XY, Li X, Guo ZB, Zou L, Wu DT (2023) Physicochemical characteristics and biological activities of soluble dietary fibers isolated from the leaves of different quinoa cultivars. Food Res Int 163:112166. https://doi.org/10.1016/j.foodres.2022.112166
9. Song Y, Lei J, Li J, Wang J, Hu JL, Zheng XQ, Hu YC, Zou L, Wu DT (2023) Structural properties and biological activities of soluble dietary fibers rich in pectic-polysaccharides from different buckwheat green leaves. Int J Biol Macromol 253:126686–126686. https://doi.org/10.1016/j.ijbiomac.2023.126686
10. McCleary BV, Sloane N, Draga A, Lazewska I (2013) Measurement of total dietary fiber using AOAC method 2009.01 (AACC International Approved Method 32-45.01): evaluation and updates. Cereal Chem J 90(4): 396–414. https://doi.org/10.1094/cchem-10-12-0135-fi
11. Megazyme (2017) Total dietary fiber assay procedure. Megazyme:1–19. https://prod-docs.megazyme.com/documents/Assay_Protocol/K-TDFR-200A_DATA.pdf. Accessed 21 Apr 2025
12. McCleary BV, McLoughlin C (2022) Determination of insoluble, soluble, and total dietary fiber in foods using a rapid integrated procedure of enzymatic-gravimetric-liquid chromatography: first action 2022.01. J AOAC Int 106(1). https://doi.org/10.1093/jaoacint/qsac098

Chapter 3

Extraction of Bioactive Compound from Leaf – Polysaccharides – An Insilico Approach

Fanting Zhao, Runyu Zhu, and Peng Chen

Abstract

Polysaccharide is an important active compound in plant leaves, and its efficient and environmentally friendly extraction technology is of great significance for the development of functional foods and drugs. In this study, the extraction process of polysaccharides from loquat leaves was systematically described by using eotolutic solvent (DES) combined with ultrasound-assisted extraction (UAE). Polysaccharides were obtained by alcohol precipitation and freeze-drying, and the relative yield was determined by phenol-sulfuric acid method.

Key words Leaf extracts, Polysaccharide extraction technology, Deep eutectic solvent, Ultrasound-assisted extraction, Phenol-sulfuric acid method, Alcohol precipitation-freeze drying

1 Introduction

Polysaccharides in plant leaves are a kind of natural macromolecular compounds with important biological activities. They exist widely in plant cell wall, vacuole, and other structures, and have many physiological functions such as anti-oxidation, immune regulation, and anti-tumor activity [1]. In recent years, leaf-derived polysaccharides with pharmacological activities have become a research hotspot of natural products [2] and have shown unique value in functional food and drug development.

The extraction methods of polysaccharides include hot water extraction (HWE), enzyme-assisted extraction (EAE), microwave-assisted extraction (MAE), ultrasonic-assisted extraction (UAE), ultrasonic-assisted enzyme extraction (UAEE), and ultrasonic microwave-assisted extraction (UMAE) [3, 4]. In recent years, ultrasonic-assisted extraction (UAE) has been widely used in order to save solvent and improve production efficiency.

As a new green solvent, eutectic solvent (DES) is a low eutectic mixture formed by Lewis/Brønsted acid and base, which has

Mitesh Kumar Dwivedi and A. Sankara Narayanan (eds.), *Leaves as Functional Foods*, Methods and Protocols in Food Science, https://doi.org/10.1007/978-1-0716-5226-8_3,

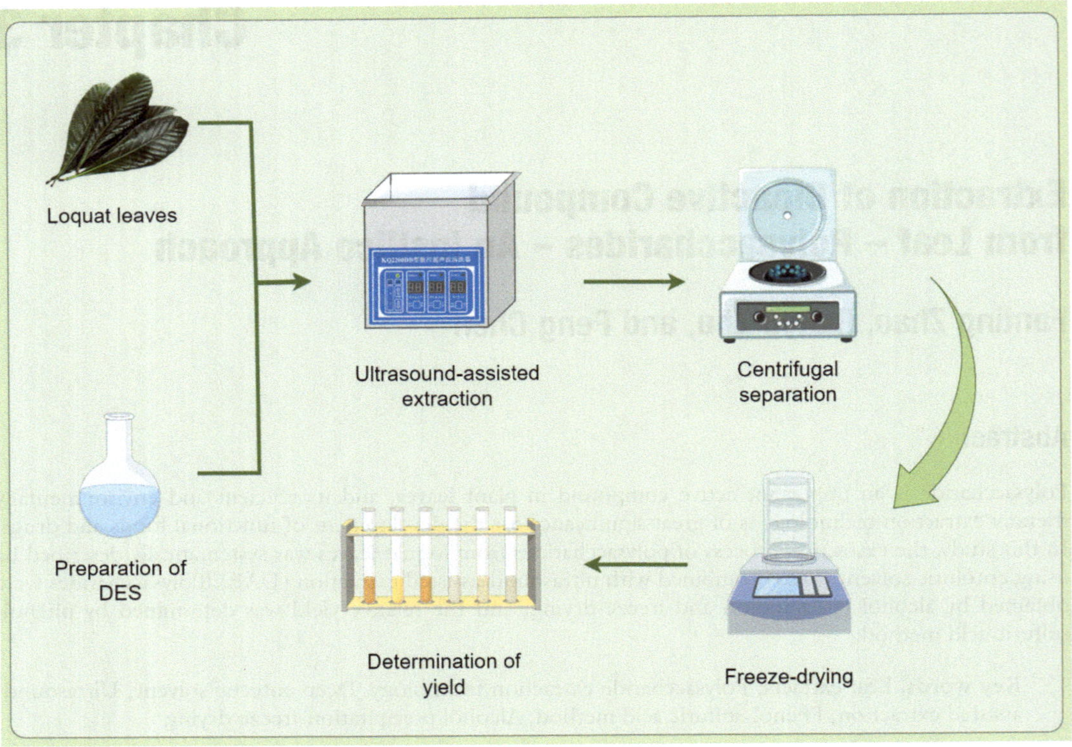

Fig. 1 Flow chart illustrating that the dried and powdered loquat leaves are subjected to polysaccharide extraction using a deep eutectic solvent (DES) via ultrasonic-assisted extraction (UAE). The resulting extract is then precipitated with ethanol, and the polysaccharides are collected via centrifugation, followed by lyophilization to obtain the crude product. The polysaccharide content is quantified by the phenol-sulfuric acid method

received extensive attention in the field of bioactive ingredient extraction. Wang et al. [5] used different extraction methods to extract polysaccharides from Qiang Huo. Compared with the hot water extraction method, the ultrasound-assisted DES extraction rate and yield were significantly improved, indicating that the extraction method has a high application potential in the field of polysaccharide preparation.

In this chapter, we chose the ultrasound-assisted extraction (DES-UAE) method of NADESs, an environmental protection solvent, as an example to extract polysaccharides from loquat leaves [6]. The extraction flowchart is shown in Fig. 1.

2 Materials

2.1 Plant Materials

1. Fresh loquat leaves (washed, oven-dried at 40 °C to constant weight, ground into 60-mesh powder).

2. Pretreatment solvents: Petroleum ether and ethanol were used to remove lipids and monosaccharides in loquat leaves [7]. The treated powder was dried and stored under seal at room temperature.

2.2 DES Components

1. Hydrogen bond acceptor (HBA): Choline chloride (ChCl, ≥98% purity).
2. Hydrogen bond donor (HBD): 1,3-Propanediol (≥99% purity).

2.3 Equipment

1. Grinder.
2. Oven.
3. Ultrasonic cleaning machine.
4. Freeze-drying machine.
5. Centrifuge.

3 Methods

3.1 DES Preparation

1. Hydrogen bond acceptor (HBA) and hydrogen bond donor (HBD), such as choline chloride and 1,3-propylene glycol, were mixed in a round-bottom flask and then mixed at a mole ratio of 1–1.5 under magnetic stirring in an oil bath.
2. Heat at 80 °C under magnetic stirring (500 rpm) in an oil bath for 120 min until a homogeneous, transparent liquid forms. Store in a desiccator to prevent moisture absorption (*see* **Note 1**).

3.2 Ultrasonic-Assisted Extraction

1. 0.25 g of loquat leaf powder was mixed with a given amount of DES in a 10 mL test tube.
2. Perform an ultrasonic cleaner operating at 200 W and 40 kHz for 30 min, maintaining the temperature at 50 °C (±2 °C) using a water bath (*see* **Note 2**).

3.3 Centrifugation

1. After extraction, the mixture was centrifuged at 4000 rpm for 10 min and the supernatant was collected (*see* **Note 3**).

3.4 Ethanol Precipitation

1. The supernatant is mixed with four times the volume of anhydrous ethanol and stored in a refrigerator at 4 °C for 12 h [8].

3.5 Freeze Drying

1. Centrifuge the precipitated polysaccharide at 4000 rpm for 5 min. Wash the pellet twice with 70% ethanol.
2. The precipitated polysaccharide is lyophilized at −80 °C for 24 h to obtain the crude polysaccharide CLLP.

3.6 Determination of the Relative Yield by Phenol-Sulfuric Acid Method

1. Preparation of sample solution: Appropriate amount of crude LLP product was diluted to prepare a stock solution of appropriate concentration.
2. Preparation of standard curve: Using a volumetric flask, prepare a 50 μg/ml solution of the glucose standard, then dilute this to prepare solutions of 0, 0.01, 0.02, 0.03, 0.04, 0.05, 0.06, 0.07, 0.08, and 0.09 mg/ml.Transfer 0.5 ml of each solution into a test tube. Add 0.3 ml of 5% phenol solution, mix thoroughly, then add 1.8 ml of concentrated sulphuric acid and mix well. An orange color appeared after 20 min at room temperature. With the first tube as the reference point, the OD value can be determined by colorimetric measurement at 490 nm, and the standard curve can be drawn [9].
3. Sample content determination: Add 0.3 mL 5% phenol solution to 0.5 mL sample solution, add 1.8 mL concentrated sulfuric acid along the tube wall immediately after mixing, shake, and mix. After reaching room temperature, the OD value at 490 nm can be measured, and the corresponding polysaccharide content can be inquired against the standard curve (*see* **Notes 4** and **5**).

$$\text{Polysaccharide content (\%)} = \frac{C \times D \times V}{W \times 10^6} \times 100$$

C: Polysaccharide concentration from the standard curve (μg/mL).

D: dilution factor.

V: Sample volume (mL).

W: Sample mass (g).

3.7 Calculation of Extraction Rate

1. The extraction yield (Y) of polysaccharides from loquat leaf can be calculated according to the following formula:

$$\text{Y (mg/g)} = \text{mass of extracted polysaccharide (mg)} / \text{mass of loquat leaf powder (g)}$$

4 Notes

1. Store DES in airtight containers with desiccants to prevent moisture absorption, which may alter its eutectic properties.
2. Strictly control UAE time and temperature (e.g., 30 min and 50 °C) to avoid polysaccharide degradation due to overheating.
3. Pre-chill ethanol to 4 °C to enhance precipitation efficiency.
4. Measure absorbance at 490 nm precisely 20 min after sulfuric acid addition; delayed readings may cause color fading and inaccuracies.
5. Handle concentrated sulfuric acid in a fume hood with PPE (gloves, goggles) due to its corrosive nature.

Acknowledgments

This work was supported by Gansu Provincial Science and Technology Major Project (Grant No. 24ZDFA001), The Lanzhou Municipal Science and Technology Program (Grant Nos. 2024-8-27, 2024-8-30, 2024-4-2), Scientific Research Project of Gansu Medical Products Administration (Grant No. 2022GSMPA008), Open Project of NMPA Key Laboratory for Quality Control of Traditional Chinese Medicine (Grant No. 2022GSMPA-KL05), Gansu Intellectual Property Project (Grant No. 22ZSCQ065), and the College Students' Innovation and Entrepreneurship Program of Lanzhou University, China (Grant Nos. 20250260006, 20250260016 and 20250260020).

References

1. Chen L, Huang G (2018) Antitumor activity of polysaccharides: an overview. Curr Drug Targets 19(1):89–96. https://doi.org/10.2174/1389450118666170704143018
2. Bhuniya S, Demina TS, Akopova TA (2024) Advances in applications of polysaccharides and polysaccharide-based materials. Int J Mol Sci 25(12):6482. https://doi.org/10.3390/ijms25126482
3. Song H, Zhang Y, Huang Q, Wang F, Wang L, Xiong L, Shen X (2025) Extraction optimization, purification, characterization, and hypolipidemic activities of polysaccharide from pumpkin. Int J Biol Macromol 307(Pt 2):141907. https://doi.org/10.1016/j.ijbiomac.2025.141907
4. Ding Z, Ying Q, Lei J et al (2025) Structural analysis and treatment of acute liver injury of polysaccharides from the *Radix Actinidiae Chinensis*. Front Pharmacol 16:1544302. https://doi.org/10.3389/fphar.2025.1544302
5. Wang C, Li Q, Qiu D, Guo Y, Ding X, Jiang K (2024) An efficient and environmentally-friendly extraction, characterization and activity prediction of polysaccharides from Rhizoma et Radix Notopterygii. Int J Biol Macromol 265 (Pt 1):130907. https://doi.org/10.1016/j.ijbiomac.2024.130907
6. Yan K, Liu X, Li L et al (2024) Ultrasound-assisted deep eutectic solvents extraction of polysaccharides from loquat leaf: process optimization and bioactivity study. Int J Biol Macromol 274(Pt 1):133308. https://doi.org/10.1016/j.ijbiomac.2024.133308
7. Xia L, Deji, Zhu M, Chen D, Lu Y (2020) Juniperus pingii var. wilsonii acidic polysaccharide: extraction, characterization and anticomplement activity. Carbohydr Polym 231:115728. https://doi.org/10.1016/j.carbpol.2019.115728
8. Lai JY, Fan XL, Zhang HB et al (2024) Polygonum cuspidatum polysaccharide: a review of its extraction and purification, structure analysis, and biological activity. J Ethnopharmacol 331:118079. https://doi.org/10.1016/j.jep.2024.118079
9. Ghosh S, Abdullah MF (2024) Extraction of polysaccharide fraction from cadamba (Neolamarckia cadamba) fruits and evaluation of its in vitro and in vivo antioxidant activities. Int J Biol Macromol 279(Pt 4):135564. https://doi.org/10.1016/j.ijbiomac.2024.135564

Chapter 4

Extraction of Bioactive Compounds: Flavonoids from *Tridax* Leaf

T. Savitha, T. Poovizhi, and A. Sankara Narayanan

Abstract

Flavonoids are chemical compounds belonging to the polyphenol family that have a basic structure consisting of two aromatic rings (A and B) linked through three carbons that usually form an oxygenated heterocycle (C ring). These molecules exhibit a number of pharmacological properties, including bacteriostasis, antioxidation, immunological modulation, free-radical scavenging, and anti-inflammatory properties. More than 8000 compounds with a flavonoid structure have been identified, many of which are responsible for the attractive colors of flowers, fruits, and leaves. They play vital roles as health-rejuvenating and disease-preventing dietary supplements. Emerging research on flavonoids focuses on novel drug development for various human ailments due to their diverse physiological activity and low toxicity. This plant possesses a number of classes of flavonoids that contribute various pharmacological activities. The present protocol needs hours for the extraction of bioactive flavonoid compounds from leaves through a traditional approach and through an easier, innovative approach.

Key words Flavonoids, Polyphenols, Pharmacological activity, Dietary supplements, Bioactive compounds

1 Introduction

Tridax procumbens, a renowned weed belonging to the Asteraceae family that is a highly potent species popular among the 30 species of the genus *Tridax*. In Ayurveda, *T. procumbens* is used as an herbal drug for wound healing, and it is found as a dominant plant in Maharashtra, Madhya Pradesh, Gujrat, Odissa, and other Indian states. It grows on roadsides and lawns and in open spaces, meadows, and croplands [1]. The phytochemistry of *T. procumbens* features tannins, flavonoids, saponins, essential oils, alkaloids, carotenoids, steroids, terpenoids, carbohydrates, and other chemical substances, according to research findings. The plant is a procumbent herb and is valued for its pharmaceutical properties. The bioactivity of plant extracts is attributed to the phytochemical constituents in the extracts. However, the phytochemical constituents

Mitesh Kumar Dwivedi and A. Sankara Narayanan (eds.), *Leaves as Functional Foods*, Methods and Protocols in Food Science, https://doi.org/10.1007/978-1-0716-5226-8_4,

of a plant extract are extremely dependent on the polarity of the solvent, the particle size of the plant material, the solvent–plant material ratio, the temperature, and the extraction method [2]. The extraction rate of flavonoids and phenolic compounds from *T. procumbens* has been found to be higher when the plant is cultivated in well-lit areas versus shaded areas [3]. Flavonoids are polyphenols found in *T. procumbens* leaves that exist as flavones and flavanones, which are responsible for antioxidant, hepatoprotective, anticancer, antibacterial, and wound-healing properties. Out of the various flavonoids present, the most popular compounds, namely quercetin, apigenin, kaempferol, and myricetin, exert radical scavenging activity, which make up the backbone of many pharmacological activities [4]. Flavonoids are the predominant class of bioactive compounds found in this plant, which exhibit many biological properties [5]. The present protocol explains the extraction of flavonoids from *Tridax* leaf through a systematic approach.

2 Materials

For extraction, use Soxhlet apparatus and solvents (acetone, ethanol, and methanol), Whatman filter paper, a rotary evaporator, and sterile containers.

For determination and purification, use 20% ammonium hydroxide, aluminum chloride, an ultraviolet-visible (UV-visible) spectrophotometer, rutin and quercetin, thin layer chromatography (TLC) plates, and CAMAG TLC equipment.

For identification, use the Fourier transform infrared spectroscopy (FTIR) model: Varian 3600, DRX-Bruker 300 MHz (Mega Hz), Switzerland.

3 Methods

3.1 Preparation of Plant Material

1. The *Tridax procumbens* leaves of fresh and sound quality should be collected. The authentication of leaves should be carried out at a reputable research institution.
2. The leaves are washed thoroughly two to three times with running tap water then once with sterile distilled water.
3. The washed leaves are then dried (shade dried/air dried at room temperature) to reduce moisture content by up to 10% and ground into a fine powder by using a domestic mixture grinder.
4. The powdered material is weighed and kept in an airtight container for further use.

3.2 Extraction of Flavonoids

1. Extraction methods such as maceration and Soxhlet extraction are commonly employed for the preliminary extraction of flavonoids from leaves.
2. In maceration, 1 g of the powdered leaves is mixed with 100 mL of suitable solvent systems (*see* **Note 1**), such as acetone, ethanol, methanol, and water, separately. Each mixture (*see* **Note 2**) is kept on the shaker for 24 hours at ambient temperature to obtain a homogenate.
3. In Soxhlet extraction, the continuous extraction of powdered material with a suitable solvent system uses Soxhlet apparatus to obtain a homogenate. These homogenates are filtered by using Whatman filter paper to separate plant debris, and the filtrates are concentrated by using a rotary evaporator to obtain a crude extract.
4. The prepared extracts are stored separately in sterile bottles at 10 °C for further screening [6].

3.3 Determination of Flavonoids

3.3.1 Alkaline Reagent Test for Qualitative Estimation [7]

First, 1 mL of extract is treated with five drops of 20% ammonium hydroxide solution. The production of an intense yellow color/fluorescence indicates the presence of flavonoids.

3.3.2 Colorimetric Method for Quantitative Estimation

1. The total flavonoid content (TFC) of *Tridax procumbens* leaf extracts is quantified by using the aluminum chloride colorimetric method [8, 9].
2. 1 mL of the extract is mixed with 1 mL of 2% aluminum chloride.
3. The mixture is vortexed, and the reaction is kept at room temperature for 10 min in dark conditions, and then the absorbance of the reaction mixture is measured at 510 nm by using a UV-visible spectrophotometer.
4. The standard compound—i.e., quercetin (10–200 mg/L)—is used for calibration, and the amount of total flavonoid is calculated as mg/100 g of dry powder to determine quercetin equivalent (QUE).

3.4 Purification of Flavonoids

The purification of the crude extract requires subjecting it to thin layer chromatography (TLC) and high-performance thin layer chromatography (HPTLC). In TLC, an optimized eluent system and the stationary phase are employed to obtain efficient extraction. In the HPTLC method, individual flavonoid compounds are identified by using rutin and quercetin as standards.

3.4.1 Purification via Thin Layer Chromatography [10]

1. In TLC, extracted samples (2 mg/mL) are dissolved in suitable solvents.
2. As the standard, quercetin (1 mg) is dissolved in methanol.
3. Different mobile phases are optimized with varying concentrations {n-butanol:acetic acid:water (2:2:6)} for the separation of flavonoids.
4. A sample and the standard are spotted on the TLC plate. Next, they should be dried.
5. The plates are exposed to the solvent system until it reaches the other end.
6. Following drying with UV light, all the plates are visualized directly at 254 nm and 366 nm in a UV TLC viewer.
7. The various *Rf* values of the respective spots are observed and calculated.
8. The dark-brown bands that are produced after keeping the plates in the iodine chamber are scratched and suspended in the mobile phase.
9. After the mobile phase, the spots are sucked out and subjected to vacuum evaporation, and the residue left is collected and subjected to HPTLC and infrared (IR) spectral analysis.

3.4.2 HPTLC Method of Estimating Flavonoids

1. The complete CAMAG TLC equipment consists of a fully automatic Linomat 5 sample applicator, a developing chamber, a CAMAG TLC scanner (allowing densitometry evaluation of chromatograms), and CATS 4 software for the interpretation of data.
2. Weigh 10 mg of the purified extracts and dispense in the relevant solvent system, making a volume of 10 mL in a standard flask (1000 μg/mL).
3. The standard flavonoid compound (10 mg of quercetin) is dissolved in methanol, making a volume of up to 100 mL in a standard flask (100 μg/mL).
4. HPTLC aluminum sheets and silica gel 60 F254 are employed as adsorbents (the stationary phase).
5. The extracts, from 10 μL to 1000 μg/mL, are applied pointwise on HPTLC aluminum sheets in different tracks in the form of 6 mm wide bands by using a CAMAG semiautomatic Linomat 5 spotter at a distance of 12 mm.
6. Nitrogen gas is supplied for the simultaneous drying of bands.
7. In the mobile phase, n-butanol:acetic acid:water (2:2:6) is used as a solvent system in the HPTLC of different extracts.

8. In the mobile phase, 10 mL is placed in each flat-bottomed CAMAG twin trough TLC chamber, for saturation, 30 min before the development of the HPTLC plate.
9. The chamber should be sealed with parafilm and covered with a steel lid.
10. By using a TLC scanner 3 with Wincats software under 364 nm, the dried developed plates are scanned.
11. All plates are visualized directly after drying, and a fingerprint profile is documented through photos by using a CAMAG Reprostar 3 under 254 nm and 366 nm in UV and visible light.
12. The digital images of chromatograms are evaluated with the CAMAG Video Scan program.
13. The captured image is subjected to a visual inspection on the computer screen.
14. The differences found are specified by the HPTLC system, in which the differences are detected and the *Rf* value (and color) of a compound in the system is determined.

3.5 Identification of Flavonoids

The spectral identification of isolated compounds is carried out by using IR, nuclear magnetic resonance (NMR), and mass spectrometry (MS) methods through structural elucidations. According to the differences in the functional groups and their relative positions in the 15-carbon skeleton, flavonoids are classified into several subgroups: flavones, flavanone, isoflavonoid, anthocyanidin, and chalcones.

1. For IR, the sample (1–2 mg) is crushed into KBr pellets (3–4 mg) using an FTIR (model: Varian 3600) with the help of mechanical pressure.
2. They are recorded within the wave number range of 4000–400 cm^{-1} in an FTIR instrument. Also, 1H NMR spectra are recorded on DRX-Bruker 300 MHz (megahertz), from Switzerland.
3. The nuclear magnetic field is obtained for the isolated flavonoid. The isolated compound is dissolved in the relevant solvent (CDCl3), and about 600 μL is poured into the NMR tube and observed on the applied magnetic field.
4. The mass was obtained via TOF MS ES + 5.18e4, TOF MS ES + 477, and TOF MS ES + 174.80ES + 3 (ABHAY_BIOTECH 28).

4 Notes

Optimization procedures for solvent selection

1. In solvent selection, the choice of the solvent and its concentration is crucial for the effective and efficient extraction of plant materials, which are optimized by evaluating the total phenolic content and the total flavonoid content of the individual extracts.
2. Simultaneously, extraction parameters such as extraction time, the solvent–sample ratio, and temperature can be optimized to increase flavonoid yield.

References

1. Mir SA, Jan Z, Mir S, Dar AM, Chitale G (2017) A concise review on biological activity of *Tridax procumbens* linn. Organic Chem Curr Res 6(177):2161–0401. https://doi.org/10.4172/2161-0401.1000177
2. Kannamba B, Teja WD, Surekha M, Lavanya B (2017) Effect of extraction methods and solvent on phytochemical composition of medicinal plant extracts. Der Pharma Chemica 9(9): 152–156. ISSN 0975-413X
3. Idris AC, Linatoc A, Mohammad SM, Aliyu AM, Abu Bakar MF (2018) Effect of light intensity on the total flavonoid and total phenolic contents of *Mikania Micrantha* and *Tridax Procumbens*. J Sci Technol 10(4). https://penerbit.uthm.edu.my/ojs/index.php/JST/article/view/3645
4. Hemalatha R (2016) Sar study of flavonoids isolated from *Tridax Procumbens* L. for their anti-oxidant activity. Eur J Biomed Pharm Sci 3(3):519; ISSN: 2349-8870
5. Ingole VV, Mhaske PC, Katade SR (2022) Phytochemistry and pharmacological aspects of *Tridax procumbens* (L.): a systematic and comprehensive review. Phytomed Plus 2(1): 100199. https://doi.org/10.1016/j.phyplu.2021.100199
6. Patil AM, Pawar SG (2018) Biochemical Investigation, Quantitative Estimation and Antioxidant activity of *Tridax procumbens* L. J Emerg Technol Innov Res 5(7):712–717. ISSN-2349-5162
7. Roy D (2018) Qualitative and quantitative screening of total phenolic and total flavonoids for different extracts of *Tridax procumbens* L. and comparative study of their anti-oxidant properties. J Med Sci Clin Res 6(9):340–345. https://doi.org/10.18535/jmscr/v6i9.59
8. Deshmukh MH, Pai SR, Nimbalkar MS, Patil RP (2009) Biochemical characterization of banana cultivars from Southern India. Int J Fruit Sci 9(4):305–322. https://doi.org/10.1080/15538360903241336
9. Ghosh P, Das C, Biswas S, Nag SK, Dutta A, Biswas M, Sil S, Hazra L, Ghosh C, Das S, Saha M, Mondal N, Mandal S, Ghosh A, Karmakar S, Chatterjee S (2020) Phytochemical composition analysis and evaluation of *in vitro* medicinal properties and cytotoxicity of five wild weeds: a comparative study. F1000Research 2:493. https://doi.org/10.12688/f1000research.22966.1
10. Sharma V, Janmeda P (2017) Extraction, isolation and identification of flavonoid from *Euphorbia neriifolia* leaves. Arab J Chem 10(4):509–514. https://doi.org/10.1016/j.arabjc.2014.08.019

Chapter 5

Comparative Extraction and Fractionation of Alkaloids Using Modern and Conventional Solvent Systems

Abhishek Sharma, Roshani Patel, Ruchika Upadhyay, Deepa Nautiyal, Dhara Patel, and Abhishek Sharma

Abstract

Alkaloids constitute a structurally heterogeneous group of nitrogenous secondary metabolites with widespread distribution in fungi and plants, and numerous members of this group are bioactive and of pharmacological interest. Conventional extraction techniques, like acid-base fractionation, have been traditional starting reagents but are plagued by key drawbacks in the form of the use of harmful solvents, low selectivity, degradation of the analytes, and high environmental load. Advances in green chemistry and extraction methodology over the recent past have helped usher in more efficient and eco-friendly alternatives. This chapter examines both traditional and novel alkaloid extraction approaches and highlights two contrasting processes. The first utilizes ultrasound-assisted complete extraction with an ionic liquid non-aqueous solvent system (UACE-ILNAS) to extract isoquinoline alkaloids from *Phellodendron amurense.* The combination of the ionic liquid [C4mim][OAc] with DMSO delivers improved penetration through lignified plant material and high yields, improving viscosity and waste solvent reduction. The second protocol describes a conventional methanol-extraction of indole and oxindole alkaloids from *Mitragyna speciosa* (kratom). The protocol involves argon-assisted methanol reflux, acid-base manipulation, solvent partitioning, and column chromatography to extract both major and minor alkaloids such as mitragynine and isopaynantheine. Collectively, these methods demonstrate how the integration of traditional principles and contemporary advances can enhance efficiency, lessen environmental footprint, and facilitate the isolation of pharmacologically active alkaloids from otherwise difficult-to-handle plant matrices.

Key words Ultrasound-assisted complete extraction (UACE), Ionic liquid non-aqueous solvent (ILNAS), Methanol reflux, *Mitragyna speciosa*, *Phellodendron amurense*, Alkaloid extraction, HPLC

1 Introduction

Alkaloids are products of alkaline nature found in fungi and plants naturally. The isolation and extraction of alkaloid compound was first achieved in 1805. Isolation of morphine was an important pharmacological discovery, which led to a series of experiments leading to isolation of caffeine, brucine, and quinine [1]. Alkaloids

Mitesh Kumar Dwivedi and A. Sankara Narayanan (eds.), *Leaves as Functional Foods*, Methods and Protocols in Food Science,
https://doi.org/10.1007/978-1-0716-5226-8_5,

are classified into three types: true, pseudo, and proto alkaloids. True alkaloids possess nitrogen-incorporated heterocyclic rings. These are synthesized from amino acid synthesis in plants and other organisms. Pseudo alkaloids have heterocyclic ring structure derived from alternate pathways. In proto-alkaloids biogenesis, nitrogen is not part of initial ring structure and is incorporated through methylation and decarboxylation [2–5].

Isolation and purification of alkaloids from plant components started in the nineteenth century with strychnine and cinchona from *Strychnos* and *Cinchona*. Further studies and advancements in chemical isolation and analysis led to discoveries of morphine, codeine, thebaine, and narcotine from poppy (*Papaver somniferum*) [6–9]. Recent uses of alkaloids include the treatment of cancer; vincristine and vinblastine alkaloids isolated from *Catharanthus roseus* are used in cancer treatments [10, 11].

Alkaloids exist in two forms naturally: ionic salts and neutral-free base. The acid-base extraction method allows the conventional extraction of these alkaloids in polar and non-polar solvents [12, 13]. Apart from replacing toxic solvent used in extraction alkaloids by using ionic liquids and deep eutectic solvents other modern methods for alkaloid extraction include Microwave Assisted Extraction (MAE), Supercritical fluid extraction (SFE) [14] and Ultrasound Assisted Extraction (UAE).

Acid-base method is traditionally used for alkaloid extraction, but the disadvantage of alkaloid degradation and erroneous analysis due to artifact generation makes this method not suitable for alkaloid extraction. MAE-based alkaloid extraction involves irradiation of sample with microwaves that are absorbed by polar analytes according to their dielectric constant, which causes heating of the analyte thereby increasing the dissolution in the solvent. Similar to MAE, UAE also increases analyte dissolution and purity of the sample with lower temperature requirement compared to MAE [15, 16].

In UAE, organic solvents are used for extraction. Large amount of solvent waste, non-alkaloid impurity, hazardous volatility, and environmental contamination are some of the limitations of UAE method. Ionic liquid aqueous solvent (ILAS) used for UAE-based extraction has multiple issues. Low inter-miscibility of ionic solvents in water, less extraction of low-polarity analytes, and less penetration of ILAS solvent in plant material make ILAS solvents in ultrasound-based alkaloid extraction inefficient [17]. Development of a suitable solvent and application of suitable physical techniques are necessary to improve the extraction process and to address the aforementioned issues. This chapter discusses UACE-ILNAS (ultrasound-assisted extraction–ionic liquid non-aqueous solvent) as well as traditional solvent-based methods for the extraction of alkaloids.

2 Materials

2.1 Extraction of Alkaloids Using UACE-ILNAS Method [17]

1. Leaves and/or bark of plant.
2. Mesh sieve, 60/80 microns.
3. Polyethylene zipper bags.
4. Incubator/Storage, 4 °C.
5. Alkaloid Standards, ≥98%.
6. Acetonitrile, Chromatographic grade.
7. Phosphoric acid, chromatographic grade.
8. Filter, 0.45 μm.
9. 1-butyl-3-methylimidazole acetate ([C4mim][OAc]) (*see* **Note 1**).
10. Dimethyl sulfoxide (DMSO) (*see* **Note 1**).
11. Methanol (*see* **Note 1**).
12. Ultrapure water, (resistivity: 18.2 MΩ·cm).
13. Kromasil-C18 RP column (4.6 mm × 250 mm, 5 μm).
14. HPLC Setup.
15. Ultrasound generator assembly.

2.2 Extraction of Alkaloids Using Traditional Solvent-Based Methods [18]

1. Kratom powder or dry kratom leaves (500 g).
2. N-Hexane.
3. Methanol.
4. Ethyl acetate.
5. Dichloromethane (DCM).
6. Ultrapure water (resistivity: 18.2 MΩ·cm).
7. Acetic acid.
8. Sodium hydroxide pellets.
9. Sodium chloride.
10. Sodium sulfate.
11. Argon gas.
12. Silica gel column and HPLC setup.
13. Rotary evaporator.
14. High vacuum drying setup.
15. Ice bath.

3 Methods

3.1 Ultrasonic-Assisted Complete Extraction in Ionic Liquid Non-aqueous Solvent (UACE-ILNAS)

1. Samples (dried leaves) are mixed with 2.01 mol/L ionic liquid ([C4mim][OAc]/DMSO) in 25 mL/g (liquid/solid) ratio.
2. Ultrasonic irradiation extraction of alkaloids with following ultrasonic cleaner parameters:
   - Temperature: 79 °C.
   - Time: 30 min.
   - Voltage: 80 W.
   - Mesh Size: 60 microns.
3. Sample is filtered using 0.45 μm size filter for further extraction of alkaloids in HPLC system (*see* **Note 2**).

3.2 HPLC for Separation of Alkaloid Extracts

1. Phosphoric acid (0.1%) as solvent A and acetonitrile as solvent B are used as mobile phase for separation of alkaloids. Input conditions for solvents are as follows: 0–8 min, 15% solvent B; 8–20 min, 15–30% solvent B; 20–35 min, 30% solvent B; 35–40 min, 30–15% solvent B.
2. Runtime is set to 40 min, with input/injection volume of 10 μL. Flow rate of eluent is set to 1 mL/min with a column temperature of 25 °C.
3. Alkaloid standards can be prepared using methanol as a solvent.

3.3 Extraction of Alkaloids from Mitragyna speciosa *[18]*

Extraction of alkaloids from Mitragyna speciosa:

1. Dry powdered kratom leaves (Kratom powder), 500 g, is refluxed in 500 mL methanol for 30 min. Reflux is performed in inert condition by using argon (*see* **Note 3**).
2. Suspension is filtered and methanol-based reflux is repeated twice. Solvent from triplicate is combined and evaporated under low pressure condition.
3. Residue is resuspended in 1 L 10% acetic acid and washed with hexane (500 mL) (*see* **Note 4**). Hexane wash is repeated three more times before the aqueous phase is cooled to 0 °C and its pH is increased to 10 slowly using 2.5 M NaOH (~1 L).
4. Alkaloids are extracted from aqueous layer by using ethyl acetate (400 mL, repeated thrice). The combined layer is washed with highly concentrated NaCl solution (250 mL). The sample is subsequently dried using sodium sulfate.

   Remaining solvent is evaporated in low pressure and extract is subsequently dried in high vacuum condition.
5. Alkaloid extract is analyzed using silica gel column chromatography (0–15% methanol in dichloromethane) to isolate fractions of mitragynine, paynantheine, speciogynine, and

speciociliatine. For minor alkaloids mitraciliatine and isopaynantheine, 5–20% methanol in dichloromethane is used as elution buffer.

4 Notes

1. Methanol, DMSO, and ethyl acetate are toxic and volatile; therefore, use under fume hoods with appropriate PPE. Low-volatility ionic liquids such as [C4mim][OAc] are highly viscous. Use after thorough mixing.
2. Precondition the column and optimize the gradient on TLC prior to fractionation by HPLC. Reverse-phase C18 column is employed for UACE-ILNAS-based HPLC detection at 229 nm. Utilize freshly prepared mobile phase and pre-filtered samples to avoid column clogging.
3. Reflux is performed under argon to prevent oxidation. If an inert gas setup is unavailable, ensure minimal air contact to protect sensitive alkaloids.
4. Multiple n-hexane and ethyl acetate washes should be performed to clean alkaloid layers. Ensure separation is complete and emulsions are avoided by using brine or centrifugation as necessary.

References

1. Aniszewski T (2007) Alkaloids—secrets of life: alkaloid chemistry, biological significance, applications and ecological role. Elsevier, Amsterdam
2. Dewick PM (2009) Medicinal natural products: a biosynthetic approach, 3rd edn. Wiley, Chichester
3. Buckingham J, Baggaley KH, Roberts AD, Szabo LF (eds) (2010) Dictionary of alkaloids, 2nd edn. CRC Press, Boca Raton
4. Hegnauer R (1988) Biochemistry, distribution and taxonomic relevance of higher plant alkaloids. Phytochemistry 27:2423–2427. https://doi.org/10.1016/0031-9422(88)87006-7
5. Khaden S, Marles RJ (2011) Chromone and flavonoid alkaloids: occurrence and bioactivity. Molecules 17:191–206. https://doi.org/10.3390/molecules17010191
6. Viegas C, Bolzani VS, Barreto EJ (2006) Os produtos naturais e a química medicinal moderna. Química Nova 29(2):326–337. https://doi.org/10.1590/S0100-40422006000200025
7. Barreiro EJ (1990) Produtos naturais bioativos de origem vegetal e o desenvolvimento de fármacos. Química Nova 13:29–39. https://doi.org/10.1590/S0100-40422009000300012
8. Concheiro M, Newmeyer MN, Costa JL, Flegel R, Gorelick DA, Huestis MA (2014) Morphine and codeine in oral fluid after controlled poppy seed administration. Drug Test Anal 7:586–591. https://doi.org/10.1002/dta.1742
9. Gulland JM, Robinson R (1923) The morphine group. Part I. A discussion of the constitutional problem. J Chem Soc 123:980–998. https://doi.org/10.1039/CT9232300980
10. Maciel SC, Silva RF, Reis MS, Oliveira CD, Ventura JA, Lima MF (2011) Characterization of a new potyvirus causing mosaic and flower variegation in *Catharanthus roseus* in Brazil. Sci Agric 68(6):698–705. https://doi.org/10.1590/S0103-90162011000600013
11. Dhyani P, Quispe C, Sharma E, Saurav K, Shukla R, Pandey AK, Bhardwaj P, Parihar

RD, Singh H (2022) Anticancer potential of alkaloids: a key emphasis to colchicine, vinblastine, vincristine, vindesine, vinorelbine and vincamine. Cancer Cell Int 22:206. https://doi.org/10.1186/s12935-022-02624-9

12. Dey P, Kundu A, Kumar A, Gupta M, Lee BM, Bhakta T (2020) Analysis of alkaloids (indole alkaloids, isoquinoline alkaloids, tropane alkaloids). In: Silva AS, Nabavi SF, Saeedi M, Nabavi SM (eds) Recent advances in natural products analysis. Elsevier, Amsterdam, pp 505–567. https://doi.org/10.1016/B978-0-12-816455-6.00015-9
13. Dostál J (2000) Two faces of alkaloids. J Chem Educ 77:993. https://doi.org/10.1021/ed077p993
14. Marillán C, Uquiche E (2023) Extraction of bioactive compounds from *Leptocarpha rivularis* stems by three-stage sequential supercritical extraction in fixed bed extractor using CO_2 and ethanol-modified CO_2. J Supercrit Fluids 197:105903. https://doi.org/10.1016/j.supflu.2023.105903
15. Belwal T, Pandey A, Bhatt ID, Rawal RS, Luo Z (2020) Optimized microwave-assisted extraction (MAE) of alkaloids and polyphenols from *Berberis* roots using multiple-component analysis. Sci Rep 10(1). https://doi.org/10.1038/s41598-020-57585-8
16. Confortin TC, Todero I, Luft L, Silva LP, Zanetti GD, Mercali GD, Marczak LDF (2021) Extraction of bioactive compounds from *Senecio brasiliensis* using emergent technologies. 3 Biotech 11(6):284. https://doi.org/10.1007/s13205-021-02845-1
17. Peng X, Sui X, Li J, Liu T, Yang L (2021) Development of a novel functionality for a highly efficient imidazole-based ionic liquid non-aqueous solvent system for the complete extraction of target alkaloids from *Phellodendron amurense* Rupr. under ultrasound-assisted conditions. Ind Crop Prod 168:113596. https://doi.org/10.1016/j.indcrop.2021.113596
18. Chakraborty S, Uprety R, Daibani AE, Rouzic VL, Hunkele A, Appourchaux K, Eans SO, Nuthikattu N, Jilakara R, Thammavong L, Pasternak GW, Pan YX, McLaughlin JP, Che T, Majumdar S (2021) Kratom alkaloids as probes for opioid receptor function: pharmacological characterization of minor indole and oxindole alkaloids from kratom. ACS Chem Neurosci 12(14):2661–2678. https://doi.org/10.1021/acschemneuro.1c00149

Chapter 6

Extraction of Bioactive Components from Leaf by Microwave-Assisted Extraction

M. Veerapagu, K. R. Jeya, A. Sankara Narayanan, and K. Arun

Abstract

Edible leaves are significant sources of bioactive chemicals vital for human health and nutrition. They operate not only as major sources of macronutrients and micronutrients but also as vehicles for the production of functional foods, owing to their diverse phytochemical composition. Leafy greens like *Moringa oleifera*, *Murraya koenigii*, *Ocimum sanctum*, and *Morus alba* are abundant in antioxidants, vitamins, minerals, and secondary metabolites, including flavonoids, phenolic acids, alkaloids, terpenoids, and saponins. These chemicals demonstrate many biological actions, including antioxidant, anti-inflammatory, antibacterial, antidiabetic, and anticancer properties. Microwave-assisted extraction (MAE) has evolved as a quick, environmentally friendly, and efficient technology for separating bioactive compounds from plant sources. This chapter delineates a detailed laboratory protocol for the extraction of phytochemicals from edible leaves via MAE, highlighting its benefits regarding yield, time efficiency, and the preservation of thermolabile compounds, thus enhancing research and applications in nutraceuticals and functional food formulations.

Key words Functional foods, Edible leaves, Bioactive compounds, *Moringa oleifera*, Microwave-assisted extraction (MAE), Antioxidants, Nutraceuticals

1 Introduction

Natural sources of primary and secondary metabolites, leaves are the most photosynthetically active and metabolically active parts of a plant. Plant leaves comprise phytochemicals with a variety of biological activities and, therefore, they have long served as the foundation for the formulation of medicines, supplements, and functional foods [1]. The emerging realization of the connectivity between health and food led to a growing demand for functional foods, which not only serve basic nutrition but also provide some physiological benefits in modulating health. Consumption of edible leaves such as curry leaf (*Murraya koenigii*), holy basil (*Ocimum*

Mitesh Kumar Dwivedi and A. Sankara Narayanan (eds.), *Leaves as Functional Foods*, Methods and Protocols in Food Science, https://doi.org/10.1007/978-1-0716-5226-8_6,

sanctum), spinach (*Spinacia oleracea*), and moringa (*Moringa oleifera*) may provide a positive health effects due to their functional food properties. [2]. Functional foods have benefits beyond basic nutrition, such as promoting general health and preventing disease. As interest in health and wellness continues internationally, the market for these healthy foods is ever increasing. Leaves of plants are an important contribution to household food security and dietary diversity. They are rich in important nutrients—including zinc, calcium, iron, magnesium, potassium, provitamin A, vitamin C, and nutritional fiber while being low in fat and carbohydrates, making them highly nutritious, nutrient-dense nutritional alternatives [3].

Moringa oleifera, generally called the drumstick tree, is an exceptional example of a nutraceutical leafy greens. Its leaves are rich in proteins, carbohydrates, carotene (provitamin A), vitamin C, and essential minerals such as calcium, potassium, iron, and phosphorus making them a valuable source of nutrition [4]. Dry moringa leaf materials contain protein levels (30.3%) on par with leguminous crops like soybeans and sort beans and also contain critical amino acids like methionine, cysteine, tryptophan, and lysine [5]. Moreover, the leaves are rich in antioxidants including flavonoids, phenolic compounds, ascorbic acid, and carotenoids. Moringa's minimal anti-nutritional factor level guarantees its safety and edibility for consumption. Due to its fat content, it retains more flavour during processing and improves texture and handling properties. It's far widely formulated with infant complementary diet, biscuits, bread and cakes [6, 7].

Edible leaves are rich sources of compounds vital for human life and health. Of these, polyphenols such as flavonoids, phenolic acids, and tannins are effective antioxidants that enhance cellular defense mechanism from oxidative damage [8]. Alkaloids are recognized for their analgesic and antimicrobial properties, even as terpenoids showcase immunomodulatory and anticarcinogenic functions [9, 10]. Saponins, another crucial group, are recognized for their ability to lower the levels of cholesterol and facilitate immune homeostasis. Olive leaf extracts, which are high in oleuropein, a substance that supports cardiovascular health by lowering blood pressure and improving lipid profiles, are an example of the advantages of these substances [11]. Further, sweet potato leaves contain polyphenols and anthocyanins that can inhibit the growth of most cancers cell [12].

Morus alba (white mulberry) is another example of a leafy vegetable rich in bioactive compounds such as phenolic acids, flavonoids, anthocyanins, nutrients, and minerals. Its leaves contain notable health-promoting molecules like quercetin, rutin, apigenin, and chlorogenic and ferulic acids. The leaves have also been shown to contain bioactive substances, such as moracin furols and flavans. Higher levels of 1-deoxynojirimycin (DNJ), a substance with antidiabetic properties, are seen in mature leaves [13].

Sweet potato leaves are also emerging as a fairly resilient and nutritious option. Their resistance to pests, diseases, and extra moisture makes them a dependable food supply throughout natural calamities like floods or droughts. These leaves are rich in polyphenols and anthocyanins, known for their antioxidant properties and potential anti- cancer effects, positioning them as a healthier alternative to synthetic dietary supplements [14]. Wild garlic (*Allium ursinum*), broadly used as conventional medicinal drug, possesses strong antimicrobial efficacy and alleviates inflammation [15].

The health-promoting capacity of edible leaves is primarily attributed to their numerous bioactive compounds. Their antioxidant properties allow for the neutralization of free radicals, thereby decreasing the likelihood of chronic non-transmissible illnesses like diabetes, carcinoma, and heart disease. The immunomodulatory properties of compounds like oleuropein and L-theanine may also alleviate conditions such as arthritis and improve coronary heart health. Furthermore, many leaf extracts exhibit antimicrobial activities, inhibiting the growth of harmful pathogens and contributing to both food shelf-life extension and immune defense.

The effectiveness of bioactive compounds in functional foods and nutraceuticals largely depends on the extraction process. Numerous techniques are commonly used, including solvent extraction with ethanol or methanol, Soxhlet extraction involving continuous solvent reflux, and ultrasound-assisted extraction (UAE), which uses sound waves to rupture plant cells and release bioactive compounds[16, 17]. Of these techniques, microwave-assisted extraction (MAE) sticks out as a fast, efficient, and eco-friendly method. It is a progressive extraction approach that makes use of microwave power to warm the plant material and solvents. This results in rapid dsirutpion of plant cell wall, enhancing the release and yield of bioactive compounds like polyphenols, flavonoids, alkaloids, terpenoids and saponins etc. into the surrounding solvent. MAE offers several advantages over traditional extraction techniques, such as reduced extraction time, lower solvent intake, minimal thermal degradation of thermolabile compounds, and advanced extraction performance. Successful application of MAE has been reported in various edible leaves like *Murraya koenigii*, moringa leaves (*Moringa oleifera* L), spinach and cabbage, which yielded higher amounts of bioactive compounds and improved antioxidant activity [17–20]. This chapter provides a detailed laboratory procedure for the extraction of bioactive compounds from leaves using the microwave-assisted extraction (MAE) method, offering a practical and reproducible protocol for experimental and research purposes.

2 Materials

- Plant material: Fresh *Moringa oleifera*, *Spinacia oleracea* (spinach), or *Murraya koenigii* (curry leaves) leaves, etc.
- Microwave extraction system—Milestone ETHOS X (Model: ETHOS X, power: 1000 W, frequency: 2450 MHz; Milestone Srl, Italy).
- Solvents: 70% Ethanol, 70% Methanol, or desired solvent (*see* **Note 1**).
- Beakers (250 mL).
- Measuring cylinders.
- Microwave-safe extraction flasks or glass containers.
- Centrifuge (REMI R-8 K).
- MCE Membrane Filter, 0.45 μm Pore Size (HAWP04700. MF-Millipore™).
- Rotary evaporator (Buchi Rotavapor R-200, Vac V-500 pump, Switzerland).
- Heating Bath (*see* **Note 2**).
- Vertical glass condenser.
- Round bottom flask.
- Receiving flask.
- Analytical balance.
- Mortar and pestle or blender.
- 60-mesh sieve.

3 Materials

3.1 Preparation of Plant Material

1. Collect fresh leaves of plant (*Moringa oleifera* or *Spinacia oleracea*).
2. Wash thoroughly with distilled water to remove any dirt, debris, or dust.
3. Air dry the leaves at room temperature (25–30 °C) in a dust-free area until a constant weight is achieved, indicating complete drying [21].
4. Grind the dried leaves into a coarse powder using a clean blender or a mortar and pestle.
5. Pass the powder through 60-mesh sieve to obtain uniform particle size for efficient extraction [22].
6. Store the powdered sample in a clean, airtight container protected from light and moisture until use (*see* **Note 3**).

3.2 Microwave-Assisted Extraction (MAE) [23]

1. Weigh 10 g of the sieved leaf powder (usually depend upon plant and solvent volume).
2. Place it into a microwave-safe borosilicate glass container (preferably 250–500 mL capacity).
3. Choose an appropriate solvent based on the bioactive compounds of interest and solubility characteristics.
4. Add 100 mL of solvent (70% ethanol) to the container, maintaining a 10:1 solvent-to-solid ratio (mL:g) to facilitate sufficient diffusion.
5. Cover the container loosely using microwave-permeable film or a vented lid to minimize solvent loss while allowing pressure release.
6. Place the container in a microwave extraction system.
7. Set the microwave parameters: time, power (watts), and temperature based on the sample and desired extraction efficiency.
   Power can range from 500 to 750 W.
   Extraction times typically vary from 5 to 30 min.
   Temperature can range from 40 to 80 °C.
8. After microwave treatment, allow the mixture to cool to room temperature (25–30 °C).
9. Open the vessel carefully (to avoid pressure release) after cooling.

3.3 Clarification of Extract [24]

1. Centrifuge the extract at 365× *g* for 10 min to sediment coarse plant debris and insoluble materials.
2. Carefully separate the supernatant without disturbing the pellet.
3. Filter the supernatant under vacuum through a 0.45 μm cellulose membrane filter to obtain a clear extract.
4. Collect the filtrate in a clean, dry container.

3.4 Concentration and Drying of Extract Rotary Evaporation [25]

1. Transfer the filtered extract into a clean evaporator round-bottom flask.
2. Attach the flask to a rotary evaporator fitted with.
   - A condenser.
   - A heating bath set to 40–60 °C depending on the solvent.
   - A vacuum pump.
3. Turn ON the vacuum pump and begin rotation at a speed of 100–120 rpm.
4. Slowly lower the flask into the heating bath using the lift.
5. Monitor vacuum pressure (optional: vacuum controller) to prevent solvent bumping.

6. Solvent vapors should condense and collect in the receiving flask.
7. Adjust vacuum and temperature as needed for smooth evaporation.
8. Monitor until most of the solvent (ethanol) is removed.
9. Once evaporation is complete which usually takes 15–30 min, depending on volume and ambient humidity, raise the flask out of the bath.
10. Turn off rotation, heating bath, and vacuum pump.
11. Gradually release vacuum pressure using the vent valve.
12. Carefully remove the flask with concentrated extract.
13. Transfer the thick semi-solid extract from the flask using a spatula or glass rod to a sterile preweighed glass vial.

3.5 Extraction Yield

1. Weigh the dried extract using an analytical balance and record final weight.
2. Calculate the extraction yield using the following formula [26].

Extraction Yield (%)

$$\text{Extraction Yield }(\%) = \frac{\text{Weight of Dried Extract (g)}}{\text{Weight of Initial Plant Material (g)}} \times 100$$

3.6 Storage of Extract

1. Clearly label all extract vials with plant name, extraction method, date, and sample code.
2. Store all samples at 4 °C to preserve bioactive compound stability until further analysis (Fig. 1).

4 Notes

1. All the chemicals and solvents should be of analytical grade which can be procured from certified suppliers.
2. Ensure the function of temperature control before starting the processing of sample.
3. Store the sample in a cool and dark place for further analysis.

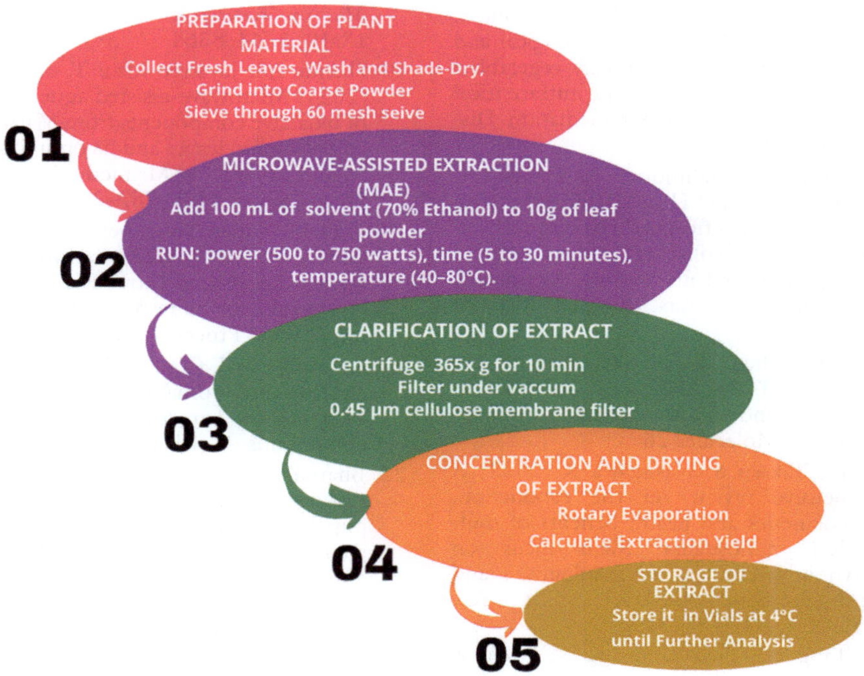

Fig. 1 Schematic representation of the microwave-assisted extraction (MAE) methodology for the extraction of bioactive components from plant leaves. The figure delineates each phase of the process, from the preliminary collection and preparation of fresh leaves to the ultimate storage of the extract and determination of extraction yield. It emphasizes extraction using the microwave-assisted technique, clarity of the extract through centrifugation and then filtration, and concentration utilizing rotary evaporation

References

1. Chen X (2019) A review on coffee leaves: phytochemicals, bioactivities and applications. Crit Rev Food Sci Nutr 59:1008–10025
2. Sarkar T, Salauddin M, Roy S, Chakraborty R, Rebezov M, Shariati MA et al (2023) Underutilized green leafy vegetables: frontier in fortified food development and nutrition. Crit Rev Food Sci Nutr 63:11679–11733
3. Ebabhi A, Adebayo R (2022) Nutritional values of vegetables. In: Vegetable crops – health benefits and cultivation. IntechOpen, London
4. Yang M, Tao L, Kang XR, Wang ZL, Su LY, Li LF et al (2023) *Moringa oleifera* Lam. leaves as new raw food material: a review of its nutritional composition, functional properties, and comprehensive application. Trends Food Sci Technol 138:399–416
5. Adewumi OO, Felix-Minnaar JV, Jideani VA (2022) Functional properties and amino acid profile of Bambara groundnut and *Moringa oleifera* leaf protein complex. Processes 10(2):205
6. Ferreira T, Gomes SM, Santos L (2023) Elevating cereal-based nutrition: *Moringa oleifera* supplemented bread and biscuits. Antioxidants 12(12):2069
7. Agba TD, Yahaya-Akor NO, Kaur A, Ledbetter M, Templeman J, Wilkin JD et al (2024) Flour functionality, nutritional composition, and in vitro protein digestibility of wheat cookies enriched with Decolourised *Moringa oleifera* leaf powder. Foods 13(11):1654
8. Masiala A, Vingadassalon A, Aurore G (2024) Polyphenols in edible plant leaves: an overview of their occurrence and health properties. Food Funct 15(13):6847–6882
9. Huda-Faujan N, Zubairi SI, Baker AAA (2023) Nutritional and bioactive constituents of antioxidant and antimicrobial properties in *Spinacia oleracea*: a review. Sains Malays 52:2571–2585

10. Patra S, Shand H, Manna A, Bose D, Some S, Mondal R (2024) Ethnopharmacological and nutritional insights into wild leafy vegetables: implications for food security, nutraceutical potential, and environmental remediation. Discover Plants 1(1):56
11. Fki I, Sayadi S, Mahmoudi A, Daoued I, Marrekchi R, Ghorbel H (2020) Comparative study on beneficial effects of Hydroxytyrosol- and Oleuropein-rich olive leaf extracts on high-fat diet-induced lipid metabolism disturbance and liver injury in rats. Biomed Res Int 2020(1):1315202
12. Nguyen HC, Chen CC, Lin KH, Chao PY, Lin HH, Huang MY (2021) Bioactive compounds, antioxidants, and health benefits of sweet potato leaves. Molecules 26
13. Parida IS, Takasu S, Nakagawa K (2023) A comprehensive review on the production, pharmacokinetics and health benefits of mulberry leaf iminosugars: Main focus on 1-deoxynojirimycin, d-fagomine, and 2-O-α-d-galactopyranosyl-DNJ. Crit Rev Food Sci Nutr 63:3468–3496
14. Jia R, Tang C, Chen J, Zhang X, Wang Z (2022) Total phenolics and anthocyanins contents and antioxidant activity in four different aerial parts of leafy sweet potato (*Ipomoea batatas* L.). Molecules 27(10):3117
15. Petropoulos SA, Di Gioia F, Polyzos N, Tzortzakis N (2020) Natural antioxidants, health effects and bioactive properties of wild allium species. Curr Pharm Des 26(16):1816–1837
16. Rodríguez García SL, Raghavan V (2022) Green extraction techniques from fruit and vegetable waste to obtain bioactive compounds—a review. Crit Rev Food Sci Nutr 62:6446–6466
17. Sandeep G, Arumugam T, Janavi GJ, Anitha T, Senthil K, Lakshmanan A (2023) A comparative study on conventional and non-conventional extraction methodologies for extraction yield, quality and antibacterial properties of moringa (*Moringa oleifera* Lam.). J Appl Hortic 25(1)
18. Mondal P, Meeran SM (2024) Effect of extraction methods on bioactives profile, antioxidant potential and anticancerous properties of *Murraya koenigii* leaves. J Food Meas Charact 18(10):8344–8364
19. Shende AS, Jayasree Joshi T, Srinivasa Rao P (2024) Microwave-assisted aqueous extraction of bioactive components from Malabar spinach (*Basella alba*) leaves and its optimization using ANN-GA and RSM methodology. J Food Meas Charact 18(1):287–298
20. Yigit U, Turabi Yolaçaner E, Hamzalıoglu A, Gökmen V (2022) Optimization of microwave-assisted extraction of anthocyanins in red cabbage by response surface methodology. J Food Process Preserv 46(1):e16120
21. Segwatibe MK, Cosa S, Bassey K (2023) Antioxidant and antimicrobial evaluations of *Moringa oleifera* lam leaves extract and isolated compounds. Molecules 28(2):899
22. Bhattarai S, Katekhong W, Klinkesorn U, Peanparkdee M (2024) Valorization of *Bombyx mori* pupae and *Moringa oleifera* leaf as sources of alternative protein: structural, functional and antioxidant properties. Future Foods 10: 100408
23. Alara OR, Abdurahman NH, Ukaegbu CI, Kabbashi NA (2019) Extraction and characterization of bioactive compounds in *Vernonia amygdalina* leaf ethanolic extract comparing Soxhlet and microwave-assisted extraction techniques. J Taibah Univ Sci 13(1):414–422
24. Boli E, Prinos N, Louli V, Pappa G, Stamatis H, Magoulas K et al (2022) Recovery of bioactive extracts from olive leaves using conventional and microwave-assisted extraction with classical and deep eutectic solvents. Separations 9(9):255
25. Alara OR, Abdurahman NH, Olalere OA (2020) Ethanolic extraction of flavonoids, phenolics and antioxidants from *Vernonia amygdalina* leaf using two-level factorial design. J King Saud Univ Sci 32(1):7–16
26. Royani A, Hanafi M, Lotulung PDN, Julistiono H, Dinoto A, Manaf A (2023) Analysis of the antibacterial activity and the total phenolic and flavonoid contents of the *Moringa oleifera* leaf extract as an antimicrobial agent against *Pseudomonas aeruginosa*. Scientifica (Cairo) 2023(1):5782063

Chapter 7

Extraction of Bioactive Components from Leaf by Pulse Electric Field Extraction

Simran Rani, Priyanka Dahiya, Chetna Rathi, K. Ananthi, Amita Suneja Dang, and Pooja Suneja

Abstract

Plant contains numerous types of bioactive compounds also referred to as "phytoconstituents" including alkaloids, terpenes, glycosides, saponins, polysaccharides, and phenolics. Leaves tend to have high concentration of these compounds which find their applications in pharmaceuticals (drug development and delivery), food (preservatives and additives), and agriculture industry (pesticides and biofertilizers). Limitations like lower yield, usage of more solvent, and longer duration along with thermal degradation of the products are associated with the conventional modes of extracting these biologically active substances. On the contrary, pulse electric field (PEF) being a non-thermal technology facilitates more convenient and efficient extraction of bioactive compounds besides other advantages like economic viability and yield improvement. This chapter provides a comprehensive description of the use of PEF technique in order to obtain economically important bioactive components from plant leaves. The precautions to be taken for the prevention of degradation of bioactive substances have also been discussed at the end to minimize the loss during the process.

Key words Bioactive compounds, Phytoconstituents, Extraction, PEF, Phenolic compounds

1 Introduction

A wide array of bioactive compounds are formed in plants primarily as a defense mechanism against pathogens and herbivores. Leaves are an important source of bioactive compounds, such as phenolic compounds, phytoestrogens, polyunsaturated fatty acids, sulfur compounds, terpenes, and vitamins, well known for their antioxidant, anticancer, antidiabetic, and antimicrobial activities [1]. The continous developement has been observed in the field of techniques used for extraction of bioactive compounds from plants. While many researchers still use some traditional techniques such as maceration, hydrodistillation, and Soxhlet extraction for the extraction of bioactive compounds, these methods have certain drawbacks,

Mitesh Kumar Dwivedi and A. Sankara Narayanan (eds.), *Leaves as Functional Foods*, Methods and Protocols in Food Science,
https://doi.org/10.1007/978-1-0716-5226-8_7,

viz., longer extraction time, nutrient loss, reduced yield, and higher energy consumption. In recent times, several ecofriendly techniques characterized by higher yield and efficacy, ultrasound-assisted extraction, supercritical fluid extraction, instant controlled pressure drop extraction, enzymatic extraction, etc., are being used to extract bioactive compounds from a variety of leaves [2, 3].

Pulse electric field (PEF) is a non-thermal, environment friendly, rapid, and highly effective technique that involves the application of short bursts of electricity to inactivate the microbes and is commonly used for food preservation. Mostly, pulses ranging between 100 ns and 1 ms with a voltage between 1 and 3 kV are used for short duration [4]. PEF also finds its application in food processing (texture modification, drying, dehydration, starch modification), water treatment, and wound disinfection [5–7]. In addition, this technique has also been used to extract biologically active compounds, polyphenols, carotenoids, and proteins from the leaves of a variety of plants viz., *Moringa oleifera* (drumstick tree) *Annona squamosa* (custard apple), *Cannabis sativa* (Indian hemp), and *Crataegus monogyna* (hawthorn) [8–11]. To isolate the bioactive components from plant samples, the cells are exposed to external electrical field for a brief period. Cell membranes undergo disruption as a result of electroporation by the high voltage short pulses, and the intracellular components including desired bioactive compounds are released [9].

2 Materials

1. Plant specimens.
2. PEF system.
3. Double distilled water.
4. High voltage power generator.

3 Methods

3.1 Sample Collection and Processing [11]

- Collect the leaves from mature plants and bring them to the laboratory.
- Wash thoroughly under water and freeze-dry for 12 h.
- Mill the leaves to particle sizes of about 0.05 cm using a blender and store in the vacuum-sealed polyethylene bags at 25 °C in dark until used (maximum 3 days).
- Mix 2.5 g of freeze-dried leaves with 50 mL of double distilled water.

- Filter this mixture through filter paper after each extraction treatment.
- Rinse the filter with double distilled water to a final volume of 50 mL.

3.2 PEF Apparatus [11, 12]

The equipment layout consists of:

- High voltage power generator (to provide high voltage for creating strong electric field).
- Digital oscilloscope (to monitor voltage, current, frequency, and pulse waveform).
- Pulse generator (to shape high-voltage output into specific pulses and also control their width, frequency, and intensity).
- Treatment chamber (to hold the product/leaf and expose it to the high-voltage electrical pulses generated by the PEF system).

3.3 Extraction Procedure [11]

- Attach positive and negative electrodes on stainless steel cylinder.
- Pour the sample containing freeze-dried leaves and double distilled water in the treatment chamber and screw in the negative electrode.
- Calculate electric field strength (E) as the ratio of applied voltage (U) and distance between electrode and cylinder (d) and set it between 0.1 and 10 kV/cm.
- Optimize pulse duration (period of application of electric field) and pulse interval (period between two pulse applications) depending upon the application/type of sample (*see* **Note 1**).
- Check the temperature of the contents placed in treatment chamber post-treatment with an infrared thermometer (*see* **Note 2**).
- Filter the mixture through filter paper and rinse the residue with double distilled water to replenish the water absorbed by the plant material during the procedure.

4 Notes

1. Several parameters such as electric field strength, duration of treatment, frequency and width of pulse, and temperature influence the efficacy of PEF. Optimization of these parameters will improve the results significantly while dealing with different plant samples [13].
2. Though PEF is a non-thermal process, heat generated by the electric current might raise the sample temperature. Therefore, it is very important to control the temperature effectively [14].

Acknowledgments

The authors acknowledge the infrastructural support from the Department of Microbiology and Centre for Medical Biotechnology (CMBT), Maharshi Dayanand University (MDU), Rohtak, India. SR and CR acknowledge the Council of Scientific & Industrial Research (CSIR), New Delhi, India, for the CSIR fellowship. PD is thankful to the University Grants Commission (UGC), New Delhi, India, for fellowship support. The authors are thankful to the Department of Science and Technology, Govt. of Haryana, India (Grant No. HSCSIT/R&D/2022/2968) for providing infrastructural and financial support to undertake this research work.

References

1. Samtiya M, Aluko RE, Dhewa T, Moreno-Rojas JM (2021) Potential health benefits of plant food-derived bioactive components: an overview. Foods 10(4):839
2. Usman I, Hussain M, Imran A, Afzaal M, Saeed F, Javed M et al (2022) Traditional and innovative approaches for the extraction of bioactive compounds. Int J Food Prop 25(1): 1215–1233
3. Azmir J, Zaidul ISM, Rahman MM, Sharif KM, Mohamed A, Sahena F et al (2013) Techniques for extraction of bioactive compounds from plant materials: a review. J Food Eng 117(4): 426–436
4. Gao X, Wang Z, Sun G, Zhao Y, Tang S, Zhu H, Li Z (2025) Pulsed electric field (PEF) technology for preserving fruits and vegetables: applications, benefits, and comparisons. Food Rev Int:1–26
5. Capodaglio AG (2021) Pulse electric field technology for wastewater and biomass residues' improved valorization. Processes 9(5): 736
6. Rubin AE, Usta OB, Schloss R, Yarmush M, Golberg A (2019) Selective inactivation of Pseudomonas aeruginosa and Staphylococcus epidermidis with pulsed electric fields and antibiotics. Adv Wound Care 8(4):136–148
7. Saulis G (2010) Electroporation of cell membranes: the fundamental effects of pulsed electric fields in food processing. Food Eng Rev 2(2):52–73
8. Bozinou E, Karageorgou I, Batra G, Dourtoglou VG, Lalas SI (2019) Pulsed electric field extraction and antioxidant activity determination of Moringa oleifera dry leaves: a comparative study with other extraction techniques. Beverages 5(1):8
9. Shiekh KA, Olatunde OO, Zhang B, Huda N, Benjakul S (2021) Pulsed electric field assisted process for extraction of bioactive compounds from custard apple (Annona squamosa) leaves. Food Chem 359:129976
10. Mpakos D, Chatzimitakos T, Athanasiadis V, Mantiniotou M, Bozinou E, Lalas SI (2024) Optimization of pulsed electric field-based extraction of bioactive compounds from cannabis sativa leaves. Analytica 5(1):90–106
11. Papazidou V, Makrygiannis I, Mantiniotou M, Athanasiadis V, Bozinou E, Lalas SI (2025) Isolation of bioactive compounds and antioxidant activity evaluation of Crataegus monogyna leaves via pulsed electric field-assisted extraction. Plants 14(15):2262
12. Kasri NF, Piah MAM, Adzis Z (2020) Compact high-voltage pulse generator for pulsed electric field applications: lab-scale development. J Electr Comput Eng 2020(1):6525483
13. Su W, Wang Q, Li J, Qiu Z, Qiu Y (2024) Effects of pulsed electric field technology on the nutritional value and biological function of plant food. Front Sustain Food Syst 8: 1385533
14. Maniglia BC, Castanha N, Rojas ML, Augusto PE (2021) Emerging technologies to enhance starch performance. Curr Opin Food Sci 37: 26–36

Chapter 8

Extraction of Bioactive Components from Leaf by Ultrasound-Assisted Extraction

Roshani Patel, Abhishek Sharma, Ruchika Upadhyay, Deepa Nautiyal, and Abhishek Sharma

Abstract

Plant-based materials serve as important sources of various bioactive compounds (BCs) with industrial applications. To maximize their yields and recover them, various experimental strategies are being deployed. Ultrasound-assisted extraction (UAE), for example, is a sophisticated, economical, environmentally friendly, and sustainable alternative for recovering BCs (polyphenols, flavonoids, anthocyanins, and carotenoids) from plant sources with high yields. Other techniques for extracting BCs include maceration, Soxhlet extraction, hydrodistillation, pulsed electric fields, enzymes, microwaves, high hydrostatic pressure, and supercritical fluids. Ultrasound-assisted extraction (UAE) is a green and efficient technique for isolating bioactive compounds from plant materials. This method has broad applications using a wide variety of plants leaves, is suitable for thermolabile substances and low concentrations, and is useful in research and industry.

Key words Bioactive compounds (BCs), Ultrasound-assisted extraction (UAE), Green extraction, Polyphenolic compounds, Thermolabile, Myrtle alcoholic extracts

1 Introduction

The food applications of plant-derived bioactive components have increased in recent years [1, 2]. Their natural components have inspired new strategies, requiring the extraction of these bioactive from natural sources [3–6]. Human health is also benefited by the bioactive compounds found in products containing antimicrobial or antioxidant properties.

Bioactive compounds (BCs) are the natural compounds that are found primarily in plants. BCs also have beneficial impacts on human health when they are consumed. BCs such as phytochemicals and secondary metabolites are present in foods such as plants, fruits, vegetables, nuts, and whole grains. The extraction of BCs occurs through methods such as maceration, pulsed electric fields,

Mitesh Kumar Dwivedi and A. Sankara Narayanan (eds.), *Leaves as Functional Foods*, Methods and Protocols in Food Science, https://doi.org/10.1007/978-1-0716-5226-8_8,

Soxhlet extraction, and hydrodistillation, but these methods are more expensive, less effective, and more time-consuming than ultrasound-assisted extraction (UAE), which can also heat thermolabile compounds. UAE is a process that uses sound waves to improve the extraction of BCs from plant materials. It facilitates the degradation of cell walls, which releases intracellular contents.

UAE involves the interaction of ultrasonic power and solvents for the extraction of bioactive components from plants. It has improved the extraction yield of bioactive compounds when compared to conventional methods [7]. UAE treatment has gained consideration as an effective food extraction method. Its high sound-wave frequency can break the low-energy chemical bonds of plant tissue through cavitation, allowing the extraction of bioactive compounds [8]. This equipment promotes cavitation where bubble collapse occurs at the interface of immiscible liquids [9]. These sound waves use frequencies audible to human ear, greater than 20 MHz, but in a clinical trial, lower sound frequencies are being used. Several ultrasound waves are used depending on the frequency. Sound waves with frequencies of 20–400 kHz are categorized as the frequencies used for the extraction process applied to food. In addition to the ultrasound frequency, the ultrasound amplitude also affects extraction efficiency: A greater amplitude causes chemical bonds or cell walls to break [10]. This method is used to reduce processing time and solvent consumption while increasing extraction yield. Ultrasounds also have positive effects and negative effects on targeting bioactive compounds depending on their intrinsic and extrinsic parameters. Using different chemicals and plants, UAE is applied to various natural products.

2 Materials

1. Quercetin, dihydrate, Folin–Ciocalteu reagent, gallic acid, 2,2-diphenyl-1-picrylhydrazyl (DPPH), and other chemicals obtained from Hi-Media Mumbai.
2. Glassware from Borosil Glass Works Limited, Mumbai, India.
3. *Allium sativum* plant collected from Haryana garlic (HG-17).
4. PID-controlled cabinet drier.
5. 20 kHz probe ultrasonicator (for UAE).
6. Ice bath.
7. Ethanol–water mixture.
8. Whatman filter paper 42.
9. One factor at a time (OFAT).
10. Sodium carbonate.
11. Spectrophotometer.

12. Aluminum chloride.
13. Sodium nitrite (5%).
14. Aluminum chloride (10%).
15. Sodium hydroxide (1 M).
16. DPPH solution (1 mM).
17. Water's HPLC 2489 equipped with Water's Foods 2023.
18. For the estimation of organosulfur and phenolic compounds: the Luna LC column.
19. Acetonitrile 50% used for mobile phase A and water 50% used for mobile phase B.
20. 1.0 mL/min constant flow rate provided at a detection wavelength of 254 nm.
21. At a flow rate of 1.0 mL/min, the determination of the phenolic compounds at different concentrations: 280 nm for solvent A (water/acetonitrile/formic acid = 96/2.9/1; v/v) and for solvent B (acetonitrile/formic acid = 99.9/0.1; v/v).
22. 0.45 nylon filter.
23. For obtaining the different compositions of input variables and analyzing the rotatable central composite design (RCCD): Expert Software.
24. For Fourier transform infrared (FTIR) analysis: a Bruker Alpha FTIR system (Germany).
25. Field emission scanning electron microscope (FEI Novanano SEM 450, Nebraska Center for Materials and Nanoscience, Lincoln, NE, USA).

3 Methods

3.1 Reagents

All the required samples and chemicals were obtained from Hi-Media Mumbai. The glassware used was from Borosil Glass Works Limited, Mumbai, India.

3.2 Raw Materials

1. Two matured green leaves from the *A. sativum* plant, a variety of Haryana garlic (HG-17), were harvested from Punjab.
2. Before use, the distal ends of the harvested green leaves, which were dried until hay-like, were removed, and then washing was carried out under running tap water twice.
3. The leaves were cut lengthwise into approximately 1.5 cm strips after water washing and were dried by using a PID-controlled cabinet drier maintained at 50 ± 1 °C.
4. To obtain garlic leaf powder, the dried garlic leaves were ground and sieved using a number 100 mesh sieve.

3.3 Ultrasound-Assisted Extraction (UAE)

1. UAE was carried out by using a 20 kHz probe ultrasonicater with a 16 mm probe.
2. A frequency of 20 kHz with a constant 700 W is required for the extraction of the target compounds.
3. The samples were kept in an ice bath to ensure that the sample temperature did not exceed 40 °C.
4. The garlic leaf powder was mixed with an ethanol–water mixture by maintaining the same distance from the probe for each run.
5. The applied energy treatment was recorded with an instrument's energy reading (J).
6. Calorimetric energy was calculated as Kikuchi and Uchida for each experiment.
7. The control of ultrasound amplitude and treatment was carried out on a digital panel on the instrument.
8. The extract was immediately filtered by using Whatman filter paper 42.
9. The extract was then kept at 4 °C for further studies.

3.4 One Factor at a Time (OFAT)

1. OFAT is used to analyze and select critical factors affecting the extraction yield.
2. Different parameters of OFAT are used: ultrasonic amplitude (20%, 30%, 40%, 50%, and 60%), ethanol concentration (20%, 40%, 50%, 60%, 80%, and 100%), and treatment time (5, 10, 15, and 20 min).
3. Solvent-to-solid ratios: 10:1, 20:1, 30:1, 40:1, and 50:1.

3.5 Analysis of Composition

1. The moisture, protein, fat, and ash contents and the pH of garlic leaves were estimated by using the standard method, and the carbohydrate content was estimated by using a different method.

3.6 Analysis of the Total Phenol Content (TPC)

1. The Folin–Ciocalteu method was followed for the determination of the total phenol content.
2. In 0.1 mL of aliquot extract, 2.5 mL of Folin–Ciocalteu reagent was added.
3. Sodium carbonate (10% w/v) was added to the mixture, which was incubated for 30 min.
4. Absorbance was analyzed by using a spectrophotometer (*see* **Note 1**).

3.7 Determining the Total Flavonoid Content (TFC)

1. An aluminum chloride colorimetry assay was performed to determine the total flavonoid content.

2. Quercetin was utilized as the standard procedure: 0.5 mL of supernatant and 4 mL of distilled water.
3. Next, 0.3 mL of sodium nitrite (5%) was added to the mixture.
4. After incubation for 5 min, 0.3 mL of aluminum chloride (10%) was added.
5. Afterward, 1 M of NaOH (2 mL) was added and mixed with 2.4 mL d/w.
6. The mixture was incubated at room temperature for 5 min, and the absorbance was measured at 510 nm by using a spectrophotometer.
7. The calibration curve of quercetin was drawn from 0–1 mg/mL and expressed as quercetin equivalent.

3.8 DPPH Assay

1. DPPH scavenging assay was proposed to determine the antioxidant activity.
2. 0.1 mL of garlic leaf extract was mixed with 3.9 mL of DPPH solution at a concentration of 1 mM.
3. Absorbance was evaluated with a spectrophotometer at 515 nm.
4. For the control, 1 mL of distilled water was mixed with a DPPH solution of 3.9 mL.
5. Absorbance was determined at the same wavelength.
6. An equation was used to calculate DPPH activity (*see* **Note 2**).

3.9 Estimation of Organosulfur Compounds (OSCs) and Phenolic Compounds via HPLC 2489

1. To estimate organosulfur and phenolic compounds, the chromatographic technique was used.
2. The experiment was performed using HPLC 2489 equipped with Water's ultraviolet (UV) detector (*see* **Note 3**).
3. OSCs were detected by using a gradient mobile phase A of 50% (acetonitrile) and a mobile phase B of 50% (water).
4. A constant flow rate of 1.0 mL/min at a detection wavelength of 254 nm was used.
5. Phenolic compounds were detected at the 280 nm wavelength with varying concentrations of solvent A and solvent B.
6. The gradient program was performed with varying percentage proportions of solvent A and solvent B (*see* **Note 4**).
7. Ultrasonication was performed for 7–10 min to remove the air.
8. Filter with a 0.45 μ nylon filter to eliminate the contaminants from sample extracts.
9. Compounds were identified with standard compound retention times.
10. OSCs and phenolic compounds were calculated by using an equation (*see* **Notes 5** and **6**).

3.10 Fourier Transform Infrared (FTIR) Spectroscopy Analysis

1. FTIR spectroscopy analysis is an analytical technique used to determine the functional groups in the extract, which helps in identifying the components in the sample.
2. The leaf extract obtained via UAE was analyzed by using Bruker Alpha FTIR system (*see* **Note 7**).

3.11 Field Emission Scanning Electron Microscopy (FE-SEM)

1. FE-SEM is a powerful imaging technique that visualizes pores or micro-cracks on the surface structure of samples at a high resolution.
2. The garlic leaf powder's structure was observed with and without ultrasound treatment (FE-SEM).
3. A magnification of 2500× was used to analyze the garlic leaf powder.

4 Notes

1. A UV-1800 Shimadzu spectrophotometer (Shimadzu, Kyoto, Japan) at 725 nm was used.
2. DPPH scavenging (%) $= \frac{(As-Ac)}{Ac} \times 100$.
3. For the estimation of organosulfur compounds (OSCs) and phenolic compounds, the Luna LC column (150 4.6 mm, 5 particle size) and C18 column (Phenomix 100 4.6 mm, 5 particle size) were used.
4. At the initial stage, the concentration of solvent A:solvent B was 100:0, at 5 min 95:5, at 25 min 85:15, at 30 min 80:20, at 39 min 75:25, at 43 min 55:45, at 48 min 5:95, at 50 min 5:95, at 55 min 80:20, and at 60 min 100:0.
5. The concentration of individual OSCs (ppm) $= \frac{Area_{sample}}{Area_{std}} \times Conc_{std}$.
6. The concentration of individual phenolic compound (ppm) $= \frac{Area_{sample}}{Area_{std}} \times Conc_{std}$.
7. A range of 4000–5000 cm^{-1} at a resolution of 4 $cm-^{1}$ was used for 64 scans per sample.

References

1. Awad AM, Kumar P, Ismail-Fitry MR, Jusoh S, Ab Aziz MF, Sazili AQ (2022) Overview of plant extracts as natural preservatives in meat. J Food Process Preserv 46:e16796. [Google Scholar] [CrossRef]
2. Awad AM, Kumar P, Ismail-Fitry MR, Jusoh S, Aziz MFA, Sazili AQ (2021) Green extraction of bioactive compounds from plant biomass and their application in meat as natural antioxidant. Antioxidants 10:1465. [Google Scholar] [CrossRef] [PubMed]
3. Freitas de Oliveira C, Giordani D, Lutckemier R, Gurak PD, Cladera-Olivera F, Ferreira Marczak LD (2016) Extraction of pectin from passion fruit peel assisted by ultrasound. LWT—Food Sci Technol 71:110–115
4. Keven Silva E, Zabot GL, Toledo Hijo AAC, Meireles MAA (2017) Encapsulation of

bioactive compounds using ultrasonic technology. In: Ultrasound: advances for food processing and preservation. Academic Press, Cambridge, MA, pp 323–350

5. Kumar P, Abubakar AA, Verma AK, Umaraw P, Nizam MH, Mehta N, Ahmed MA, Kaka U, Sazili AQ (2022) New insights in improving sustainability in meat production: opportunities and challenges. Crit Rev Food Sci Nutr. [Google Scholar] [CrossRef] [PubMed]
6. Kumar P, Verma AK, Umaraw P, Mehta N, Malav OP (2020) Plant phenolics as natural preservatives in food system. In: Plant phenolics in sustainable agriculture. Springer, Singapore, pp 367–406. [Google Scholar]
7. Mehta N, Ahlawat SS, Sharma DP, Dabur RS (2015) Novel trends in development of dietary fiber rich meat products—a critical review. J Food Sci Technol 52:633–647. [Google Scholar] [CrossRef] [PubMed]
8. Mehta N, Chatli MK, Kumar P, Malav OP, Verma AK, Kumar Y, Kumar D (2019) Development of dietary fiber-rich meat products: technological advancements and functional significance. In: Bioactive molecules in food. Springer, Cham, pp 763–795. [Google Scholar]
9. Senrayan J, Venkatachalam S (2018) A short extraction time of vegetable oil from Carica papaya L. seeds using continuous ultrasound acoustic cavitation: analysis of fatty acid profile and thermal behavior. J Food Process Eng 42(1):1–9
10. Shekhar S, Prakash P, Singha P, Prasad K, Singh SK (2023) Modeling and optimization of ultrasound-assisted extraction of bioactive compounds from *Allium sativum* leaves using response surface methodology and artificial neural network coupled with genetic algorithm. Foods 12(9):1925

Chapter 9

Extraction of Bioactive Components from Leaf by Super Critical Fluid Extraction

Priyanka Dahiya, Nitin Chawaria, Simran Rani, Chetna Rathi, Bhupender Yadav, Pooja Suneja, and Amita Suneja Dang

Abstract

The search and extraction of bioactive compounds from natural sources is largely motivated by their significant applications across various commercial sectors, including pharmaceuticals, food, and chemicals. This highlights the necessity for developing suitable and standardized method to extract the maximum amount of these active components from plant materials. An ideal extraction method should be sustainable, nontoxic, and less time consuming and should ensure quantitative as well as qualitative recovery and easy separation of extracts from the solvent. Therefore, advancement and application of alternative green technologies substituting traditional methods are crucial. Supercritical fluid technology for bioactive compound extraction is one such sustainable method addressing the limitations associated with conventional extraction methods. The advantages it offers include reduced solvent consumption and substitution of lethal chemicals with safer options while improving the yield and extract quality.

Key words Bioactive compounds, Non-conventional extraction, Green method, Supercritical fluid extraction

1 Introduction

Humans have relied on plants since the beginning of civilization. Initially, they were used for their nutritional benefits. Later, the discovery of their medicinal properties made them vital for treating sicknesses and improvement of human health after. Plants produce an array of secondary metabolites for their defence, and many of them are bioactive in nature influencing biological systems [1]. These bioactive compounds have been documented for possessing numerous health advantages for both humans and animals, including antibacterial, antimicrobial, anti-inflammatory, anti-aging, and anti-cancer effects [2]. These compounds are in demand as they are natural and therefore are safe for use in different sectors, including cosmetics, food, feed, agriculture, and pharmaceuticals.

Mitesh Kumar Dwivedi and A. Sankara Narayanan (eds.), *Leaves as Functional Foods*, Methods and Protocols in Food Science, https://doi.org/10.1007/978-1-0716-5226-8_9,

Bioactive compounds such as essential oils, carotenoids, phenolic acids, and flavonoids have been extracted from different plant parts such as leaves, stems, roots, and flowers using various extraction methods [3]. Extraction is a multi-step process. The quantity and quality of the extract depend on many factors including the type of extraction method, time and temperature used, the nature of the solvent, and the solvent-to-sample ratio [1].

Traditional techniques like Soxhlet extraction, maceration, and hydro distillation have been commonly used to extract bioactive compounds. However, these methods have several pitfalls, such as long processing times taking hours or even days and the use of large quantities of costly and harmful solvents. Compounds that are sensitive to heat or volatile are at risk of damage as prolonged exposure to high temperatures can lead to their degradation or activity loss. The extract obtained is usually very dilute and need further concentration steps and may have residual toxic solvents [4, 5].

In recent times, transition has occurred towards more environmentally friendly and efficient extraction methods. These modern techniques intent is to lower energy consumption, reduce solvent usage, and replace toxic reagents with safer options, all while enhancing yield and extract quality. Examples include ultrasound-assisted, microwave-assisted, pulsed electric field-assisted, and supercritical fluid extraction. These methods not only offer efficiency but also comes under the broader principles of green chemistry: safer chemicals, renewable resources, and processes that minimize waste and hazards [6]. Among the modern extraction techniques, supercritical fluid extraction (SFE) has attracted significant interest as a sustainable alternative to conventional methods. In this, the fluid becomes supercritical when it is maintained above its critical temperature and pressure, where it exhibits properties of both liquids and gases. In this state, it diffuses easily through solid materials like a gas while dissolving compounds effectively like a liquid. Supercritical fluids penetrate the sample matrix more effectively, significantly reducing extraction time. Other advantage is the complete and simple solute recovery by adjusting temperature and/or pressure and depressurization respectively [7]. The process operates at relatively low temperatures, preventing degradation of heat-sensitive molecules. The samples and solvent are required in less amount compared to conventional methods and solvent can be reused therefore additionally reduce environmental damage [8]. Carbon dioxide (CO_2) is the most widely used supercritical fluid due to its easy availability, low cost, chemical inertness, and food-grade safety and can be easily removed from the final extract. Carbon dioxide (CO_2) is particularly suitable as it has relatively low critical temperature (31.2 °C) protecting heat labile bioactive compounds and the extractions is done under vacuum so reduces the risk of oxidative degradation [9]. This method is particularly effective for non-polar compounds, and addition of co-solvents like ethanol, methanol, or water often improves the yield of polar and larger molecules [3]. This chapter provides a brief overview of protocol for extracting bioactive components from leaf using super critical fluid extraction.

2 Materials

Plant leaves, drying chamber, super critical fluid extraction apparatus, separator, conical flask, beaker, double distilled water, liquid nitrogen.

3 Methods

3.1 Plant Material and Sample Preparation [10–15]

1. Collect fresh leaves and wash thoroughly under running water. Blot excess water using tissue paper.
2. The leaves can be dried using different methods like sun drying, drying chamber (55 °C) and a dry air flow, freeze-drying or liquid nitrogen freezing followed by lyophilization. Determine moisture content using a moisture analyzer (*see* **Notes 1** and **2**).
3. Grind dried leaves using a mortar or cutting mill and sieve to obtain particle size 0.2–0.75 mm (*see* **Note 3**).
4. Store the powdered material in vacuum-laminated polyethylene bags at −20 °C until further use.

3.2 Supercritical Fluid Extraction Apparatus

Liquid CO_2 is cooled in a water bath and compressed to the desired pressure. The compressed CO_2 is heated in an extraction vessel (stainless steel tubular cell) placed inside temperature-controlled oven. Supercritical CO_2 flows through the packed vessel where separation takes place.

3.3 Extraction Procedure

1. Weigh the stored material (generally 5–10 g depending on vessel size) and place into the extraction vessel. Add co-solvent (e.g., ethanol) to enhance the extraction of target compounds such as phenols (*see* **Note 4**).
2. Fill remaining vessel space with inert glass spheres to improve mass transfer and ensure uniform solvent distribution.
3. Seal the extraction vessel and run with intermediate conditions: 150 bar and 65 °C or under the required conditions as per compound. Set temperature between 35 and 50 °C to maintain supercritical conditions and avoid degradation of thermolabile compounds. For flavonoids, a wider range of 35–80 °C can be used as they remain stable up to 80 °C.

4. Adjust pressure within 85–250 bar, with 100–200 bar generally suitable for phenolic compounds. Select a processing time between 30 and 180 min, depending on the extraction requirements (*see* **Note 5**).
5. Finally, collect the extracts in amber-colored flasks using a cyclone separator.
6. Store the flasks at −20 °C in the dark to prevent degradation of bioactive compounds until further analysis by high-performance liquid chromatography (HPLC), ultra-high-performance liquid chromatography (UHPLC), and spectrophotometer.

4 Notes

1. The water in the sample can compete with the solute and lower extraction yield and therefore the moisture content should be controlled between 4% and 14% [3].
2. Freeze drying causes lesser damage as compared to high oven tempertaures that reduce the yield of bioactive compounds as some are heat labile. Air flow drying improves the impact of solvent on the surface structure, so the selection should be based on the property of bioactive compound [5].
3. The recommended mean particle size for SFE of natural products ranges between 0.25 and 2.0 mm. The smaller particle size enhances extraction by increasing surface area and reducing diffusion resistance, but excessively fine particles may agglomerate, limiting CO_2 flow and reducing yield [3].
4. Among co-solvents, methanol and ethanol are frequently used at concentrations around 15% of the quantity of CO_2 employed for the extraction [5].
5. Extraction time must be optimized with solvent flow rate. The short extraction time may cause incomplete recovery, while extended time can waste solvent and degrade compound of interest. A preliminary test is recommended to determine ideal conditions [16].

Acknowledgments

The authors acknowledge the infrastructural support from the Department of Microbiology and Centre for Medical Biotechnology (CMBT), Maharshi Dayanand University (MDU), Rohtak, India. PD is grateful to the University Grants Commission (UGC), New Delhi, India, for fellowship support. NC, SR, CR, and BY acknowledge the Council of Scientific and Industrial Research (CSIR), New Delhi, India, for the fellowship.

References

1. Azmir J, Zaidul ISM, Rahman MM, Sharif KM, Mohamed A, Sahena F et al (2013) Techniques for extraction of bioactive compounds from plant materials: a review. J Food Eng 117(4): 426–436. https://doi.org/10.1016/j.jfoodeng.2013.01.014
2. Da Silva RP, Rocha-Santos TA, Duarte AC (2016) Supercritical fluid extraction of bioactive compounds. TrAC Trends Anal Chem 76: 40–51. https://doi.org/10.1016/j.trac.2015.11.013
3. Uwineza PA, Waśkiewicz A (2020) Recent advances in supercritical fluid extraction of natural bioactive compounds from natural plant materials. Molecules 25(17):3847. https://doi.org/10.3390/molecules25173847
4. De Castro ML, Garcıa-Ayuso LE (1998) Soxhlet extraction of solid materials: an outdated technique with a promising innovative future. Anal Chim Acta 369(1–2):1–10. https://doi.org/10.1016/S0003-2670(98)00233-5
5. Khaw KY, Parat MO, Shaw PN, Falconer JR (2017) Solvent supercritical fluid technologies to extract bioactive compounds from natural sources: a review. Molecules 22(7):1186. https://doi.org/10.3390/molecules22071186
6. Bitwell C, Indra SS, Luke C, Kakoma MK (2023) A review of modern and conventional extraction techniques and their applications for extracting phytochemicals from plants. Sci Afr 19:e01585. https://doi.org/10.1016/j.sciaf.2023.e01585
7. Camel V (2001) Recent extraction techniques for solid matrices—supercritical fluid extraction, pressurized fluid extraction and microwave-assisted extraction: their potential and pitfalls. Analyst 126(7):1182–1193
8. Lang Q, Wai CM (2001) Supercritical fluid extraction in herbal and natural product studies—a practical review. Talanta 53(4): 771–782. https://doi.org/10.1016/S0039-9140(00)00557-9
9. Lopez-Hortas L, Rodriguez P, Diaz-Reinoso B, Gaspar MC, de Sousa HC, Braga ME, Dominguez H (2022) Supercritical fluid extraction as a suitable technology to recover bioactive compounds from flowers. J Supercrit Fluids 188:105652. https://doi.org/10.1016/j.supflu.2022.105652
10. Pereira P, Cebola MJ, Oliveira MC, Bernardo-Gil MG (2016) Supercritical fluid extraction vs conventional extraction of myrtle leaves and berries: comparison of antioxidant activity and identification of bioactive compounds. J Supercrit Fluids 113:1–9. https://doi.org/10.1016/j.supflu.2015.09.006
11. Goyeneche R, Fanovich A, Rodrigues CR, Nicolao MC, Di Scala K (2018) Supercritical CO_2 extraction of bioactive compounds from radish leaves: yield, antioxidant capacity and cytotoxicity. J Supercrit Fluids 135:78–83. https://doi.org/10.1016/j.supflu.2018.01.004
12. Khor BK, Chear NJY, Azizi J, Khaw KY (2021) Chemical composition, antioxidant and cytoprotective potentials of Carica papaya leaf extracts: a comparison of supercritical fluid and conventional extraction methods. Molecules 26(5):1489. https://doi.org/10.3390/molecules26051489
13. Park HS, Choi HK, Lee SJ, Park KW, Choi SG, Kim KH (2007) Effect of mass transfer on the removal of caffeine from green tea by supercritical carbon dioxide. J Supercrit Fluids 42(2): 205–211. https://doi.org/10.1016/j.supflu.2007.03.002
14. Yerena-Prieto BJ, Gonzalez-Gonzalez M, García-Alvarado MÁ, Casas L, Palma M, Rodríguez-Jimenes GDC et al (2022) Evaluation of the effect of different co-solvent mixtures on the supercritical CO_2 extraction of the phenolic compounds present in Moringa oleifera Lam. leaves. Agronomy *12*(6):1450. https://doi.org/10.3390/agronomy12061450
15. Reverchon E, De Marco I (2006) Supercritical fluid extraction and fractionation of natural matter. J Supercrit Fluids 38(2):146–166. https://doi.org/10.1016/j.supflu.2006.03.020
16. Pourmortazavi SM, Hajimirsadeghi SS (2007) Supercritical fluid extraction in plant essential and volatile oil analysis. J Chromatogr A 1163(1–2):2–24. https://doi.org/10.1016/j.chroma.2007.06.021

Chapter 10

Extraction of Plant Proteins from Leaf and Its Functional Properties

Felix Kwashie Madilo and Emmanuel Letsyo

Abstract

Plant proteins are affordable and sustainable alternatives to animal proteins, which usually meet the nutritional needs of consumers. Hence, this study is designed to extract proteins with maximum functional properties from plant leaves. Proteins are compounds mainly used as food additives and ingredients in the food system. They have several functional properties and attributes, including foaming and gelling agents, and as emulsifiers, which could contribute significantly to the quality and sensory properties of the food. However, the extraction and purification techniques often used affect the quality and functionality. Hence, the extraction techniques employed should be carefully selected so that the functionality of the proteins, nutritional compositions, and the sensory properties could be protected and maximized.

Key words Proteins, Plant leaf, Extraction technique, Quantification, Functionality, Labile additives

1 Introduction

Proteins are very important nutrients that are required in large amounts by the body since they are the primary building blocks of the body. Plant proteins are abundant, affordable, and sustainable alternatives to animal proteins [1], usually meeting the body's protein demands. These proteins protect and sustain the environment and are considered a renewable alternative to animal-based ones [2]. A study by [3] reported that plant-based protein alternatives are healthier and more environmentally sustainable than animal proteins. Hence, they are considered more ethically sound than animal-derived proteins [4]. They can also be extracted from sustainable and cheap sources, such as by-products of crops, reducing food waste [2].

In general, proteins are compounds mainly used as food additives and ingredients in the food system. They have several functional properties and attributes such as gelation, foaming and gelling agents, emulsion properties, and water and oil-holding

Mitesh Kumar Dwivedi and A. Sankara Narayanan (eds.), *Leaves as Functional Foods*, Methods and Protocols in Food Science, https://doi.org/10.1007/978-1-0716-5226-8_10,

capacity, contributing significantly to the food's quality and sensory properties [4, 5]. Plant protein could be considered a functional ingredient for food formulations, stabilizers of emulsions and foams, and binding agents for fat and water [2]. Plant-based proteins also have antioxidant and antimicrobial properties [6] and can prevent cancer and diabetic disorders [7].

Several techniques are used to extract proteins from plant parts. Plant proteins extracted and purified using different extraction techniques have usually resulted in different protein profiles, quality, and functionality. Therefore, these techniques should be carefully selected to maximize the functionality, nutritional compositions, and sensory properties of the proteins [8, 9]. Extraction of proteins with different qualities and functionalities from plants such as pea, lentil, almond, and lupins has recently been reported [10–12]. Some of the available methods include chemical extraction [13]; ultrasound-assisted extraction [14]; microwave-assisted extraction [15, 16]; pulsed electric field extraction [8]; high hydrostatic pressure [16]; and enzymatic extraction [17, 18]. These techniques are also used to determine the quality and functionality of the protein. Hence, this study designs a standard procedure for extracting proteins from plant leaves with maximum functional properties.

2 Materials

1. −80 °C freezer.
2. Blender.
3. Centrifuge.
4. Digital balance.
5. Eppendorf tube rack.
6. Eppendorf tubes.
7. Fine-tipped black sharpie.
8. Mortars.
9. Parafilm.
10. Pencil.
11. Pestles.
12. Whatman 3MM paper.
13. Coomassie staining solution (325 mL water, 125 mL isopropanol, 50 mL acetic acid, 250 mg Coomassie Blue R-250).
14. Homogenization buffer.
15. Homogenization Buffer Base Stock.
16. Ice.

17. Labile additives (ß-mercaptoethanol, PMSF, leupeptin, and pepstatin).
18. Leaves.
19. Liquid nitrogen.
20. Protein extraction buffer.
21. Water.

3 Methods

1. Make a list of samples to be extracted.
   (a) Label 1.4 mL Eppendorf tubes with fine-tipped black sharpie (*see* **Note 1**).
   (b) Put materials together: mortars and pestles, protein extraction buffer, Eppendorf tube rack with ice, Whatman paper.
   (c) Label with a pencil the positions where each sample will be spotted.
2. Prepare homogenization buffer.
   (a) Add labile additives (*see* **Note 2**) such as ß-ME, PMSF, leupeptin (*see* **Note 3**) and pepstatin (*see* **Note 4**) to a 5 mL aliquot of the homogenization buffer base stock. This will be sufficient to prep ~20 samples.
   (b) Store the complete buffer on ice and discard unused buffer after 30 min.
3. Prepare protein homogenization buffer base stock and store it at 4 °C.
   (a) For 100 mL: prepare 1 M Tris–HCl (*see* **Note 5**) pH 7.5 up to 10 mL; 50% glycerol (*see* **Note 6**) up to 20 mL; 0.5 M EDTA (*see* **Note 7**) up to 1 mL; 0.1 M EGTA (*see* **Note 8**) up to 2 mL; ddH20 (*see* **Note 9**) up to 67 mL.
4. Before use, aliquot sufficient buffer for 30 min of extractions (about 5 mL) and add:
   (a) PMSF to 2 mM (1/20th volume of a 40 mM stock in isopropanol and store at −20 °C) (*see* **Note 10**).
   (b) Mix and heat enough to put any crystals into solution before pipetting.
   (c) ß-mercaptoethanol (ß-ME) to 40 mM (27 μL of the pure liquid stock per 10 mL).
   (d) Pepstatin to 2 μg/mL (1/500th volume of a 1 mg/mL stock and store at −20 °C).

   (e) Leupeptin to 2 μg/mL (1/5000th volume of a 10 mg/mL stock and store at −20 °C).

5. Add 13 μL Β-ME, 10 μL 1 mg/mL pepstatin, 250 μL 40 mM PMSF and 1 μL 10 mg/mL leupeptin for 5 mL of complete buffer and discard after 30 min.
6. Grind the tissues.
   (a) Weigh about 100 g of leaves using a digital balance.
   (b) Grind in a small mortar and pestle or blend in a 1 L blender.
   (c) Grind the unfrozen tissue directly in ice-cold homogenization buffer (200 μL) in a pre-chilled mortar and pestle if the tissues are fresh.
   (d) Grind or blend tissue to a fine powder in liquid nitrogen if the tissues are frozen and then add 200 μL homogenization buffer to the frozen powder and continue grinding/blending to a paste. It is essential to keep the tissue frozen until the buffer is added.
   (e) Grind hard/blend at a high speed to break bundle sheath cells. Failure to do this will reduce the yield of Rubisco and other bundle sheath-specific proteins.
7. For total leaf extract, pipette the extract into the labeled Eppendorf tube, on ice.
   (a) If quantifying by Coomassie spotting (i.e., if the amount of tissue harvested was not similar for all samples), spot 2 μL onto Whatman 3MM paper and label the spot with a pencil.
   (b) For the analysis of separated membrane and soluble proteins:
      (i) Pellet the membranes by microfuging for 3 min at a higher speed in the microfuge in the cold.
      (ii) Pipette off the supernatant, avoiding the green pellet, and place in a new microfuge tube. This is the "soluble" fraction.
   (c) Rinse the membranes to be sure they are free of soluble proteins:
      (i) Add 0.2 mL homogenization buffer to each membrane sample.
      (ii) Flick gently with your finger, and microfuge for 2 min.
      (iii) Discard the supernatant and resuspend the membrane pellets in 50–200 μL homogenization buffer, by pipetting up and down gently.

   (iv) Preparing a series of samples on a gel, resuspend the smallest green pellet in 50 μL and the remaining pellets in whatever volume it takes to give an equal density of greenness.

8. Store the protein samples in freezer at −80 °C. Freeze the samples in small quantities to avoid repeated freeze-thaw cycles as this could denature or degrade the protein.
9. Quantify the protein. If samples contained uneven amounts of leaf tissue, the relative protein concentrations can be estimated by the following procedure:
   (a) Spot 2 μL of each sample onto Whatman 3MM (with samples labeled).
   (b) Spot 1:2 dilutions of several samples that are concentrated to enable quantification.
   (c) Each spot should have the same volume, i.e., on a piece of parafilm, mix 2 μL of sample to 2 μL water, and then spot 2 μL of this onto the Whatman paper. This will enable you to see 1/2 as much protein looks like.
   (d) Spot 1, 3, and 10 μg of bovine serum albumin (BSA) in a 2 μL volume for better quantification.
   (e) Place the Whatman 3 mm paper with the protein spots in a container that has previously been used for Coomassie staining (*see* **Note 11**).
   (f) Add the Coomassie staining solution (325 mL water, 125 mL isopropanol, 50 mL acetic acid, 250 mg Coomassie Blue R-250).
   (g) Rock briefly until you can see the spots turn blue (10–20 s).
   (h) Rinse with water until the background is fairly white.
   (i) Compare signal intensities while the paper is either fully wet or fully dry. If it is partially dry, results may be misleading.
   (j) Gels-flick tube to suspend stuff.
   (k) Pipette out what you want to put in the sample, heat to 70 °C, 1 min spin.

4 Notes

1. It is important to number on top and side because numbers can rub off.
2. Labile additives are substances (polymers) that degrade or change their forms when subjected to heat or specific chemical processes.

3. Leupeptin is a naturally occurring protease inhibitor (a peptide aldehyde) that inhibits serine, cysteine, and theanine proteases.
4. Pepstatin is a protein and a specific inhibitor of aspartic proteases, a class of enzymes that cleave peptide bonds in proteins.
5. Tris–HCl, also called Tris–hydrochloride, is a common buffering agent used in biochemistry and molecular biology. It is a white crystalline powder that is highly soluble in water. It maintains a stable pH range (7.0–9.0) in various biological applications, including DNA and RNA extraction, protein purification, enzyme assays, and electrophoresis.
6. Glycerol (glycerin) is a colorless, odorless, sweet-tasting, viscous liquid that is a simple triol compound. It is a polyol, a polyhydric alcohol, with three hydroxyl groups, making it hydroscopic and water soluble.
7. EDTA acid (ethylenediaminetetraacetic acid) is an aminopolycarboxylic acid. It is a white, slightly water-soluble solid and is used to bind to iron and calcium ions, forming water-soluble complexes even at neutral pH.
8. EGTA (ethylene glycol tetraacetic acid) is a chelating agent. It has a higher affinity for calcium ions but a lower affinity for magnesium ions than EDTA. It protects against cell death caused by nitric oxide-induced calcium influx into nerve cells.
9. ddH_2O stands for double distilled water. It is a type of highly purified water (repeatedly distilled tap water) used in the laboratories.
10. PMSF (phenylmethanesulfonyl fluoride) is toxic and inactivates within 30 min after being added to aqueous solution. It is a serine protease inhibitor mostly used in biochemistry when cell lysates are prepared. It is added to lysis buffers to prevent the degradation of proteins by proteases.
11. Coomassie staining is a blue staining technique used to observe proteins separated by electrophoresis in a polyacrylamide gel. It is a quick, simple, and affordable technique used for detecting proteins on gels.

References

1. Estell M, Hughes J, Grafenauer S (2021) Plant protein and plant-based meat alternatives: consumer and nutrition professional attitudes and perceptions. Sustainability 13:1478
2. Claire MD (2024) A review of alternative plant protein sources, their extraction, functional characterization, application, nutritional value and pinch points to being the solution to sustainable food production. Int J Food Sci Technol 2024(59):462–472. https://doi.org/10.1111/ijfs.16467
3. Bryant CJ (2022) Plant-based animal product alternatives are healthier and more environmentally sustainable than animal products. Future Foods 6:100174
4. Navaf M, Sunooj KV, Aaliya B, Sudheesh C, Akhila PP, Mir SA, Nemtanu MR, George J, Lackner M, Mousavi Khaneghah A (2023)

Contemporary insights into the extraction, functional properties, and therapeutic applications of plant proteins. J Agric Food Res 14: 100861. https://doi.org/10.1016/j.jafr.2023.100861

5. Tan M, Nawaz MA, Buckow R (2021) Functional and food application of plant proteins–a review. Food Rev Int 39(2021):2428–2456. https://doi.org/10.1080/87559129.2021.1955918
6. Jafari SM, Doost AS, Nasrabadi MN, Boostani S, van der Meeren P (2020) Phytoparticles for the stabilization of Pickering emulsions in the formulation of novel food colloidal dispersions. Trends Food Sci Technol 98:117–128
7. Islam MS, Wang H, Admassu H, Sulieman AA, Wei FA (2022) Health benefits of bioactive peptides produced from muscle proteins: antioxidant, anti-cancer, and anti-diabetic activities. Process Biochem 116:116–125. https://doi.org/10.1016/j.procbio.2022.03.007
8. Maningat CC, Jeradechachai T, Buttshaw MR (2022) Textured wheat and pea proteins for meat alternative applications. Cereal Chem 99:37–66
9. Zhang T, Dou W, Zhang X et al (2021) The development history and recent updates on soy protein-based meat alternatives. Trends Food Sci Technol 109:702–710
10. Maykish A, Nishisaka MM, Talbott CK, Reaves SK, Kristo AS, Sikalidis AK (2021) Comparison of whey versus almond protein powder on nitrogen balance in female college students; the California almond protein powder project (CAlmond-P3). Int J Environ Res Public Health 18:11939
11. Jarpa-Parra M (2018) Lentil protein: a review of functional properties and food application. An overview of lentil protein functionality. Int J Food Sci Technol 53:892–903
12. Munialo CD, Martin AH, van der Linden E, de Jongh HH (2014a) Fibril formation from pea protein and subsequent gel formation. J Agric Food Chem 62:2418–2427
13. Kumar M, Tomar M, Potkule J, Verma R, Punia S, Mahapatra A, Belwal T, Dahuja A, Joshi S, Berwal MK, Satankar V, Bhoite AG, Amarowicz R, Kaur C, Kennedy JF (2021) Advances in the plant protein extraction: mechanism and recommendations. Food Hydrocoll 115:106595. https://doi.org/10.1016/j.foodhyd.2021.106595
14. Wang F, Zhang Y, Xu L, Ma H (2020) An efficient ultrasound-assisted extraction method of pea protein and its effect on protein functional properties and biological activities. LWT Food Sci Underw Technol 127:109348. https://doi.org/10.1016/j.lwt.2020.109348
15. Görgüç A, Özer P, Yılmaz FM (2020) Microwave-assisted enzymatic extraction of plant protein with antioxidant compounds from the food waste sesame bran: comparative optimization study and identification of metabolomics using LC/QTOF/MS. J Food Process Preserv 44. https://doi.org/10.1111/jfpp.14304
16. Barrios C, Fernández-Delgado M, López-Linares JC, García-Cubero MT, Coca M, Lucas S (2022) A techno-economic perspective on a microwave extraction process for efficient protein recovery from agri-food wastes. Ind Crop Prod 186:115166. https://doi.org/10.1016/j.indcrop.2022.115166
17. Pojić M, Mišan A, Tiwari B (2018) Eco-innovative technologies for extraction of proteins for human consumption from renewable protein sources of plant origin. Trends Food Sci Technol 75:93–104. https://doi.org/10.1016/j.tifs.2018.03.010
18. Ochoa-Rivas A, Nava-Valdez Y, Serna-Saldívar SO, Chuck-Hernández C (2017) Microwave and ultrasound to enhance protein extraction from peanut flour under alkaline conditions: effects in yield and functional properties of protein isolates. Food Bioprocess Technol 10: 543–555. https://doi.org/10.1007/s11947-016-1838-3

Chapter 11

Extraction of Bioactive Compounds from Leaf: Essential Oils

Naznin Shaikh, Vaibhavi Bhatia, and Mayank Sharma

Abstract

Essential oils (EOs) are aromatic compounds valued for their therapeutic, antioxidant, and antimicrobial properties. Growing demand for natural, eco-friendly products has driven interest in sustainable extraction methods. This review highlights bioactive EOs from plant leaves, comparing traditional and emerging green extraction techniques. Emphasis is placed on advanced techniques like supercritical fluid, enzyme, and microwave extraction for their efficiency, reduced processing time, and preservation of bioactive compounds. Comparative analyses offer guidance for optimizing EO extraction in industrial and pharmaceutical applications.

Key words Bioactive, Essential oil, Leaf, Extraction

1 Introduction

Essential oils (EOs) are intricate blends of over 200 compounds derived from various plant parts, with leaves being especially rich sources due to their high concentration of volatile phytochemicals stored in specialized structures [1–4]. Comprising a volatile fraction (90–95%) and a non-volatile fraction (5–10%), EOs possess potent antimicrobial, antioxidant, anti-inflammatory, and anticancer properties, making them valuable across pharmaceuticals, cosmetics, food preservation, and perfumery [2, 5]. Notable examples include Citrus aurantifolia and Cordia verbenacea leaf oils, known for their strong bioactivities [6]. As demand for natural products grows, key EO contributors like rosemary, basil, and lavender dominate the global market [7]. Extraction methods critically influence EO quality, with traditional techniques often compromising sensitive compounds, prompting the rise of green technologies such as MAE, UAE, DES, SFE, and solar-assisted distillation that enhance yield, preserve bioactivity, and support sustainability goals [1, 6]. This chapter focuses on evaluating these extraction methods for leaf-derived EOs in terms of efficiency, environmental impact, and preservation of bioactive constituents.

Mitesh Kumar Dwivedi and A. Sankara Narayanan (eds.), *Leaves as Functional Foods*, Methods and Protocols in Food Science, https://doi.org/10.1007/978-1-0716-5226-8_11,

2 Materials

1. Plant Materials (Rose, Basil, Coriander or Eucalyptus).
2. Solvents.
3. Solar Steam Distillation unit.
4. Clevenger Apparatus.

3 Methods

3.1 Extraction Techniques for Essential Oils from Plant Leaves

1. Conventional/Classical techniques.
2. Hydro-Distillation (HD).
   1. *Selection of Plant Material* (*see* **Note 1**).

      Choose aromatic or medicinal plants such as rose [6], basil [8], coriander [9], or eucalyptus [10] for essential oil extraction.
   2. *Preparation of the Mixture.*

      Immerse the plant material in water or a solvent. For example, 150 g of basil leaves were mixed with 6 L of water in a referenced study [8].
   3. *Heating the Mixture* (*see* **Note 2**).

      Heat the water-plant mixture to release volatile compounds. This causes the essential oils to evaporate along with water vapor [6].
   4. *Condensation of Vapors.*

      Direct the vapors through a condenser where they cool down and condense into liquid form [1].
   5. *Separation of Essential Oil.*

      Use a decanter to separate the essential oil from the condensed water (distillate) [1].
   6. *Collection and Analysis.*

      Collect the essential oil. In the basil study, 65 compounds were identified, with linalool (48.4%) and methyl chavicol (14.3%) being the most abundant [8].
   7. *Application and Evaluation.*

      Evaluate the oil's bioactivity. Basil oil is known for antimicrobial and antioxidant properties [8], coriander oil for nutraceutical value [9], and eucalyptus oil for insect-repellent and antimicrobial effects [10].

8. *Consider Limitations.*

   Be aware of HD drawbacks: long extraction times, high energy consumption, and thermal degradation of heat-sensitive compounds [6].

3. Steam Distillation (SD).
   1. *Harvesting and Drying* Fully matured aromatic plants are harvested and left to dry in the field for several days.
   2. *Preparation of Plant Material* Once dried, the plant material (referred to as "hay") is chopped and transferred to the distillation unit using a wagon, forming a packed bed of leaves.
   3. *Steam Application* Steam from the boiler passes through the plant material, causing essential oil components to evaporate at temperatures close to the boiling points of water and the oils (typically 250–350 °C) [6].
   4. *Condensation and Separation* The vapor mixture is condensed, and the essential oil is separated from the water in the flask [6].
   5. *Consider Limitations* Conventional SD relies on fossil fuels, contributing to greenhouse gas emissions and environmental pollution.
   6. *Sustainable Alternative* Solar steam distillation (SSD) uses solar energy to meet energy demands while reducing pollution [6].
   7. *Applications and Examples.*
      - Ocimum basilicum: Extracted using SD, produces high-grade essential oil with a strong fragrance, used in aromatherapy massages and for alleviating digestive issues, headaches, muscle tension, and nervous exhaustion [11].
      - Tobacco leaves: SD used by Xianzhong Zhang et al. [12].
      - Tea leaves: SD applied by Gary R. Baker et al. [13].
   8. *Commercial Relevance* SD is widely adopted for commercial essential oil extraction due to its efficiency and ability to preserve volatile compounds.

3.2 Non-conventional Techniques

3.2.1 Super Critical Fluid Extraction (SFE)

1. *Overview of SFE Technique* Supercritical Fluid Extraction (SFE) is an advanced and eco-friendly method used to extract essential oils from plant materials. It offers high efficiency and preserves heat-sensitive compounds.
2. *Use of Supercritical CO_2* Carbon dioxide (CO_2) is commonly used as the supercritical fluid due to its dual liquid-gas properties, enabling effective penetration and dissolution of target compounds.

3. *Operating Conditions* SFE operates under moderate temperatures and tunable pressures, minimizing thermal degradation and maintaining the integrity of volatile constituents [14–16].
4. *Optimization Parameters* Key parameters such as pressure, temperature, and extraction time influence yield and selectivity. Co-solvents like ethanol can enhance phytochemical solubility.
5. *Advantages Over Traditional Methods* Compared to Soxhlet extraction and hydro-distillation, SFE provides faster processing, solvent-free extracts, and better preservation of bioactive compounds.
6. *Green Chemistry Benefits* The use of non-toxic, easily removable CO_2 supports SFE's status as a green and safe extraction technique [17].
7. *System Configuration* Liquid CO_2 is supplied from a cylinder with a siphon tube and pumped into the system via a dual-piston pump (up to 68 MPa). The system includes four parallel extraction vessels—two for supercritical CO_2 extraction and two for supercritical water oxidation.
8. *Thermal Control and Flow Regulation* CO_2 is preheated using a preheater and heating jacket. A variable restrictor valve regulates flow rate and maintains system pressure. An electrical heating mechanism prevents sample blockage, and a flow meter monitors CO_2 flow [17].
9. *Optimization Example* Shazana Azfar Radzali et al. optimized SFE of phenolic compounds from Labisia pumila using Response Surface Methodology. Using supercritical CO_2 with ethanol-water (78% v/v) as co-solvent at 283 bar, 32 °C, and 16% co-solvent concentration, they achieved 14.05 ± 0.76% (g/g) total phenolics. Key compounds were identified via HPLC [18].
10. *Applications and Selectivity.*
    1. Curry flowers, sage leaves, and their mixture: Extracted using conventional SFE by Svetolik Maksimović et al. [19].
    2. Moringa oleifera: Extracted by Suwei Zhao and Dongke Zhang at 50 MPa, 60 °C, 120 min, yielding 6.34% oil. Only 12 compounds were identified compared to 42 from Soxhlet extraction, indicating higher selectivity [17].

3.2.2 Microwave-Assisted Extraction (MAE)

1. *Overview of MAE Technique* Microwave-Assisted Extraction (MAE) is an advanced and efficient method for extracting essential oils and bioactive compounds from plant materials. It uses microwave energy to rapidly heat polar solvents and plant tissues, promoting cell wall rupture and phytoconstituent release.

2. *Mechanism of Action MAE* operates through ionic conduction and dipole rotation, enabling fast and uniform heating that enhances extraction efficiency.
3. *Advantages Over Traditional Methods* Compared to hydro-distillation and steam distillation, MAE offers shorter extraction times, lower energy consumption, and better preservation of heat-sensitive compounds [4].
4. *Green Chemistry Benefits MAE* is recognized as a green and sustainable technology due to reduced solvent and energy requirements, aligning with eco-friendly industrial practices.
5. *System Configuration and Parameters MAE* typically involves a modified microwave system combined with extraction setups. Key parameters include microwave power, extraction time, and solvent-to-material ratio, which are optimized for maximum yield.
6. *Solvent-Free Extraction* MAE can be performed with or without solvents. Solvent-free approaches are especially beneficial for essential oil extraction, offering higher purity and eliminating the need for solvent recovery.
7. *ApplApplications and Examples.*
   1. *Patchouli leaves (Pogostemon cablin Benth.):* Extracted using SFME and MHD by Heri Septya Kusuma et al. SFME used a modified domestic microwave oven (800 W, 2450 MHz) with a Clevenger apparatus. SFME achieved higher yields and met ISO 3757:2002(E) standards [20].
   2. *Magnolia sieboldii leaves:* Extracted using MSDE by Sarah Hamid et al., yielding 0.94 ± 0.05% oil. MSDE showed superior efficiency, sustainability, and cost- effectiveness.
   3. *Liquidambar formosana leaves and fruits:* Extracted using a deep eutectic solvent-assisted microwave technique by Ke Zhang et al., enabling recovery of essential oils, gallic acid, and ellagic acid [4].
      - *Industrial Relevance MAE* is a rapid, efficient, and eco-friendly alternative with promising industrial applications for essential oil extraction and multifunctional phytochemical recovery.

4 Notes

1. Follow the standard procedure to identify the plant specimens.
2. Maintain the proper temperature in the heating process, or otherwise it will affect heat sensitive bioactive components present in plants.

References

1. Katekar VP, Rao AB, Sardeshpande VR (2023) A hydrodistillation-based essential oils extraction: a quest for the most effective and cleaner technology. Sustain Chem Pharm 36:101270. https://doi.org/10.1016/j.scp.2023.101270
2. Luque De Castro MD, Jiménez-Carmona MM, Fernández-Pérez V (1999) Towards more rational techniques for the isolation of valuable essential oils from plants. TrAC Trends Anal Chem 18:708–716. https://doi.org/10.1016/S0165-9936(99)00177-6
3. Visakh NU, Pathrose B, Chellappan M, Ranjith MT, Sindhu PV, Mathew D (2023) Extraction and chemical characterisation of agro-waste from turmeric leaves as a source of bioactive essential oils with insecticidal and antioxidant activities. Waste Manag 169:1–10. https://doi.org/10.1016/j.wasman.2023.06.030
4. Zhang K, Gao H, Yang Y, Ping Y, Tian H, Gu H, Yang L (2025) Natural acidic deep eutectic solvent-mediated microwave-assisted simultaneous hydrodistillation, hydrolysis, and extraction for obtaining essential oils, gallic acid and ellagic acid from Liquidambar formosana leaves and fruits. Sustain Chem Pharm 43: 101870. https://doi.org/10.1016/j.scp.2024.101870
5. Khalid S, Chaudhary K, Amin S, Raana S, Zahid M, Naeem M, Mousavi A, Muhammad R (2024) Recent advances in the implementation of ultrasound technology for the extraction of essential oils from terrestrial plant materials : a comprehensive review. Ultrason Sonochem 107:106914. https://doi.org/10.1016/j.ultsonch.2024.106914
6. Kant R, Kumar A (2022) Review on essential oil extraction from aromatic and medicinal plants: techniques, performance and economic analysis. Sustain Chem Pharm 30:100829. https://doi.org/10.1016/j.scp.2022.100829
7. Cassel E, Vargas RMF, Martinez N, Lorenzo D, Dellacassa E (2009) Steam distillation modeling for essential oil extraction process. Ind Crop Prod 29:171–176. https://doi.org/10.1016/j.indcrop.2008.04.017
8. Chenni M, El Abed D, Rakotomanomana N, Fernandez X, Chemat F (2016) Comparative study of essential oils extracted from egyptian basil leaves (ocimum basilicum l.) using hydrodistillation and solvent-free microwave extraction. Molecules 21:113. https://doi.org/10.3390/molecules21010113
9. Abbas A, Anwar F, Ahmad N, Shahid M, Al-Mijalli SH, Yaseen M, Farooq S, Iqbal M (2022) Characterization of bioactives and nutra-pharmaceutical potential of supercritical fluid and hydro-distilled extracted coriander leaves essential oil. Dose-Response 20:1–10. https://doi.org/10.1177/15593258221130749
10. Ngo TCQ, Tran TH, Le XT (2020) The effects of influencing parameters on the Eucalyptus globulus leaves essential oil extraction by hydrodistillation method. IOP Conference Series: Materials Science and Engineering 991:012126. https://doi.org/10.1088/1757-899X/991/1/012126
11. Mindaryani A, Rahayu SS (2007) Essential oil from extraction and steam distillation of ocimum basillicum. World Congr Eng Comput. Sci, pp 4–8
12. Zhang X, Gao H, Zhang L, Liu D, Ye X (2012) Extraction of essential oil from discarded tobacco leaves by solvent extraction and steam distillation, and identification of its chemical composition. Ind Crop Prod 39:162–169. https://doi.org/10.1016/j.indcrop.2012.02.029
13. Baker GR, Lowe RF, Southwell IA (2000) Comparison of oil recovered from tea tree leaf by ethanol extraction and steam distillation. J Agric Food Chem 48:4041–4043. https://doi.org/10.1021/jf0004356
14. Pourmortazavi SM, Hajimirsadeghi SS (2007) Supercritical fluid extraction in plant essential and volatile oil analysis. J Chromatogr A 1163: 2–24. https://doi.org/10.1016/j.chroma.2007.06.021
15. Yousefi M, Rahimi-Nasrabadi M, Pourmortazavi SM, Wysokowski M, Jesionowski T, Ehrlich H, Mirsadeghi S (2019) Supercritical fluid extraction of essential oils. TrAC—Trends Anal Chem 118:182–193. https://doi.org/10.1016/j.trac.2019.05.038
16. Dashtian K, Kamalabadi M, Ghoorchian A, Ganjali MR, Rahimi- Nasrabadi M (2024) Integrated supercritical fluid extraction of essential oils. J Chromatogr A 1733:465240. https://doi.org/10.1016/j.chroma.2024.465240
17. Zhao S, Zhang D (2013) Supercritical fluid extraction and characterisation of Moringa oleifera leaves oil. Sep Purif Technol 118: 497–502. https://doi.org/10.1016/j.seppur.2013.07.046
18. Radzali SA, Markom M, Md Saleh N (2022) Parameter effects and optimisation in supercritical fluid extraction of phenolic compounds from Labisia pumila. Separations 9:385. https://doi.org/10.3390/separations9120385

19. Maksimovic S, Kesic Z, Lukic I, Milovanovic S, Ristic M, Skala D (2013) Supercritical fluid extraction of curry flowers, sage leaves, and their mixture. J Supercrit Fluids 84:1–12. https://doi.org/10.1016/j.supflu.2013.09.003

20. Kusuma HS, Altway A, Mahfud M (2018) Solvent-free microwave extraction of essential oil from dried patchouli (Pogostemon cablin Benth) leaves. J Ind Eng Chem 58:343–348. https://doi.org/10.1016/j.jiec.2017.09.047

Part II

Immuno Modulatory Effect of Leaf

Chapter 12

Determination of Immunomodulatory Activities of Leaf

Manjarima Ganguli and Tamalika Chakraborty

Abstract

The immune system of our body plays an important role in maintaining the physiological activities of our body by protecting it from various foreign microorganisms and pathogens. Some recent studies reveals that the exploration of plant obtained from natural products having immunomodulatory potential have gained therapeutic significance. T-cell proliferation assay is one of the reliable in vitro methods by which the immunomodulation of leaves can be determined. This chapter focuses on the isolation and preparation of cytotoxic T cells from murine spleen; preparation of processed extracts; cell culture and stimulation conditions with mitogens; and quantification with colorimetric and fluorescence-based assays (ELISA). It also discusses key aspects such as optimization of extraction concentration of natural products, cytotoxicity control, and selection of adequate mitogens (Concanavalin A) to achieve reliable and reproducible results.

Key words Immunomodulation, T-Cell proliferation assay, Leaf extracts, Mitogen stimulation

1 Introduction

Medicinal plants have been known for their therapeutic activities with least adverse effects where the leaves of those plants having rich source of bioactive compounds. These compounds comprise alkaloids, glycosides, terpenoids, and polysaccharides that can modulate immune cells such as NK cells, T cells, and B cells by their stimulating or inhibitory effects. Analysis of the immunomodulatory effects of leaf extracts includes a series of systematic steps—preparation of leaf extracts, isolation of T cells and cell culture media preparation, activation of T cells and treatment with leaf extracts and standardized modulator, and assessment of T-cell proliferation by cytokine profiling with ELISA [1–5].

Recent studies have discussed about the immunomodulatory role of several plant species. Aqueous extracts of *Moringa oleifera* leaves have been shown to improve the proliferation of $CD4^+$ and $CD8^+$ T cells in vitro, showcasing their immunostimulatory effects [6]. Similarly, *Tinospora cordifolia* extracts have been described to

Mitesh Kumar Dwivedi and A. Sankara Narayanan (eds.), *Leaves as Functional Foods*, Methods and Protocols in Food Science, https://doi.org/10.1007/978-1-0716-5226-8_12,

regulate Th17 cell responses, highlighting their impact in immune regulation [1, 7]. *Limonium gmelinii* stimulated the secretion of TNF-α and IL-6 in non-activated macrophages and lymphocytes, while inhibiting their production in activated immune cells [8]. *Moringa oleifera* flower and leaf extract can suppress the activity potential of inflammatory mediators in RAW 264.7 macrophages via NF-κB pathway [9, 10]. These key findings emphasize the role of standardized approaches to examine the immunomodulatory actions of plant extracts..

This chapter provides systematic guidelines to evaluate the immunomodulatory effects of leaf extracts of holy basil on cytotoxic T (Tc) cells. Basically, the dried leaves of *Ocimum sanctum* L. (synonym: *Ocimum tenuiflorum*), belonging to the Lamiaceae family, contain various proinflammatory cytokines with immunomodulatory effects. Researchers [11, 12] have studied the activation of macrophages and proliferation of T cells by lowering the proinflammatory cytokines like TNF-α/IL-6. Tulsi, the Holy basil is mainly found in places like India, Bangladesh, Nepal and is cultivated largely in in Southeast Asian countries like Thailand and Malaysia [11]. The content of the protocol ranges from the preparation of its leaf extracts to the isolation, culture of Tc cells, activation procedures, treatment of the extracts, and the evaluation phase of Tc cell proliferation. Following the advanced biotechnological methods, researchers will assess the functionality of various plant extracts in altering the immune functions which can aid into formulations of new immunotherapeutic drugs.

2 Materials

2.1 Plant Material and Extraction

- *Fresh or dried leaves* of holy basil or tulsi.
- *Solvents* for extraction: *Ethanolic extract*: 70% Ethanol (*see* **Note 1**).
- *Soxhlet apparatus.*
- *Rotary evaporator.*
- *Filter paper/Whatman No. 1.*
- *Centrifuge and centrifuge tubes.*

2.2 Isolation of T Cells

- *Primary T cells are mainly CD8+ Cytotoxic T Cells (Tc)* (from spleen or lymph nodes of mice/rats).

2.3 Cell Culture Reagents and Equipment

- *Fetal Bovine Serum (FBS) (10% v/v):* Thaw the FBS by placing it in the refrigerator at 4 °C temperature and mix gently to form a uniform mixture. After ensuring that the water bath is preheated to 56 °C, place the FBS bottle or aliquots in a water bath for 30 minutes to incubate. Gentle swirling is done during the heating every 5–10 minutes to maintain the heating evenly. Lastly, cool the aliquots in −4 °C temperature and store in −20 °C to −80 °C temperature until use.

- *RPMI-1640 medium:* To prepare RPMI-1640 medium, dissolve RPMI-1640 into 10% FBS with 1% Pen-Strep and 2 mM L-glutamine. Lastly, store it at 4 °C; heat the solution to 37 °C before use.
- *Phosphate Buffered Saline (PBS):* Dissolve commercial PBS tablet in distilled water or prepare by mixing NaCl 137 mM, KCl 2.7 mM, Na_2HPO_4 10 mM, KH_2PO_4 1.8 mM thoroughly and adjust pH to 7.4.
- *Trypan Blue solution (0.4%):* Accurately weigh 0.4 grams of Trypan Blue powder and dissolve the powder in distilled water or PBS with magnetic stirring. Transfer the solution to a volumetric flask to make up the volume up to 100 ml and store at 4 °C for further use.
- *T-25 or T-75 flasks, 96-well flat-bottom tissue culture plate.*
- *CO_2 incubator (5% CO_2 at 37 °C).*
- *Hemocytometer or automated cell counter.*

2.4 T-Cell Activation

- Murine spleen cells.
- *RBC lysis buffer:* Add NH_4Cl 0.155 M, $KHCO_3$ 0.01 M, EDTA 0.1 mM accordingly and adjust the pH to 7.4. After filtering the solution, sterilize it and keep it at 4 °C.
- *CD3/CD28 antibodies.*
- *Concanavalin A (Con A)* (5 μg/ml): Dissolve in PBS to a final concentration of 1 mg/ml to prepare the standard solution. Use 5 μg/ml in the assay.

2.5 Cytokine Profiling Assay Components

- *Pre-coated ELISA plates.*
- *Cytokine standards:* Reconstitute lyophilized cytokine standard in sterile PBS with 0.1% BSA.
- *Detection antibodies:* Dilute in antibody diluent (PBS 1% BSA) (typically 1:100–1:1000).
- *Streptavidin-HRP (Horseradish Peroxidase):* Mixing 0.1 gm of Tween 20 and 1 gm bovine serum albumin (BSA) in 100 mL PBS.
- *TMB (3,3′,5,5′-tetramethylbenzidine) substrate.*
- *Stop solution* (e.g., *1 M H_2SO_4 for TMB):* By adding 55.5 mL of concentrated H_2SO_4 (18 M) into 944.5 mL of distilled water (with stirring in ice bath).
- *Wash buffer:* Mix 0.5 mL Tween 20 per 1 L PBS (0.05% v/v).

- *Blocking buffer:* 5 gm of bovine serum albumin (BSA) is dissolved in 100 ml of PBS.
- *Sample diluent:* This is a mixture of 0.05 gm Tween 20 and 1 gm of BSA in 100 mL of PBS.

3 Methods

3.1 Preparation of Leaf Extracts

1. Collect the fresh leaves of holy basil (*Ocimum sanctum*/tulsi). Wash it thoroughly by air drying and grind it to form coarse powder.
2. Weigh 50–100 gm of dried leaf powder and placed in Soxhlet apparatus. Then add 500 mL of solvent (e.g., ethanol for polar compounds, hexane for non-polar) into the flask. Run the procedure for 6–8 ho until the solvent in the flask become clear.
3. After extraction, filter the mixture by using Filter paper/Whatman No. 1 for the removal of solid residues of the plants (*see* **Note 2**).
4. Centrifuge the filtrate at 3000–5000 rpm for 10 min and lastly the filtrate is concentrated using a rotary evaporator under reduced pressure and at a controlled temperature.

3.2 Isolation of T Cells

1. The experimental use of animals must adhere to national and institutional animal ethics guidelines. Euthanasia should be conducted using approved humane methods, such as carbon dioxide asphyxiation or cervical dislocation, following institutional animal ethics committee (IAEC). Strict aseptic techniques are to be maintained throughout the procedure to prevent contamination.
2. Following euthanasia, the abdominal cavity is dissected under sterile conditions to expose and carefully excise the spleen. The spleen is then placed into a sterile Petri dish containing cold RPMI-1640 medium.
3. To obtain individual splenocytes, the spleen is gently ground using a plunger or sterile glass slide over a mesh filter in the medium.
4. Spleen suspensions contain a considerable proportion of erythrocytes, which are selectively removed using a red blood cell lysis buffer solution. The suspension is incubated with the lysis buffer for 2–5 min at room temperature.
5. The cell suspension is then filtered through a 70 μm sterile cell strainer to eliminate tissue debris and cell aggregates.
6. Add 5–10 mL of cold PBS or RPMI-1640 to neutralize lysis buffer. Centrifuge at 300–400× *g* for 5–7 min at 4 °C. Discard the supernatant carefully without disturbing the pellet.

7. Repeat wash once or twice to completely remove lysis buffer.
8. To isolate Tc cells, magnetic-activated cell sorting (MACS) is employed. This process includes:

   Labeling cells with magnetic microbeads conjugated to anti-CD8 antibodies; passing the labeled cell suspension through a magnetic separation column; and retaining positively selected Tc cells in the column.
9. After labeling cells with magnetic beads (e.g., anti- CD8), wash with MACS buffer. Centrifuge at 300× *g* for 10 min. Resuspend in fresh buffer before applying to magnetic columns.

3.3 Cell Culture

1. Wash isolated Tc cells once in complete RPMI medium by centrifuging at 300× *g* for 5 min, discard supernatant.
2. Place the necessary T-25 or T-75 flasks or a 96-well flat-bottom tissue culture plate and pre-warm the RPMI-1640 medium to 37 °C.
3. Next, seed cells at 1×10^5 to 2×10^5 cells per well into the flask or dish with the pre-warmed medium.
4. Then, culture the cells at 37 °C and atmospheric CO_2 for incubation.
5. Microcentrifuge the collected cells and resuspend it with freshly prepared PBS.
6. Then, mix an aliquot of the cell suspension with 0.4% Trypan Blue solution to load the mixture onto a hemocytometer or automated cell counter.

3.4 T-Cell Activation

1. Freshly prepared Con A is thawed from stock and diluted to working concentrations of 5 μg/mL in complete RPMI-1640 medium.
2. Isolated T cells are then resuspended in complete RPMI-1640 medium and plated at a density of 1×10^5 to 2×10^5 cells per well in a sterile 96-well flat-bottom tissue culture plate.
3. 100 μL of the Con A solution is added to the designated wells, and the plate is gently mixed before being incubated at 37 °C in a humidified 5% CO_2 incubator for a duration of 48 and 72 h.

3.5 Treatment with Leaf Extract

1. The leaf extract is subsequently added to the wells at various concentrations, typically 1 μg/ml, 50 μg/ml, 100 μg/ml, and 200 μg/ml [13, 14].
2. The negative control group does not receive the plant extract or any activating agent. The vehicle control group is treated with the same volume of solvent used to dissolve the extract (e.g., ethanol as polar solvent), but without the extract itself.

3. A positive control group must also be included, wherein Tc cells are stimulated using Con A without any extracts.
4. The plates are transferred to a 37 °C, 5% CO_2 humidified incubator, to incubate for 48 and 72 h.

3.6 Cytokine Profiling Assay by ELISA

1. Take 100 μL of the capture antibody solution diluted in PBS for each well and pour it into the ELISA plate. Wrap the plate and place it in the fridge overnight at 4 °C to let the antibody attach.
2. Rinse the plate with 300 μL per well of PBST (pH 7.4) three times by either using a multichannel pipette or a plate washer. Next, add 5% BSA in PBS (200 μL per well) for efficient protein binding by blocking buffer to each well. Finally, incubate the plate at room temperature for 1 h.
3. In the sample diluent, make a dilution sequence of the recombinant cytokine standard starting from 1000 pg/mL to 0 pg/mL. Then, use the ELISA plate to dispense 100 μL of every serial standard concentration in duplicate wells.
4. Carefully thaw the originally stored supernatants on ice and give them a short centrifuge to eliminate the presence of any inherent objects. After that, distribute 100 μL of every supernatant sample duplicate to the plate.
5. Use the sealing film to cover the plate and keep it at room temperature for 2 h or, alternatively, move it to the refrigerator at 4 °C and let the cytokine bind overnight.
6. Wash the plate five times with PBST (300 μL per well) to remove unbound proteins.
7. Add 100 μL per well of the biotinylated detection antibody diluted in PBS with 1% BSA and then incubate the plate for 1 h at room temperature.
8. Wash the plate five times with PBST. Add 100 μL per well of streptavidin-HRP conjugate. Incubate for 30 min at room temperature in the dark.
9. Wash the plate five times with PBST. Add 100 μL per well of TMB substrate solution and incubate for 15–20 min at room temperature until the blue color develops.
10. Add 50 μL per well of 1 M sulfuric acid stop solution to the previously made reaction mixture to turn the solution yellow.
11. Scan the plate at 450 nm using a microplate reader and use 570 nm as the reference wavelength.
12. Design a standard graph by the acquisition of the known cytokine concentration values to the absorbance the meter shows.

4 Notes

1. All the chemicals and solvents should be of analytical grade which can be procured from certified suppliers.
2. Store the extract properly to prevent degradation of heat-sensitive phytochemicals as well as take care of mold growth.

References

1. Priya TS et al (2024) Bioactive constituents from Tinospora cordifolia (willd.): isolation, synthesis and their immunomodulatory activity. Fitoterapia 176:105988
2. Uto T et al (2021) Immunological separation of bioactive natural compounds from crude drug extract and its application for cell-based studies. Antibodies 10(4):48
3. Lawal H et al (2022) Moringa oleifera leaf ethanol extract immunomodulatory activity discriminates between chronic myeloid Leukaemia cell line and Normal lymphocytes. Malays J Med Health Sci 18(s21):101–110
4. Chikkamath V et al (2024) In vitro and in vivo immunomodulatory activity of hydro Ethanolic extract of Grewia hirsuta Vahl. Int J Pharm Invest 14(2):539–546
5. Behl T et al (2021) Exploring the multifocal role of phytochemicals as immunomodulators. Biomed Pharmacother 133:110959
6. Rachmawati I, Rifa'I M (2014) In vitro immunomodulatory activity of aqueous extract of Moringa oleifera lam. Leaf to the CD4 +, CD8+ and B220+ cells in Mus musculus. J Exp Life Sci 4(1):15–20
7. Nandan A et al (2023) Deciphering the mechanism of Tinospora cordifolia extract on Th17 cells through in-depth transcriptomic profiling and in silico analysis. Front Pharmacol 13
8. Kassymova D et al (2023) Phytochemical profiles and in vitro immunomodulatory activities of extracts obtained from Limonium gmelinii using different extraction methods. Plants 12(23):4019
9. Tan WS et al (2015) Moringa oleifera flower extract suppresses the activation of inflammatory mediators in lipopolysaccharide-stimulated RAW 264.7 macrophages via NF-κB pathway. Mediators Inflamm 2015(1): 1–11
10. Luetragoon T et al (2020) Bioactive compounds in Moringa oleifera lam. Leaves inhibit the pro-inflammatory mediators in lipopolysaccharide-induced human monocyte-derived macrophages. Molecules 25(1):191
11. Pattanayak P et al (2010) Ocimum sanctum Linn. A reservoir plant for therapeutic applications: an overview. Pharmacogn Rev 4(7):95
12. Venkatachalam, V.V. B. Rajinikanth, Immunomodulatory activity of aqueous leaf extract of Ocimum sanctum. 2013, Springer India. p. 425–432
13. Amirghofran Z et al (2009) Immunomodulatory activities of various medicinal plant extracts: effects on human lymphocytes apoptosis. Immunol Investig 38(2):181–192
14. Ang HY et al (2014) Immunomodulatory effects of Potentilla indica and Dendrophthoe pentandra on mice splenocytes and thymocytes. Exp Ther Med 7(6):1733–1737

Chapter 13

Determination of Immunomodulatory Activity of Beta-Glucan from Barley and Wheat Grains

Jigisha Roy Panda and Tamalika Chakraborty

Abstract

Beta-glucans are active carbohydrates in barley and wheat that boost the immune system. These (1 → 3), (1 → 4)-β-D-glucans work with immune receptors like Dectin-1 and CR3 making the body's defense stronger by helping cells phagocytose toxins. This chapter elaborates on how to extract these hydrocarbons from natural sources, using ultrasound waves and microwaves or hot water under pressure. For purification, DEAE-52 cellulose or ethanol was used. For further characterization, FTIR was used. To check its effects on immune system, RAW 264.7 and THP-1 cells were used in labs and their immune-modulatory chemicals (TNF-α, IL-6, IL-10) were also traced. Adding beta-glucans makes immune cells work better without affecting them, which means they could be beneficial for health as foods and medicines to keep people healthy.

Key words Beta-glucan, Barley, Wheat, Immunomodulation, Extraction, Cytokines, Macrophages, Purification

1 Introduction

Beta-glucans are non-starch polysaccharides composed of glucose units linked primarily through β-(1 → 3) and β-(1 → 4) glycosidic bonds [1–3]. They are present in large amounts in cereal grains such as barley and wheat, as well as in fungi and yeast. These polysaccharides have recently gained attention for their health-promoting, immunity-enhancing effects, including cholesterol reduction, glycemic control, etc. [4, 5]. The immunomodulatory activity of beta-glucans is triggered by pattern recognition receptors on human immune cells, such as Dectin-1 and complement receptor 3 (CR3), resulting in enhanced phagocytosis, cytokine production, and activation of both innate and adaptive immunity [6, 7]. The bioactivity of beta-glucans is influenced by their molecular weight, branching structure, and solubility. In this protocol, an integrated approach has been proposed for advanced extraction of

Mitesh Kumar Dwivedi and A. Sankara Narayanan (eds.), *Leaves as Functional Foods*, Methods and Protocols in Food Science,
https://doi.org/10.1007/978-1-0716-5226-8_13,

beta-glucans through ultrasonic-microwave-assisted extraction (UME) and pressurized hot water extraction (PHW) [8, 9]. This was followed by purification and characterization steps ensuring the isolation of bioactive fractions in a structurally integrated fashion. The in vitro assay of immunomodulatory effects of beta-glucans have been further proposed in this protocol, using macrophage and monocyte cell lines through cytokine assays, nitric oxide production, and phagocytosis, providing a robust framework for evaluating beta-glucan efficacy [10].

2 Materials

2.1 Chemicals and Reagents (see Note 1)

1. Barley or wheat flour.
2. Ethanol (80%, 95%).
3. Sodium hydroxide (NaOH), hydrochloric acid (HCl).
4. DEAE-52 cellulose resin.
5. Phosphate-buffered saline (PBS).
6. Lipopolysaccharide (LPS, positive control).
7. RPMI-1640 medium or DMEM.
8. Fetal bovine serum (FBS).
9. Penicillin-streptomycin antibiotic mix.
10. ELISA kits for IL-6, TNF-α, IL-10.
11. Calcofluor White stain.
12. Trypan Blue solution.
13. MTT reagent.
14. Griess reagent (for nitric oxide assay).
15. FITC-dextran or fluorescent beads (for phagocytosis assay).

2.2 Equipment

1. Ultrasonic-microwave extractor (480 W).
2. High-pressure view chamber for PHW extraction.
3. Centrifuge (capable of 4000–15,000× g).
4. Freeze dryer (lyophilizer).
5. FT-IR spectrometer (e.g., Nicolet iS20).
6. SEC-MALLS system or HPLC-SEC with UV/Vis detector.
7. UV-Vis spectrophotometer.
8. CO_2 incubator (37 °C, 5% CO_2).
9. ELISA plate reader.
10. Inverted fluorescence microscope.
11. Flow cytometer (optional for phagocytosis assay).
12. Laminar flow hood.

13. 96-well culture plates.
14. Pipettes and sterile tips.
15. Cell counter or hemocytometer.

2.3 Cell Lines

1. RAW 264.7 murine macrophages (ATCC).
2. THP-1 human monocytes (differentiated into macrophages with PMA) (*see* **Note 2**).
3. Optional: Peripheral blood mononuclear cells (PBMCs).

3 Methods

3.1 Extraction and Purification of Beta-Glucan

3.1.1 Sample Preparation and Defatting [3]

(a) Grind barley or wheat grains into fine powder (~240 μm particle size).
(b) Screen the powder through a 60-meshsieve to obtain a fine powder.
(c) Suspend 100 g of flour with 80% ethanol (1:10 w/v) and heat it at 80 °C for 4 hours to remove lipids and deactivate enzymes.
(d) Centrifuge at 4000 g for 15 min, discard supernatant, and keep the pellet for further use.
(e) Dry the pellet at 40 °C.

3.1.2 Ultrasonic-Microwave-Assisted Extraction (UME)

(a) Mix 1 gm of flour in 250 mL distilled water.
(b) Sonicate the suspension at 480 W for 20 min.
(c) Follow with microwave treatment at 480 W for 4 min.
(d) Centrifuge at 4000 g for 20 min, collect supernatant.
(e) Concentrate the supernatant under vacuum.
(f) Precipitate beta-glucan by adding 95% ethanol (1:2 v/v), incubate at 4 °C overnight.
(g) Centrifuge at 4000 g for 15 min, wash pellet with ethanol.
(h) Freeze-dry the pellet with the help of a lyophilizer.

3.1.3 Pressurized Hot Water Extraction (PHW) (Alternative Method)

(a) Load 1 g of milled flour into a high-pressure chamber.
(b) Add distilled water to fill chamber, seal.
(c) Heat to 155 °C, maintain 20 bar pressure for 18 min with stirring.
(d) Cool and decompress chamber.
(e) Centrifuge at 10,000 g for 10 min.
(f) Precipitate beta-glucan with 3 volumes of 95% ethanol at 4 °C overnight.
(g) Collect precipitate by centrifugation, wash with ethanol, dry under vacuum.

3.1.4 Purification via DEAE-52 Cellulose Chromatography

(a) Dissolve crude beta-glucan in distilled water (20 mg/mL).
(b) Load onto DEAE-52 column.
(c) Elute with 0–0.5 M NaCl gradient.
(d) Collect fractions, dialyze (3.5 kDa cutoff), and lyophilize.

3.2 Structural Characterization [4]

3.2.1 FT-IR Spectroscopy

(a) Mix 1 mg purified beta-glucan with KBr, press into pellet.
(b) Scan 4000–400 cm^{-1}.
(c) Identify characteristic β-glycosidic bonds (~895 cm^{-1}) and hydroxyl groups (~3400 cm^{-1}).

3.2.2 Monosaccharide Composition Analysis

(a) Hydrolyze 5 mg beta-glucan in 2 M trifluoroacetic acid (TFA) at 121 °C for 2 h.
(b) Analyze by high-performance anion-exchange chromatography with pulsed amperometric detection (HPAEC-PAD).

3.2.3 Molecular Weight Determination (SEC-MALLS or HPLC-SEC)

(a) Dissolve beta-glucan in 0.1 M $NaNO_3$, filter (0.45 μm).
(b) Inject into SEC system.
(c) Determine molecular weight (Mw) and polydispersity.

3.3 Immunomodulatory Activity Assays [5]

3.3.1 Cell Culture and Treatment

(a) Culture RAW 264.7 or THP-1 cells in RPMI-1640 or DMEM supplemented with 10% FBS and 1% penicillin-streptomycin.
(b) Seed 1×10^5 cells/well in 96-well plates.
(c) Incubate overnight for adherence.
(d) Treat with beta-glucan (1–200 μg/mL), LPS (1 μg/mL, positive control), or PBS (negative control).
(e) Incubate for 24–48 h.

3.3.2 Cell Viability Assay (MTT)

(a) Add MTT reagent (0.5 mg/mL) to wells.
(b) Incubate 4 h at 37 °C.
(c) Dissolve formazan crystals in DMSO.
(d) Measure absorbance at 570 nm.

3.3.3 Cytokine Quantification by ELISA

(a) Collect culture supernatants.
(b) Quantify TNF-α, IL-6, and IL-10 using ELISA kits per manufacturer's protocol.

3.3.4 Nitric Oxide Production Assay

(a) Mix supernatant at a ratio of 1:1 with Griess reagent.
(b) Measure absorbance at 540 nm.
(c) Compare to sodium nitrite standard curve.

Phagocytosis Assay (Optional)

(a) Incubate beta-glucan-treated cells with FITC-dextran or fluorescent beads (1 mg/mL) for 1 h.

(b) Wash cells and analyze fluorescence by flow cytometry or fluorescence microscopy.

4 Notes

1. All the chemicals and solvents should be of analytical grade which can be procured from certified suppliers.
2. Before starting the work, ensure the sterility of cell line.

Acknowledgments

The authors are grateful to the honorable director Prof. (Dr) Abhijit Sengupta and respected principal Prof. (Dr) Lopamudra Datta of Guru Nanak Institute of Pharmaceutical Science and Technology for their support while preparing the manuscript.

References

1. Zhu Y et al (2016) Beta-glucan recognition by immune cells: receptors and signaling pathways. Front Immunol 7:1–13. https://doi.org/10.3389/fimmu.2016.00001
2. Chen L et al (2024) Ultrasonic-microwave-assisted extraction of beta-glucan from barley: optimization and bioactivity. Food Chem 400: 134123. https://doi.org/10.1016/j.foodchem.2023.134123
3. Smith JA et al (2022) Pressurized hot water extraction of cereal beta-glucans: Yield and molecular weight analysis. J Agric Food Chem 70(15):4567–4575. https://doi.org/10.1021/acs.jafc.2c01023
4. Brown GD et al (2003) Dectin-1 is a major beta-glucan receptor on macrophages. J Exp Med 197(9):1119–1124. https://doi.org/10.1084/jem.20021752
5. Novak M, Vetvicka V (2008) Beta-glucans, history, and the present: Immunomodulatory aspects and mechanisms of action. J Immunotoxicol 5(1):47–57. https://doi.org/10.1080/15476910701726902
6. Williams DL (2007) Immunomodulation by fungal beta-glucans: the role of the receptor Dectin-1. Microbes Infect 9(6):619–625. https://doi.org/10.1016/j.micinf.2007.03.002
7. Vetvicka V, Vetvickova J (2011) Beta-glucan as an immunomodulator. Molecules 16(3): 2346–2361. https://doi.org/10.3390/molecules16032346
8. Volman JJ et al (2008) Dietary beta-glucans and their immunomodulatory properties. Nutr Rev 66(4):237–251. https://doi.org/10.1111/j.1753-4887.2008.00032.x
9. Vetvicka V, Vetvickova J (2014) Effects of beta-glucan on immune system. Ann Transl Med 2(4):1–7. https://doi.org/10.3978/j.issn.2305-5839.2014.04.01
10. Chan GC et al (2009) The effects of beta-glucan on human immune and cancer cells. J Hematol Oncol 2:25. https://doi.org/10.1186/1756-8722-2-25

Chapter 14

Determination of Immunomodulatory Activities of Seaweed-Derived Bioactive Components

Jeenatara Begum, Tamalika Chakraborty, Rajarshi Das, and Piyali Basak

Abstract

The immune system is the primary defense mechanism that protects the body from infections, as well as it contributes to the overall well-being of organisms. Seaweeds are found to contain many bioactive compounds, such as polysaccharides, phlorotannins, fucoidans, and sulfated galactans, with immune-modulatory activity. These compounds appear to potentially enhance both innate and adaptive immune responses such as pseudopodia formation, cytokine production, and modulation of immune signaling pathways. Studying the bioactive components from seaweeds for immunomodulatory activity opens a new avenue toward the development of natural therapeutic agents, which could immunostimulate efficacy for managing immune dysfunction. The chapter aims to understand the possible mechanisms through which these effects take place and emphasize the potential of seaweed-based products for immune health.

Key words Immune modulation, Seaweed, Polysaccharides, Phlorotannins, Fucoidans, Cytokines, Immune response

1 Introduction

Marine organisms have emerged with intense and possible health and "bioactive" actions. Seaweeds contain a number of polysaccharides, polyphenols, proteins, peptides, and sulfated carbohydrates which are known to be endowed with antioxidant, anti-inflammatory, antimicrobial, and antitumoristic properties. Thus, the seaweed-derived bioactive compounds seem to bring about a promising immunomodulatory effect through the enhancement of innate immunity [1].

Innate immune system thus has a considerable role in homeostasis and the subsequent defense of the body against infection. However, homeostatic balance is disrupted by aberration of immune responses leading to chronic inflammation, autoimmune

Mitesh Kumar Dwivedi and A. Sankara Narayanan (eds.), *Leaves as Functional Foods*, Methods and Protocols in Food Science,
https://doi.org/10.1007/978-1-0716-5226-8_14,

diseases, or increased risk for infections; therefore, it has fueled a demand for natural immunomodulators. All these processes were modulated by bioactive compounds such as fucoidan, laminarin, and carrageenan, which stimulated innate and adaptive immune responses by activating macrophages, stimulating NK cells, or shaping cytokine production through Toll-like receptors and other immune pathways [1, 2].

Instead, the marine biotechnological extracts act as natural immunomodulators to fulfil natural health products amid increasing infectious disease burdens and a declining efficacy of antibiotics [3].

2 Materials

2.1 For Seaweed Extraction

- Seaweed.
- Extraction solvents—ethanol/methanol (*see* **Note 1**).

2.2 Cell Lines and Culture Media

- RAW 264.7 mouse macrophage cell line (*see* **Note 2**).
- THP-1 human monocytic cell line (*see* **Note 2**).
- RPMI-1640 medium, supplemented with 10% fetal bovine serum (FBS) & 1% penicillin-streptomycin (*see* **Note 3**).

2.3 Reagents and Kits

- MTT reagent (*see* **Note 4**).
- Griess reagent (*see* **Note 5**).
- IL-6 ELISA kit (*see* **Note 6**).

2.4 Equipment

- 96-well culture plates.
- Cell incubator (37 °C, 5% carbon dioxide (CO_2)) (*see* **Note 7**).
- Microplate reader—for absorbance readings (*see* **Note 8**).
- BD FACS analyzer (BD Biosciences)—for flow cytometric analysis.
- FlowJo software (FlowJo LLC)—for flow cytometry data analysis.

2.5 Other Minor Equipment and Accessories

- Micro pipettes (100 μL, 1000 μL).
- Centrifuge (Eppendorf, Model: 5424).
- Sterile cell culture flasks.
- Disposable gloves, lab coats, and safety goggles.

3 Methods

3.1 Preparation of Seaweed Extracts (see Note 9) [2, 3]

3.1.1 Collection and Cleaning

- Seaweed samples were sourced from the coastal areas and thoroughly rinsed with distilled water to get rid of dirt and surface impurities. Clean samples were shade-dried at room temperature for 48 h (*see* **Note 10**).

3.1.2 Extraction

- The dried seaweed was finely ground. Thereafter, 10 g of it was extracted with 100 mL 90:10 (v/v) ethanolic or methanolic solution by vigorous stirring at room temperature for 24 h (150 rpm) (*see* **Note 1**). Then, the extract was then passed through a filter and concentrated at 40 °C under a rotary evaporator under vacuum and stored at −20 °C for further use.

3.2 Cell Culture and Maintenance (see Note 11) [4]

3.2.1 Cell Lines Preparation

1. The two immune-related cell lines utilized for the experiment were RAW 264.7 and THP-1.
2. The cells were maintained in RPMI-1640, a nutrient-bearing medium.
3. The constituents of the medium comprise:
   (a) 10% Fetal bovine serum (FBS), which serves as a source of nutrients and growth factors.
   (b) 1% Penicillin-streptomycin to prevent bacterial contamination (*see* **Note 3**).
4. The cells were incubated at 37 °C (body temperature) in an incubator with humidity control and 5% carbon dioxide (CO_2) gas to enable a condition that resembles that of the human body (*see* **Note** 7).

3.2.2 Subculture

1. Grow and multiply the cells over time.
2. Cells were ready for transfer to a new container upon attaining approximately 70–80% confluency within 2–3 days after their initial seeding. This measure was taken to prevent overcrowding.
3. To detach cells from container surfaces: A 0.05% trypsin–EDTA solution was added (*see* **Note 12**). Trypsin, a protein-digesting enzyme, disintegrates protein barriers and allows cells to separate.
4. The released cells were then transferred to fresh nutrient medium to continue their growth.

3.2.3 Cell Viability Assay ("MTT" Assay) (see **Note 13**) [5]

1. Seeding the cells.
2. Cells (RAW 264.7 and THP-1) were added into a well of a 96-well culture plate at a concentration of 1×10^4 cells per well (or 10,000 cells in a well).
3. The culture well was kept for 24 h for cells to get attached and settle at the bottom of the well.

3.3 Treatment with Seaweed Extract

1. The seaweed extract concentrations were applied to cells, 10, 50, 100, and 200 μg/mL, after a period of 24 h of attachment.
2. The cells that were not treated with extract serves as control (the cells should be those exposed only to culture medium without extract).

3.4 MTT Assay (Fig. 1)

1. After 24 h of treatment, 10 μL of yellow MTT reagent was added to each culture well (*see* **Note 4**).

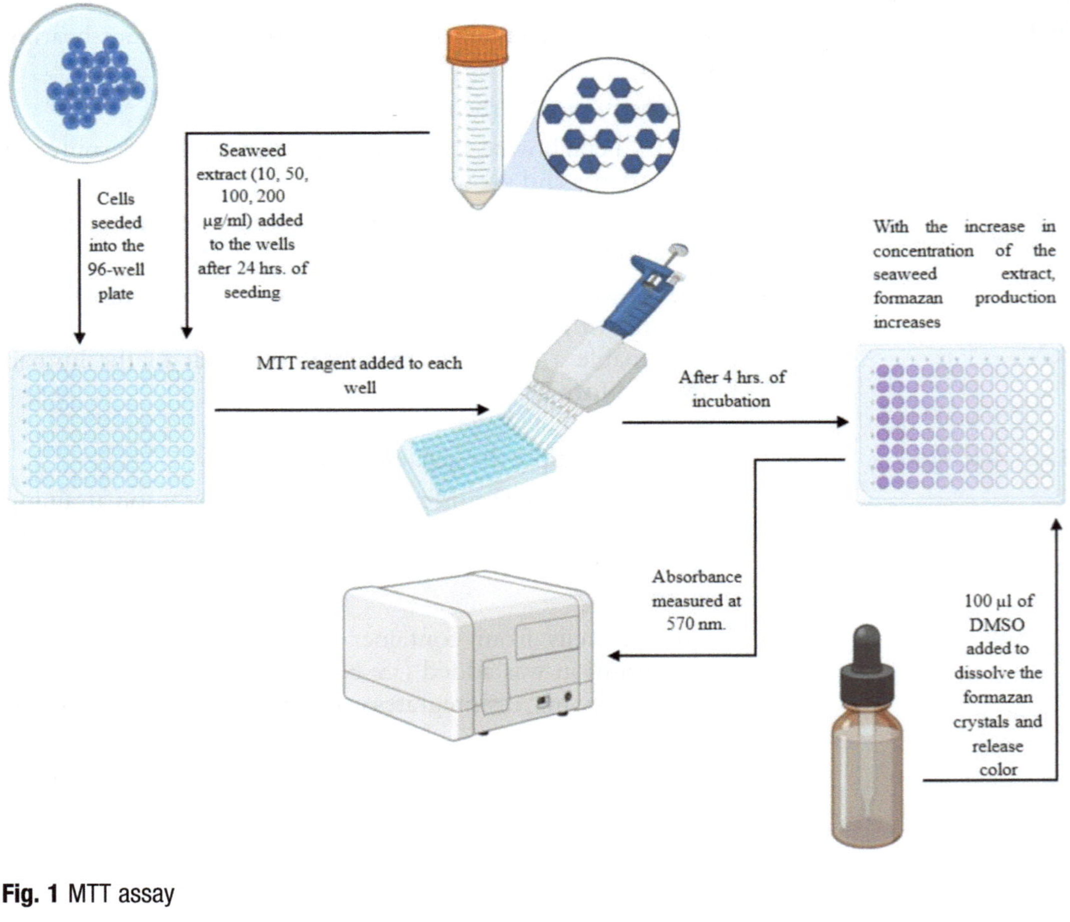

Fig. 1 MTT assay

2. This is a chemical that the viable cells can cleave into purple formazan crystals.
3. These plates were then incubated for 4 h at 37 °C to allow a proper reaction time.
4. Following 4 h of incubation, the formazan created by living cells was dissolved in 100 μL of DMSO (dimethyl sulfoxide) and distributed into each of the culture wells.
5. DMSO is used to solubilize the crystals and free the purple color into solution.
6. Colorimetric concentration was used to assess the absorbance of the samples using a microplate reader. The absorbance was recorded at a wavelength of 570 nm to quantify color intensity (which correlates with live cells) (*see* **Note 8**).
7. The more the color intensity, the more the living cells.

3.5 Calculating Cell Viability

$$\text{Cell Viability (\%)} = \frac{\text{Absorbance of treated cells}}{\text{Absorbance of control cells}} \times 100$$

3.6 Nitric Oxide (NO) Production Assay (see Note 14) [6, 7](Fig. 2)

3.6.1 Stimulation and Treatment

1. The RAW264.7 and THP-1 cell lines were plated into each plate of 96-well culture plates each at a density of 1×10^5 cells per well (i.e., 100,000 cells per well).
2. These many cells are needed to accurately measure nitric oxide production.
3. On settlement, cells were treated for an hour with the seaweed extract.
4. This was intended to allow interaction of the extract with the cells and possibly prep them for an immune response.
5. After this pretreatment period, the cells were subjected to 1 μg/mL lipopolysaccharide (LPS) for 24 h.
6. LPS is a lining surface polysaccharide found in both Gram-positive and Gram-negative bacteria that evokes an immune response through macrophages.
7. This step essentially allows one to test whether the seaweed extract potentiates or modulates the immune response in cells under stress (inflammation).

3.6.2 Griess Reagent Reaction

1. The supernatants from all wells after 24 h incubation would have contained the nitric oxide (NO) released from the cells.
2. 50 μL of the supernatant were taken for the assay.
3. The supernatant sample obtained was mixed with an equal volume of Griess reagent (50 μL) (*see* **Note 5**).

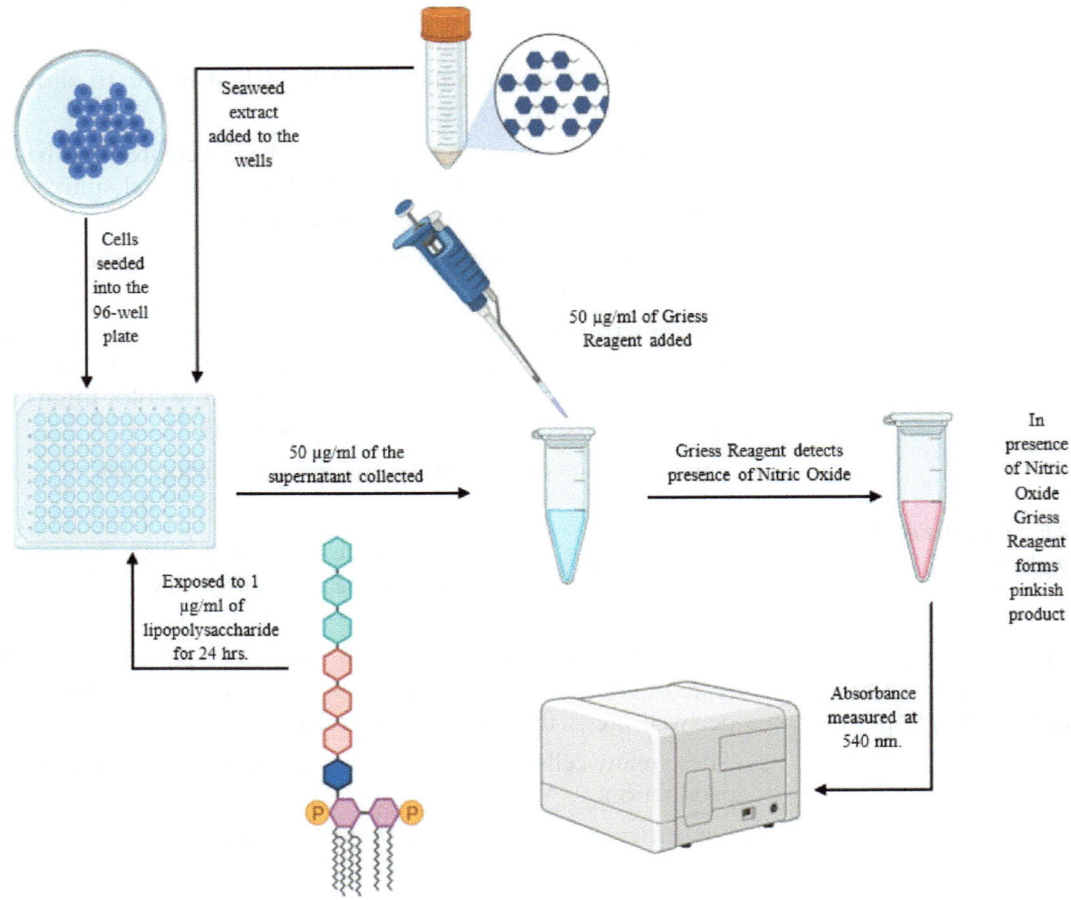

Fig. 2 NO production assay

4. The Griess reagent detects the presence of NO by means of developing a pinkish/purple product.
5. The mixture was incubated for a full 10 min at room temperature for complete development of the reaction.
6. Finally, the intensity of absorbance was read at 540 nm in a microplate reader, correlated to the NO production (*see* **Note 8**).
7. More color = more nitric oxide produced = more immunoresponse.
8. Each absorbance value was associated with the amount of nitric oxide determined from an established standard curve of sodium nitrite, while the level of nitric oxide present was determined according to the standard curve defined from sodium nitrite (*see* **Note 15**).

3.7 Cytokine Production (IL-6) Assay (see Note 16) [7, 8] (Fig. 3)

3.7.1 Stimulation and Treatment

1. The RAW264.7 and THP-1 cell lines were plated into each well of 96-well culture plate at a density of 1×10^5 cells per well (1×10^5 cells per well); this allows for the good detection of any cytokine production.
2. The cells were treated with the seaweed extract 1 h before stimulation.
3. This allows bioactive compounds from the extract to be absorbed and elicit a response from the cells.
4. After the 1-h pre-treatment time, the cells were treated with lipopolysaccharide (LPS) for the stimulation at a concentration of 1 μg/mL for 24 h.

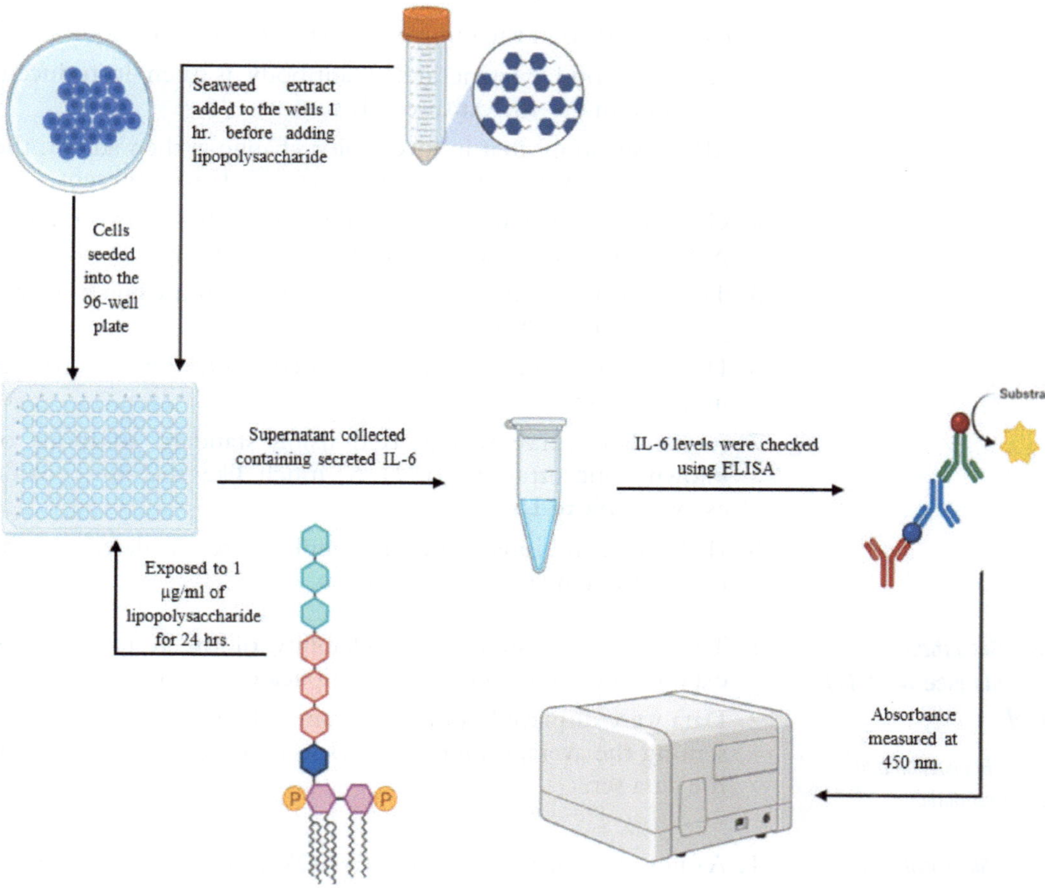

Fig. 3 Cytokine production assay

5. LPS helps to produce an immune response from macrophages, causing the production and secretion of pro-inflammatory cytokines like IL-6.
6. These tests help understand the ability of the extract to enhance or downregulate the immune responses while the LPS treatment is done.

3.7.2 ELISA Measurement

1. The supernatants were carefully withdrawn after 24 h of incubation from all wells containing secreted IL-6.
2. IL-6 levels were determined using an ELISA kit (enzyme-linked immunosorbent assay) (*see* **Note 6**).
3. The procedure involves the following steps:
   (a) The supernatants are added to plates coated with IL-6-specific antibodies.
   (b) Any IL-6 in the sample will bind to those antibodies.
   (c) A second enzyme-linked antibody is then introduced, binding to the captured IL-6.
   (d) A substrate that induces color change is then added; the enzyme will induce a color change if IL-6 is present.
4. Given this, the change in the intensity of the color is directly proportional to the amount of IL-6 in the sample.
5. The intensity of the color was read at 450 nm using a microplate reader (*see* **Note 8**).
6. Higher absorbance = higher IL-6 concentration = stronger immune response.
7. A standard curve was plotted using standard solutions of known concentrations of IL-6, much like the nitric oxide assay (*see* **Note 15**).
8. IL-6 concentrations in the test samples were calculated according to the standard curve equation.

3.8 Statistical Analysis (see Note 17) [5–8]

3.8.1 Replication and Data Presentation

1. To ensure consistency and reliability of the results, all the experiments were conducted in triplicate ($n = 3$).
2. Data were displayed as mean ± standard deviation (SD) representing the average value with respect to the variability within the data set.

3.8.2 Statistical Testing

1. A one-way analysis of variance (ANOVA) was conducted to test how various groups differ, for example, among different concentrations of seaweed extract.
2. When an ANOVA identified a significant difference, the Tukey post hoc test was applied to differentiate which specific groups differed from each other.

3. Level of Significance: The level of significance was set as $p < 0.05$, which implied that there would be a less than 5% chance that the difference observed was due to random chance (*see* **Note 18**).

4 Notes

1. *Ethanol/Methanol (90%: 10%, v/v)*: This particular solvent combination provides very good extractive capability for the majority of phytochemicals present in seaweed. Such chemicals include polyphenols, flavonoids, alkaloids, and other bioactive ones. Polar solvents like ethanol solubilize hydrophilic compounds, while methanol adds solubility for slightly less polar components, further maximizing extraction yield and effectiveness.
2. *RAW 264.7 and THP-1 cell lines*: They are immune-related cell lines for the evaluation of immunomodulatory properties. From the murine macrophage cell line, RAW 264.7 assesses endogenous NO and cytokines. On the other hand, THP-1 is a human monocytic leukemia cell line along which cells can be induced to differentiate into macrophage like by stimulant agents (such as PMA). These lines provide direct yet translational culture systems for in vitro examination of the anti-inflammatory response of plant or marine extracts.
3. *RPMI 1640 medium + supplements:* RPMI 1640 provides essential nutrients, amino acids, and vitamins for metabolism and proliferation of immune cells. Adding 10% fetal bovine serum (FBS) to the media would supply proteins, hormones, and attachment factors to the cells, while 1% penicillin-streptomycin sustains the culture to inhibit bacterial contamination for a sterile and consistent condition during any experiments with reproducible results.
4. *MTT reagent (3-(4,5-Dimethylthiazol-2-yl)-2,5-diphenyltetrazolium bromide):* A yellow dye reduced by the action of succinate dehydrogenase mitochondrial enzymes of viable cells forming crystalline purple formazan. This assay is carried out by dissolving them in either DMSO or acidified isopropanol and measuring absorbance reading at 570 nm. It is a standard assay used to find out cytotoxicity, proliferation, or viability in cells that have been exposed to test substances.
5. *Griess reagent (sulfanilamide + NED)*: This assay is a two-step colorimetric assay that quantifies nitrite, a stable and non-volatile breakdown product of nitric oxide (NO), which is one of the most important mediators in inflammation. Nitrite reacts with sulfanilamide and NED (N-(1-naphthyl)

ethylene diamine) to cause the formation of purple azo measurable at 540 nm. It provides indirect evidence of macrophage activation and oxidative stress.

6. *IL-6 ELISA kit:* It captures and detects cytokines in supernatants by an enzyme-linked immunosorbent assay using monoclonal antibodies specific to IL-6. IL-6, a pro-inflammatory cytokine released by macrophages after stimulation (e.g. by LPS), modulation of which by natural compounds can provide information on the anti-inflammatory or immunomodulatory potential of extracts.
7. *Incubator conditions (37 °C, 5% CO_2):* These conditions replicate the physiological environment inside the human body. 5% of CO_2 maintains the pH of the culture medium via the bicarbonate buffer system, ensuring cells remain healthy and functionally active throughout the experiment.
8. *Microplate reader:* A spectrophotometric device designed for high-throughput analysis of colorimetric assays. Correct selection of wavelength (570 nm for MTT, 540 nm for Griess, and 450 nm for ELISA) leads to the precise quantification of analytes based on absorbance. The data from the reader is imperative for plotting standard curves applicable to experimental outcome deduction.
9. *Seaweed extract preparation:* Involves drying under shade so as to conserve bioactivity, and solvent extraction performed under room temperature to prevent heat-sensitive compounds' degradation. Gentle stirring under solid suspension promotes diffusion of compounds into the solvent, thereby enhancing extraction efficiency without causing damage to phytochemicals.
10. *Shade drying:* Contrary to sun drying, shade drying protects the thermolabile compounds (e.g., vitamins, phenols) against photodegradation and thermal degradation. This way, biological activity of compounds pertinent to anti-inflammatory and antioxidant action is preserved.
11. *Cell culture maintenance:* Cultures must be gibberish every 2–3 days to avoid over-confluency, nutrient depletion, or contamination. Maintenance including aseptic techniques within the biosafety cabinet ensures cell health, reproducibility, and experimental validity.
12. *Trypsin-EDTA solution:* Used to detach adherent cells from the cell culture by degrading proteins involved in adhesive interactions at the cell surface. Trypsin cleaves peptide bonds; EDTA chelates divalent cations (Ca^{2+}, Mg^{2+}) needed for cell adhesion, thereby providing a model for non-damaging yet effective cell harvesting for downstream application.

13. *Principle of MTT assay (Cell viability):* It determines the metabolic activity of cells, which serves as an indirect parameter for cell viability evaluation. Accordingly, if there is a decline in the formazan formation, it means that there is a corresponding decline in cell viability, which is usually an effect due to cytotoxicity of the test substances. This assay is now extensively used in pharmacology and toxicological evaluation processes.
14. *Nitric oxide (NO) in inflammation:* Generated by the inducible nitric oxide synthase (iNOS) in activated macrophages, it serves as both a defensive weapon molecule and an inflammatory mediator. Therefore, damage done by excess NO production in oxidative insults and prolonged inflammation makes it a good candidate for anti-inflammatory screening purposes.
15. *Standard curve generation:* Is about plotting the absorbance values against known concentrations for a certain reference standard so as to form a calibration curve. Interpolation concentrations for samples can then be derived from experimental readings, enabling accurate and quantitative data in colorimetric assays like Griess and ELISA.
16. *IL-6 as a major inflammatory marker:* IL-6 plays as a key mediator for the acute-phase immune response and is involved in the pathogenesis associated with a range of chronic inflammatory diseases. Thus, natural products that lower IL-6 levels are among potential candidates for causing beneficial effects on inflammation.
17. *Triplicate analysis and statistical testing (ANOVA + Tukey's* post hoc*):* Carrying out triplicate experimentation decreases variability and increases confidence of reproducibility. ANOVA determines whether the variations between different groups are significant while Tukey's test further identifies where differences exist among the means.
18. *Use of licensed statistical software:* Eventually, all such analyses draw upon appropriate software for data analysis, data graphical presentations, and the significance test, such as GraphPad Prism, SPSS, or Origin: instruments that ensure rigorous, reproducible, and publication-worthy statistical interpretation.

References

1. Kraiem M, Ben Hamouda S, Eleroui M, Ajala M, Feki A, Dghim A, Boujhoud Z, Bouhamed M, Badraoui R, Pujo JM, Essafi-Benkhadir K, Kallel H, Ben Amara I (2024) Anti-inflammatory and immunomodulatory properties of a crude polysaccharide derived from green seaweed *Halimeda tuna*: computational and experimental evidences. Mar Drugs 22(2): 85. https://doi.org/10.3390/md22020085
2. Lomartire S, Gonçalves AMM (2022) An overview of potential seaweed-derived bioactive compounds for pharmaceutical applications. Mar Drugs 20(2):141. https://doi.org/10.3390/md20020141
3. Qin L, Xu H, He Y, Liang C, Wang K, Cao J, Qu C, Miao J (2022) Purification, chemical characterization and immunomodulatory activity of a sulfated polysaccharide from Marine

Brown Algae *Durvillaea*. Mar Drugs 20(4):223. https://doi.org/10.3390/md20040223

4. Wang L, Je J-G, Huang C, Oh J-Y, Fu X, Wang K, Ahn G, Xu J, Gao X, Jeon Y-J (2022) Anti-inflammatory effect of sulfated polysaccharides isolated from *Codium fragile* in vitro in RAW 264.7 macrophages and in vivo in zebrafish. Mar Drugs 20(6):391. https://doi.org/10.3390/md20060391
5. Mahendran S, Sankaralingam S, Sethu SM, Kathiresan D, Muthumani M, Kousalya L, Palpperumal S, Harinathan B (2022) Evaluation of antioxidant and cytotoxicity activities of polyphenol extracted from brown seaweed *Sargassum tenerrimum* biomass. Biomass Convers Biorefinery 14(2):2063–2069. https://doi.org/10.1007/s13399-022-02301-x
6. Premarathna AD, Tuvikene R, Somasiri M, De Silva M, Adhikari R, Ranahewa T, Wijesundara R, Wijesekera S, Dissanayake I, Wangchuk P, Rjabovs V, Jayasooriya AP, Rajapakse R (2023) A novel therapeutic effect of mannitol-rich extract from the brown seaweed *Sargassum ilicifolium* using in vitro and in vivo models. BMC Complementary Med Ther 23(1). https://doi.org/10.1186/s12906-023-03840-0
7. Sanjay, Yoon NY, Park E-J, Lee H-J (2024) Immune-enhancing effects of marine algae extracts: modulation of macrophage activation by Sargassum horneri, Sargassum fusiforme, and Undaria pinnatifida. Appl Sci 14(5):1794. https://doi.org/10.3390/app14051794
8. Sharma PP, Chonche MJ, Mudhol S, Muthukumar SP, Baskaran V (2023) Anti-inflammatory efficacy of brown seaweed (*Padina tetrastromatica*) in 3T3-L1 adipocytes and low-dose LPS induced inflammation in C57BL6 mice. Algal Res 71:103027. https://doi.org/10.1016/j.algal.2023.103027

Part III

Leaf and Human Health

Chapter 15

Determination of Polysaccharides from Seaweed and Its Influence on Human Gut Microbiota

Tamalika Chakraborty, Sobhanjan Bhunia, Priyanka Ray, Sumana Roy, and Sriparna Kundu Sen

Abstract

Seaweed-derived polysaccharides have attracted more attention due to their possible prebiotic qualities and positive health impacts. The extraction, identification, and measurement of polysaccharides from particular seaweed species are the main objectives of this work. An in vitro evaluation of the polysaccharide's effects on the human gut microbiota is conducted after they have been extracted, characterized, and quantified from a few chosen seaweed species. Ethanol precipitation and dialysis were used to purify the polysaccharides after they were isolated by the use of hot water extraction. A structural study was performed to determine the composition of monosaccharides using FTIR (Fourier-transform infrared spectroscopy) and HPLC (High-performance liquid chromatography). Human fecal inocula were used in in vitro fermentation experiments to evaluate changes in the structure of the microbial community by the use of 16S rRNA gene sequencing. The results showed that seaweed polysaccharides, especially sulfated galactans and fucoidans, decreased the number of calculating pathogens while highly improving the growth of *Lactobacillus* and *Bifidobacterium*, which are two types of healthy gut bacteria. According to these results, seaweed polysaccharides may act as functional prebiotics that support a more wholesome gut flora and provide a basis for dietary therapies targeted at enhancing metabolic and gastrointestinal health.

Key words Prebiotics, Gut microbiota, Polysaccharide extraction, Dialysis, FTIR, HPLC, Human fecal inocula, 16S rRNA gene sequencing, Sulfated galactans, Fucoidans, Gut bacteria

1 Introduction

The human gut microbiota is affected in immunological response, digestion, and mental health, which makes it an essential component of total health maintenance. Numerous illnesses including inflammatory bowel disease, obesity, diabetes, and colon cancer have been connected to dysbiosis, an imbalance in this complex microbial ecosystem. A lot of attention has been focused on finding natural chemicals with prebiotic potential substances that specifically support the growth and activity of beneficial gut microbes

Mitesh Kumar Dwivedi and A. Sankara Narayanan (eds.), *Leaves as Functional Foods*, Methods and Protocols in Food Science,
https://doi.org/10.1007/978-1-0716-5226-8_15,

because the diet is a significant modulator of gut microbial composition [1].

The structurally varied polysaccharides are found in seaweeds, and it is also known as marine macroalgae. It is fermentable by gut bacteria but mostly indigestible by human enzymes. Alginates, carrageenans, fucoidans, and ulvans are examples of polysaccharides that differ depending on the type of seaweed and the surrounding environment. Seaweed polysaccharides are in contrast to fibers from terrestrial plants that frequently include uncommon sugars and special sulfate groups, which may have different impacts on the gut microbiota [1].

The connection between seaweed polysaccharides and gut microbial regulation is still poorly understood even with an increased interest. The objective of this work is to use in vitro fermentation models to extract and analyze the polysaccharides from certain seaweed species and to examine their effects on the human gut flora. This work advances the development of innovative dietary therapies to promote gut health and prevent disorders linked to the microbiota by finding chemicals derived from seaweed that have prebiotic potential [1].

2 Materials

2.1 Materials for Polysaccharide Extraction and Characterization

I. Seaweed Samples.
   - Fresh or dried seaweed (e.g., *Gracilaria*, *Ulva*, *Sargassum*).

II. Chemicals and Reagents (*see* **Note 1**).
   - Distilled water.
   - Chloroform/methanol (2:1).
   - Ethanol (95% or absolute).
   - Phenol-sulfuric acid.
   - Concentrated sulfuric acid (H_2SO_4).
   - Glucose.

III. Equipment.
   - Hot plate.
   - Water bath.
   - Centrifuge.
   - Incubator.
   - UV-Vis spectrophotometer.
   - pH meter.

2.2 Materials for Microbiota Influence Study (In Vitro Fermentation)

I. Fecal Microbiota Source.
 - Fresh human fecal samples (collected from healthy volunteers under ethical approval).

II. Anaerobic Culture Media.
 - Brain heart infusion (BHI) broth or other fermentation media.
 - Yeast extract.
 - Peptone.
 - Phosphate-buffered saline (PBS).

III. Fermentation Setup.
 - Anaerobic chamber or jars with gas packs ($CO_2/H_2/N_2$).
 - Serum bottles or sealed glass vials.
 - Gas-permeable stoppers or butyl rubber septa.
 - Incubator set to 37 °C.

IV. Microbiota Analysis Tools.
 - DNA extraction kit (e.g., Qiagen QIAamp Fast DNA Stool Mini Kit).
 - PCR reagent (primer for 16S rRNA gene, Taq polymerase).
 - Next-generation sequencing platform (Illumina MiSeq or similar).
 - Bioinformatics software for microbiota analysis (QIIME2, SILVA, or Greengenes database).
 - GC or HPLC for SCFA (short-chain fatty acid) profiling.

3 Methods

3.1 Extraction and Characterization of Seaweed Polysaccharides

1. Sample Preparation [2].
   - Collect dried seaweed samples (e.g., *Gracilaria*, *Ulva,* and *Sargassum*).
   - Wash thoroughly with distilled water to remove salts, sand, and epiphytes.
   - Air dry or oven dry at 50–60 °C.
   - Grind into fine powder and store in airtight containers (*see* **Note 2**).
2. Defatting and Depigmentation [2].
   - Weigh ~10 g of powdered seaweed.
   - Soak in chloroform: methanol (2:1 v/v) overnight to remove pigments and lipids.
   - Filter and dry the residue.

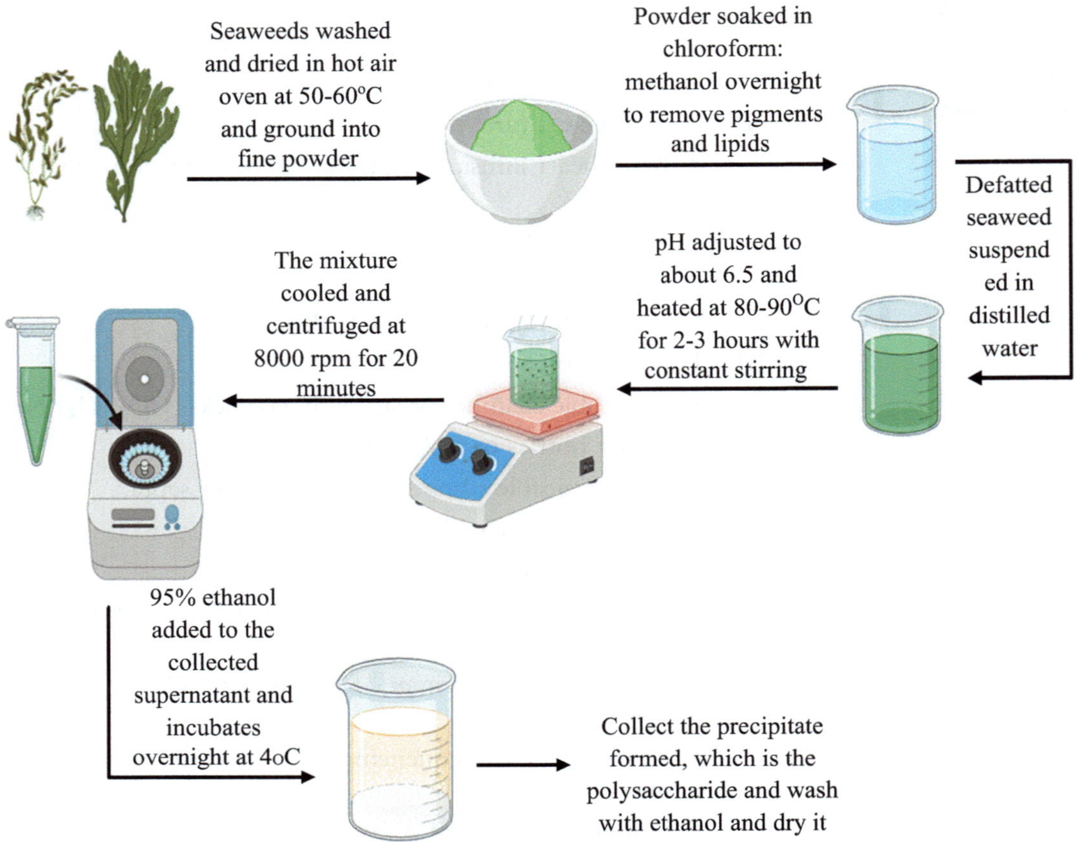

Fig. 1 Extraction of seaweed polysaccharides

3. Polysaccharide Extraction [2].
   - Suspend the defatted seaweed in distilled water (1:20 w/v).
   - Adjust pH to ~6.5 and heat at 80–90 °C for 2–3 h with constant stirring.
   - Cool and centrifuge at 8000 rpm for 20 min.
   - Collect the supernatant (Fig. 1).
4. Precipitation [2, 3].
   - Add 3 volumes of 95% ethanol to the supernatant to precipitate polysaccharides.
   - Keep the mixture at 4 °C overnight.
   - Centrifuge and collect the precipitate.
   - Wash with ethanol and dry (air-dry or freeze-dry).

5. Quantification of Total Polysaccharides [3].
   - Use the *phenol-sulfuric acid method*:
   - – Mix 1 mL sample with 1 mL of 5% phenol and 5 mL concentrated H_2SO_4.
   - – Incubate for 30 min at room temperature.
   - – Measure the absorbance at 490 nm using a UV-Vis spectrophotometer.
   - – Use glucose as standard.

3.2 In Vitro Fermentation to Assess Gut Microbiota Modulation

1. Fecal Inoculum Preparation [3].
   - Collect fresh feces from healthy volunteers in anaerobic conditions.
   - Dilute in sterile phosphate-buffered saline (1:10 w/v).
   - Homogenize and filter through sterile gauze.
   - Use immediately or store at −80 °C with 10% glycerol (Fig. 2).

2. Fermentation Setup [3].
   - Prepare anaerobic fermentation media (e.g., BHI + yeast extract + vitamins + minerals).
   - 50 mL of media was transferred into sterile serum bottles or flasks.
   - Add extracted seaweed polysaccharide (e.g., 1% w/v final concentration).

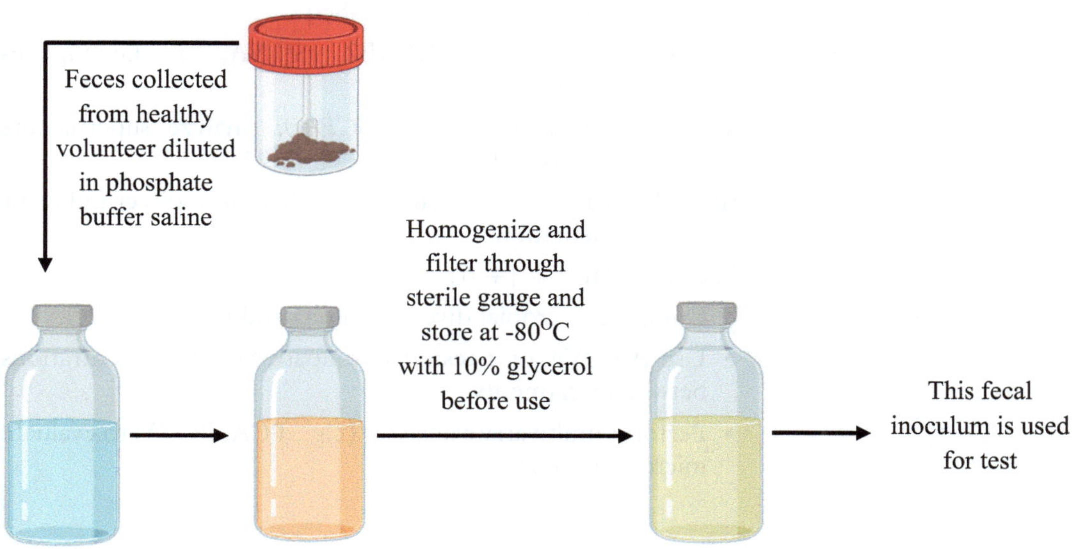

Fig. 2 Fecal inoculum preparation

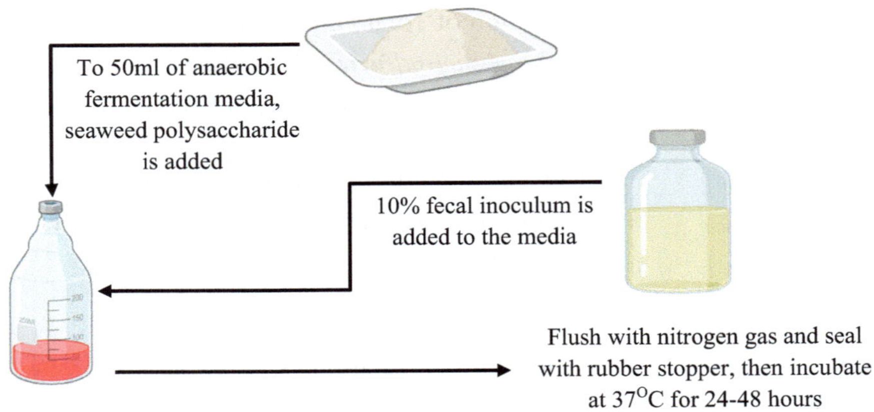

Fig. 3 In vitro fermentation setup to assess gut microbiota modulation

- Inoculate with 10% (v/v) fecal slurry.
- Flush with nitrogen gas and seal with rubber stoppers.
- Incubate at 37 °C for 24–48 h under anaerobic conditions (Fig. 3).

3. Sampling and Analysis [4].
   - Collect samples at 0, 24, and 48 h.
   - Analyze:

   (i) Microbiota composition:
   - Extract DNA using stool DNA kits.
   - Amplify 16S rRNA gene (e.g., V3–V4 region).
   - Sequence via Illumina MiSeq.
   - Analyze with QIIME2, SILVA, or Greengenes database.

   (ii) *Short-chain fatty acids (SCFAs):* Analyze supernatants using GC or HPLC.

   (iii) *pH measurement* of culture broth as an indirect indicator of fermentation.

4. Statistical Analysis [4–6].
   - Compare microbial diversity (alpha and beta diversity).
   - Use ANOVA or t-tests to compare SCFA concentrations between treatments.
   - Perform multivariate analyses (e.g., PCA, PCoA) to evaluate microbiota shifts.

4 Notes

1. All the chemicals and solvents should be analytical grade, which can be procured from certified suppliers.
2. Store the extract properly to prevent degradation of heat-sensitive phytochemicals as well as take care of mold growth.

References

1. Bannon CD, Eckenberger J, Snelling WJ, Huseyin CE, Allsopp P, Strain C, Ramnani P, Chitarrari R, Grant J, Hotchkiss S, Philp K, Campbell R, Tuohy KM, Claesson MJ, Ternan NG, Dooley JSG, Sleator RD, Rowland I, Gill CIR (2021) Low-molecular-weight seaweed-derived polysaccharides lead to increased Faecal bulk but do not alter human gut health markers. Foods 10(12):2988. https://doi.org/10.3390/foods10122988
2. El Abed N, Ben Hadj Ahmed S, Ozogul F (2024) Seaweed-derived polysaccharides and peptides and their potential benefits for human health and food applications. In: Seaweeds and Seaweed-Derived Compounds, pp 117–155. https://doi.org/10.1007/978-3-031-65529-6_5
3. Zang L, Baharlooeian M, Terasawa M, Shimada Y, Nishimura N (2023) Beneficial effects of seaweed-derived components on metabolic syndrome via gut microbiota modulation. Front Nutr 10. https://doi.org/10.3389/fnut.2023.1173225
4. Hagan M, Fungwe T (2023) Determining the effect of seaweed intake on the microbiota: a systematic review. Functional Food Sci 3(6). https://doi.org/10.31989/ffs.v3i6.1117
5. Pudlo NA, Pereira GV, Parnami J, Cid M, Markert S, Tingley JP, Unfried F, Ali A, Varghese NJ, Kim KS, Campbell A, Urs K, Xiao Y, Adams R, Martin D, Bolam DN, Becher D, Eloe-Fadrosh EA, Schmidt TM et al (2022) Diverse events have transferred genes for edible seaweed digestion from marine to human gut bacteria. Cell Host Microbe 30(3):314–328.e11. https://doi.org/10.1016/j.chom.2022.02.001
6. Gan L, Wang J, Guo Y (2022) Polysaccharides influence human health via microbiota-dependent and -independent pathways. Front Nutr 9. https://doi.org/10.3389/fnut.2022.1030063

Chapter 16

Determination of Tannins from Seaweed and Its Influence on Human Gut Microbiota

Tamalika Chakraborty, Sobhanjan Bhunia, Jeenatara Begum, Prapti Chakraborty, and Manjarima Ganguly

Abstract

Seaweeds are rich sources of polyphenolic chemicals, particularly tannins, which are becoming well-known for their possible antibacterial and prebiotic qualities. Tannins are isolated and described in this work from a few species of red and brown seaweed. Their impact on the makeup and activity of the human gut microbiota is then evaluated. The tannins were extracted by the use of aqueous and organic solvents, and their quantification was done by the use of Folin–Ciocalteu and protein precipitation procedures. Mass spectrometry and high-performance liquid chromatography (HPLC) were used to identify the tannin ingredients. In vitro fermentation models repeated the human colon, which was used to test the effects of tannin exposure on microbial diversity by the use of 16S rRNA gene sequencing. The results showed that tannins produced from seaweeds decrease aggressive pathogens like Enterobacteriaceae and upgrade beneficial bacterial species like *Bifidobacterium* and *Lactobacillus*, hence modulating the gut microbiota. Additionally, the production of short-chain fatty acids (SCFAs) was positively affected, which indicates improved colonic health. These results suggest that seaweed tannins could be useful bioactives for regulating the gut microbiota, which calls for more research into their potential application in dietary modifications and gut health treatments.

Key words Polyphenolic chemicals, Human gut microbiota, Folin–Ciocalteu method, Mass spectrometry, HPLC, Human colon, 16S rRNA gene sequencing, Short-chain fatty acids

1 Introduction

The human gut microbiota is essential for immunological regulation, digestion, and metabolism, which contribute to host health. Inflammatory bowel disease, obesity, diabetes, and colorectal cancer are among the illnesses that have been linked to dysbiosis or disturbances in the balance of gut microbes. Dietary elements that have the potential to positively alter the gut flora are therefore gaining attention. Polyphenols, especially tannins are among these substances that have drawn interest due to their antibacterial, antioxidant, and prebiotic-like qualities [1].

Mitesh Kumar Dwivedi and A. Sankara Narayanan (eds.), *Leaves as Functional Foods*, Methods and Protocols in Food Science, https://doi.org/10.1007/978-1-0716-5226-8_16,

Tannins are a class of polyphenolic substances that are typically found in terrestrial plants.

Marine algae have recently developed as a viable alternative supply, especially in red (Rhodophyceae) and brown (Phaeophyceae) seaweeds. These tannins from seaweeds are commonly found as phlorotannins, and other complex polyphenols have special chemical structures and bioactivities that are rare in land-based sources [1].

Marine elements provide molecules that differ structurally from their physical equivalents, so they contribute greatly to world biodiversity and chemical innovation. Seaweed tannins are believed to withstand upper gastrointestinal digestion and reach the colon where they may directly interact with gut microbiota. They have also shown antibacterial properties against a variety of pathogenic bacteria. This interaction may have systemic benefits and improve gut health by decreasing infections and selectively increasing beneficial bacterial populations.

The aim of this study is to quantify and extract tannins from particular seaweed species by using in vitro colonic fermentation models and to investigate their effects on the makeup of the human gut microbiota. The study investigates the potential of tannins derived from seaweed as functional agents for gut health modulation by determining the metabolite profiles including short-chain fatty acids (SCFAs) as well as microbial changes [1].

2 Materials

2.1 Materials for Tannin Extraction and Determination from Seaweed

I. Seaweed Samples.

- Brown algae (e.g., *Fucus vesiculosus*, *Sargassum* sp.).
- Red algae (e.g., *Gracilaria*, *Kappaphycus* sp.).
- Green algae (e.g., *Ulva* sp.).

2.2 Chemicals and Reagents (See Note 1)

- Acetone (70%) or methanol for extraction.
- Distilled water.
- Folin–Ciocalteu reagent.
- Sodium carbonate (Na_2CO_3).
- Tannic acid.

2.3 Equipment

- Blender or grinder.
- Conical flask.
- Soxhlet extractor or shaker incubator.
- Rotary evaporator.
- Water bath.

- Pipette.
- Test tube.
- UV-Vis spectrophotometer.

2.4 Materials for Studying Influence on Human Gut Microbiota (In Vitro Simulation)

I. Fecal Inoculum.
   - Fresh human fecal samples from healthy donors.
   - Anaerobic collection containers.

II. Media and Reagents for Microbiota Culture.
   - Brain Heart Infusion (BHI) broth or Basal Medium (like YCFA or GAM broth).
   - Anaerobic gas mixture (CO_2, H_2, N_2).
   - Phosphate-buffered saline (PBS).

III. Analytical Tools.
   - DNA extraction kits.
   - qPCR setup or 16S rRNA gene sequencing.
   - GC-MS or HPLC (*see* **Note 2**).
   - Plate count agar media (e.g., MRS, MacConkey).
   - Statistical software (e.g., GraphPad prism).

3 Method

3.1 Determination of Tannins from Seaweed

1. Sample Preparation [2].
   - Seaweed samples are collected and thoroughly washed with distilled water to remove salt, sand, and epiphytes.
   - Shade dry or oven dry at 40–50 °C.
   - Grind dried seaweed into fine powder and store it in airtight containers.
2. Extraction of Tannins [2].
   - Weigh ~5 g of seaweed powder.
   - Extract with 70% aqueous acetone or methanol (100 mL) in a conical flask.
   - Shake on an orbital shaker for 24 h at room temperature.
   - Filter through Whatman No. 1 filter paper.
   - Concentrate the extract using a rotary evaporator or water bath at 40 °C to remove the solvent.
   - Store an extract at 4 °C for further analysis (Fig. 1).

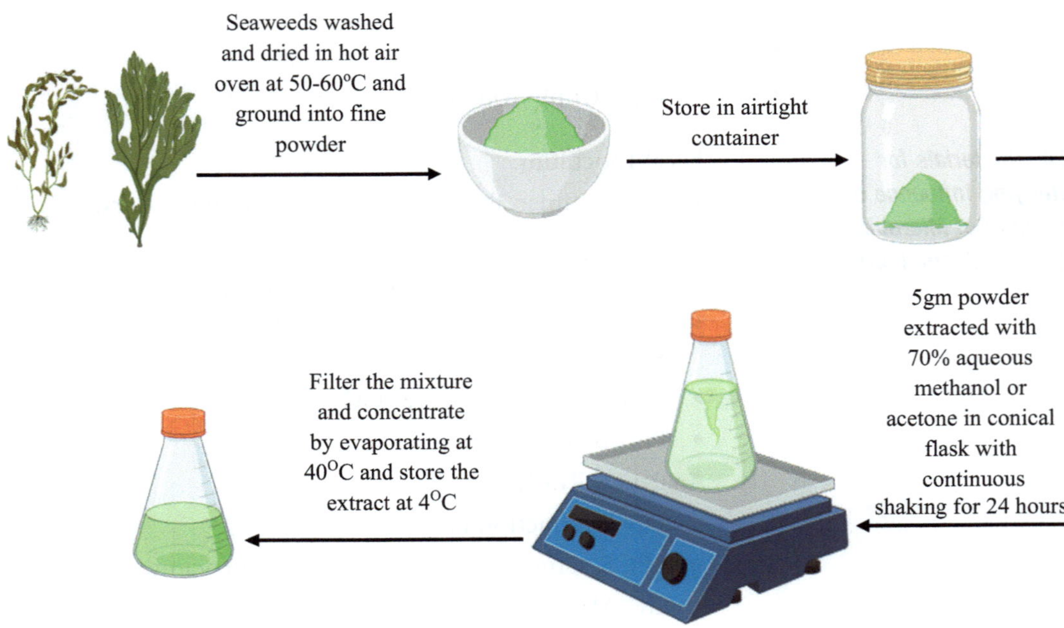

Fig. 1 Tannin extraction from seaweed

3. Quantification of Total Tannins (Folin–Ciocalteu Method) [2, 3].
   - Pipette 0.5 mL of extract into a test tube.
   - Add 3 mL of distilled water and 0.5 mL of Folin–Ciocalteu reagent.
   - Wait 3 min and then add 2 mL of 20% sodium carbonate solution.
   - Mix and incubate at room temperature for 30–40 min in the dark.
   - Measure the absorbance at 725 nm using a UV-Vis spectrophotometer.
   - Prepare a calibration curve using tannic acid standards (0–100 μg/mL).
   - Express results as mg tannic acid equivalents (TAE)/g dry weight of seaweed (Fig. 2).

3.2 Influence on Human Gut Microbiota (In Vitro Fermentation)

1. Preparation of Human Fecal Inoculum [3].
   - Collect fresh fecal samples from healthy adult donors within 2 h.
   - Homogenize 10 g of feces in 100 mL sterile anaerobic phosphate-buffered saline (PBS, pH 7.4) under anaerobic conditions.

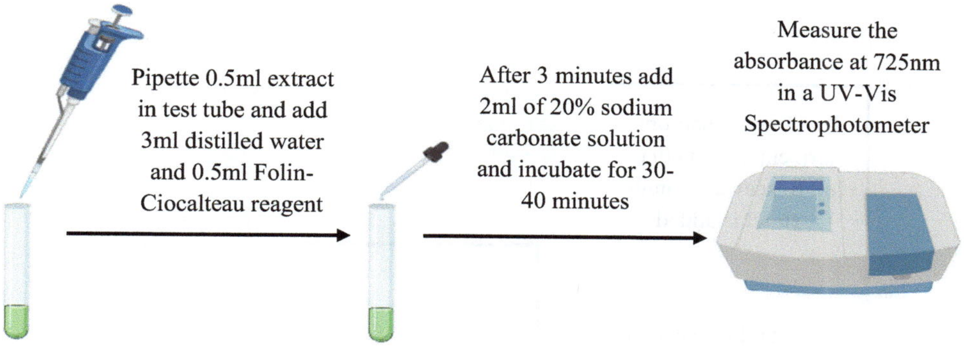

Fig. 2 Quantification of total tannins

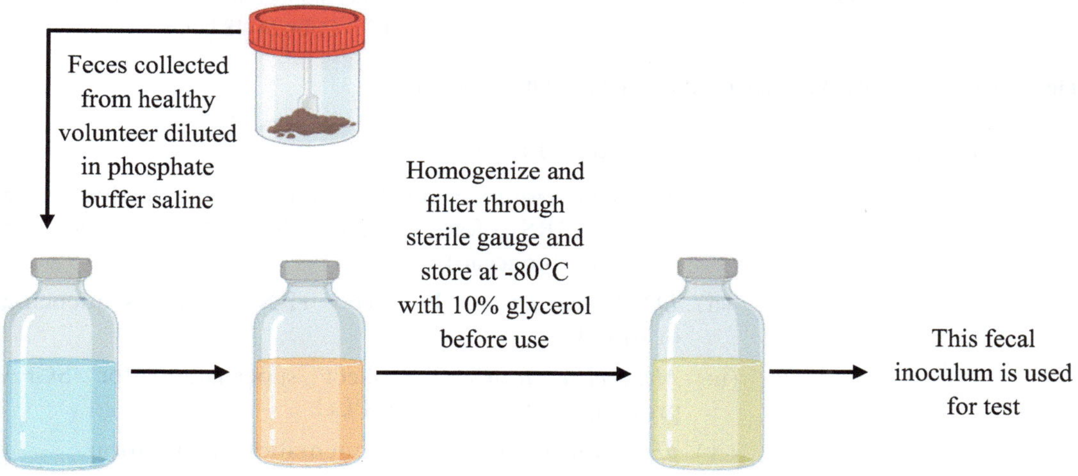

Fig. 3 Fecal inoculum preparation

- Filter through gauze to remove large particles.
- Use 10% (v/v) fecal slurry as inoculum (Fig. 3).

2. In Vitro Fermentation Setup [3].
   - Inoculate 45 mL of anaerobic growth medium (e.g., BHI or YCFA) in serum bottles with 5 mL of fecal slurry and 1 mL of seaweed tannin extract.
   - Flush headspace with anaerobic gas ($CO_2/H_2/N_2$).
   - Incubate at 37 °C for 24–48 h under static anaerobic conditions.
   - Perform experiments in triplicates (Fig. 4).

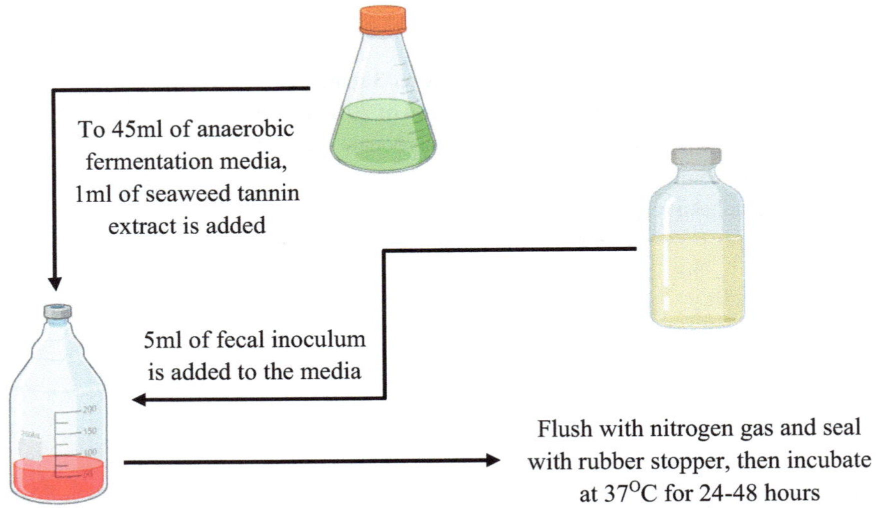

Fig. 4 In vitro fermentation setup to determine tannin from seaweed

3.2.1 Monitoring and Analysis [3–5]

- At 0, 24, and 48 h:
  - (i) *Microbial enumeration:* Plate counts on selective media (e.g., MRS for *Lactobacillus*, MacConkey for Enterobacteriaceae).
  - (ii) *DNA extraction:* For 16S rRNA gene sequencing or qPCR.
  - (iii) *Metabolite analysis:* Collect supernatant for SCFA profiling via GC-MS or HPLC.
  - (iv) *pH measurement:* Track fermentation acidification.

3.2.2 Data Analysis [3, 6]

- Compare microbial diversity (α and β diversity) between control and treated samples.
- Quantify changes in beneficial vs. pathogenic bacteria.
- Assess the correlation between tannin concentration and SCFA production or microbial shifts.
- Use statistical software (e.g., GraphPad Prism or QIIME2 for sequencing data).

4 Notes

1. All the chemicals and solvents should be analytical grade, which can be procured from certified suppliers.
2. Use HPLC grade reagents and chemicals.

References

1. Sallam IE, Abdelwareth A, Attia H, Aziz RK, Homsi MN, von Bergen M, Farag MA (2021) Effect of gut microbiota biotransformation on dietary tannins and human health implications. Microorganisms 9(5):965. https://doi.org/10.3390/microorganisms9050965
2. Shannon E, Conlon M, Hayes M (2021) Seaweed components as potential modulators of the gut microbiota. Mar Drugs 19(7):358. https://doi.org/10.3390/md19070358
3. Zang L, Baharlooeian M, Terasawa M, Shimada Y, Nishimura N (2023) Beneficial effects of seaweed-derived components on metabolic syndrome via gut microbiota modulation. Front Nutr 10. https://doi.org/10.3389/fnut.2023.1173225
4. Fraga-Corral M, Otero P, Cassani L, Echave J, Garcia-Oliveira P, Carpena M, Chamorro F, Lourenço-Lopes C, Prieto MA, Simal-Gandara J (2021) Traditional applications of tannin rich extracts supported by scientific data: chemical composition, bioavailability and bioaccessibility. Foods 10(2):251. https://doi.org/10.3390/foods10020251
5. Tap J, Ruppé E, Derrien M (2022) The human gut microbiota in all its states: from disturbance to resilience. In: Comprehensive Gut Microbiota, pp 161–178. https://doi.org/10.1016/b978-0-12-819265-8.00039-5
6. Kumar M, Sharma A, Verma A, Shrivastava R (2022) Emergence of antibiotic resistance in gut microbiota and its effect on human health. Human-Gut Microbiome:211–232. https://doi.org/10.1016/b978-0-323-91313-3.00002-7

Chapter 17

Determination of Antiviral Properties of Leaf Extracts

Tamalika Chakraborty, Sobhanjan Bhunia, Sumana Roy, Priyanka Ray, and Sriparna Kundu Sen

Abstract

Alternative treatment medicines are urgently needed as demonstrated by an increase in drug-resistant viral strains and the increasing prevalence of viral infections. This study focuses on the antiviral properties of specific leaf extracts from medicinal plants against specific viral infections. Aqueous ethanolic and methanol solvents of different intensities have been utilized to extract leaves from commonly utilized medicinal plants after they had been shade-dried. To find bioactive substances such as flavonoids, alkaloids, tannins, and saponins, the phytochemical composition of the extracts was first screened. Appropriate cell lines and viral strains were used in in vitro antiviral tests such as cytopathic effect (CPE) inhibition and plaque reduction assays. According to the results, particular leaf extracts showed strong antiviral activity, which was probably related to the presence of secondary metabolites that are known to inhibit viral entrance, replication, or assembly. As a natural source of antiviral medicines, these results suggest that plant-derived leaf extracts have potential and should be investigated further using compound isolation and mechanistic research.

Key words Aqueous ethanolic extraction, Methanol extraction, Cell lines, Cytopathic Effect (CPE) inhibition, Plaque reduction assay, Secondary metabolites, Viral replication

1 Introduction

The current outbreaks and the increasing prevalence of viral diseases like influenza, dengue, herpes simplex virus (HSV), and new coronaviruses show that viral infections continue to pose a danger to global public health. The development of drug-resistant virus strains and the limited effectiveness of current treatments and advances in antiviral therapy highlight the urgent requirement for novel, potent, and reasonably priced antiviral medicines.

The therapeutic potential of medicinal plants has contributed to their frequent usage in traditional medical systems worldwide [1]. Most common in plant leaves are secondary metabolites such as flavonoids, phenolics, alkaloids, tannins, and terpenoids. These compounds have been found to produce a variety of biological effects including antibacterial, antiviral, and anti-inflammatory

Mitesh Kumar Dwivedi and A. Sankara Narayanan (eds.), *Leaves as Functional Foods*, Methods and Protocols in Food Science,
https://doi.org/10.1007/978-1-0716-5226-8_17,

properties. Research on plant-based antivirals provides a cost-effective substitute to synthetic drugs that are less toxic, and it has fewer side effects.

Plant extract screening for antiviral activities utilizes a variety of in vitro and in vivo models that have formed the dependence of several investigations in recent years. The primary goals of this study are to investigate the antiviral properties of leaf extracts from several plant species, identify the active components, and evaluate their inhibitory effects against specific viral infections. The findings of this study may help to generate new plant-based antiviral medications and encourage the introduction of natural products into present antiviral therapy [1].

2 Materials

2.1 Plant Material

- Fresh, healthy leaves of selected medicinal plants (e.g., *Azadirachta indica*, *Ocimum sanctum*, *Psidium guajava*, etc.).

2.2 Chemicals and Reagents

- Ethanol.
- Methanol.
- Distilled water.
- DMSO (Dimethyl sulfoxide).
- MTT reagent.
- Formazan crystal.

2.3 Equipment

- Mechanical grinder.
- Laminar airflow cabinet.
- Whatman filter paper.
- CO_2 incubator.
- Inverted microscope.
- Microplate reader.
- Autoclave.
- Rotary evaporator.
- 96-well microplates.

2.4 Cell Lines and Virus

- *Cell Lines:* Vero cells.
- *Virus Strains:* Influenza A, HSV (Herpes Simplex Virus), Dengue virus.

2.5 Culture Media and Supplements

- Dulbecco's Modified Eagle's Medium (DMEM).
- Fetal bovine serum (FBS).
- Penicillin-streptomycin antibiotics.

3 Methods

3.1 Collection and Preparation of Plant Material [2]

- Fresh leaves were collected from the selected plant washed thoroughly with distilled water and shade-dried at room temperature.
- Dried leaves were ground into fine powder using a mechanical grinder.
- The powdered material was stored in airtight containers until extraction (Fig. 1).

3.2 Extraction of Leaf Compounds [2, 3]

- Approximately 50 grams of leaf powder was soaked in 500 mL of solvent (e.g., ethanol, methanol, or water) for 72 h at room temperature with occasional shaking.
- The extract was filtered using Whatman No.1 filter paper.
- The filtrate was concentrated using a rotary evaporator under reduced pressure at 40–50 °C.
- The crude extract was dried and reconstituted in DMSO or sterile water to prepare stock solutions of known concentration (e.g., 100 mg/mL).
- Extracts were stored at 4 °C until further use (Fig. 2).

3.3 Cell Line Maintenance [3]

- Vero cells or other appropriate cell lines were cultured in Dulbecco's Modified Eagle's Medium (DMEM), which was supplemented with 10% fetal bovine serum (FBS) and 1% penicillin-streptomycin.
- Cells were incubated at 37 °C in a 5% CO_2 atmosphere and subcultured every 2–3 days.

3.4 Cytotoxicity Assay (MTT Assay) [3, 4]

- Cells were seeded in a 96-well plate at a density of ~1×10^4 cells/well and incubated for 24 h.
- Serial dilutions of the extract were added to the wells in triplicate.
- After 48 hours, 20 μL of MTT (5 mg/mL) was added to each well and incubated for 4 hours.

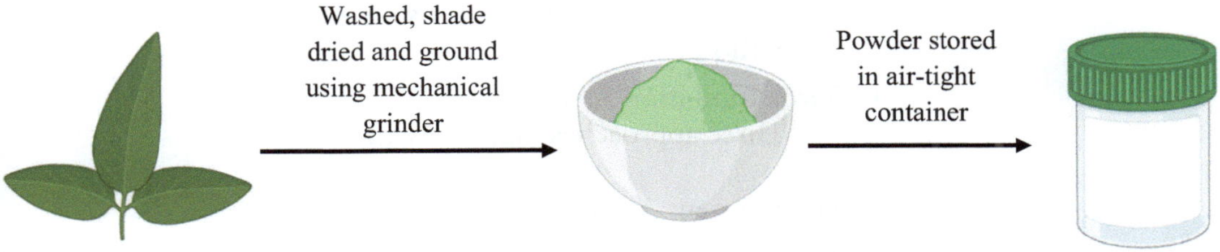

Fig. 1 Powdering of the collected leaves

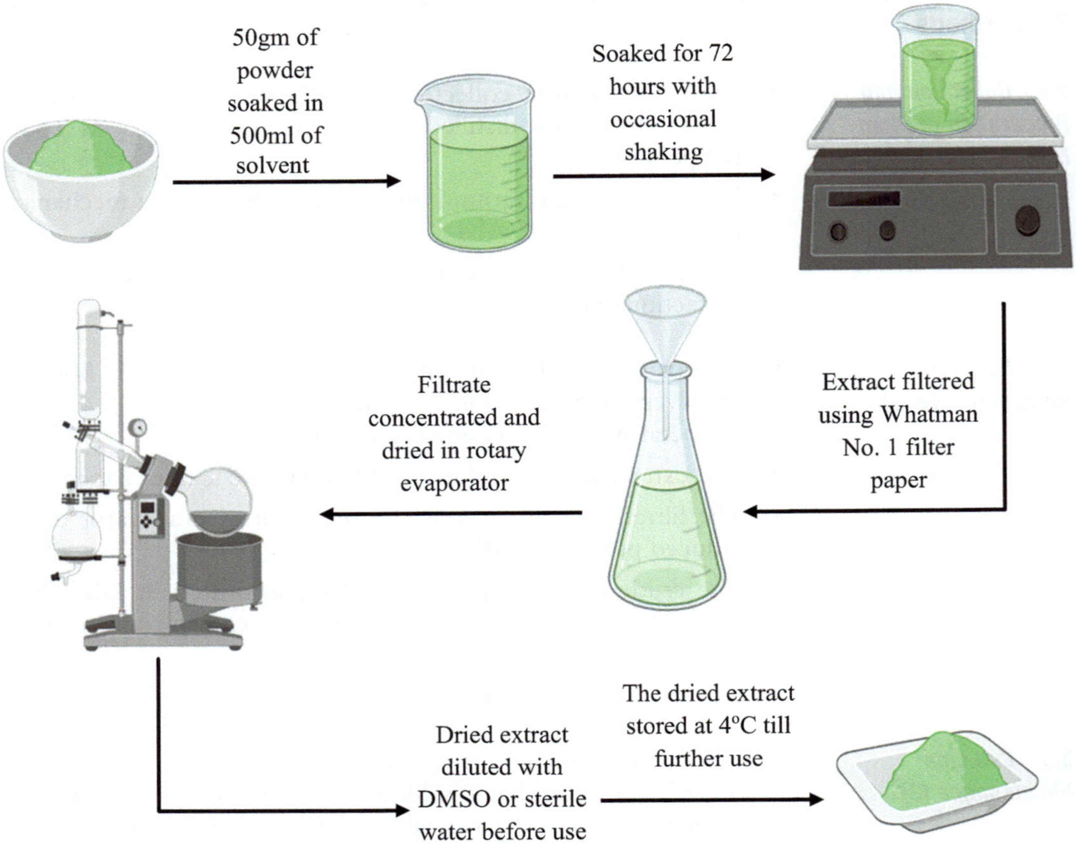

Fig. 2 Extraction done from the dried leaves powder

- The medium was removed, and formazan crystals were dissolved in 100 μL of DMSO.
- Absorbance was measured at 570 nm using a microplate reader.
- Cell viability (%) was calculated and the **CC_{50}** (concentration that kills 50% of the cells) was determined (Fig. 3).

3.5 Antiviral Assay [4]

- Virus Preparation.
  - A lab-adapted virus strain (e.g., Herpes Simplex Virus, Influenza A, Dengue Virus) was propagated in Vero cells.
  - Virus stock was titrated using plaque assay or $TCID_{50}$ method.
- Virus Inhibition Assays.

  Three approaches may be used:

(i) Pretreatment Assay.

- Cells were treated with nontoxic concentrations of extract for 2 hours.

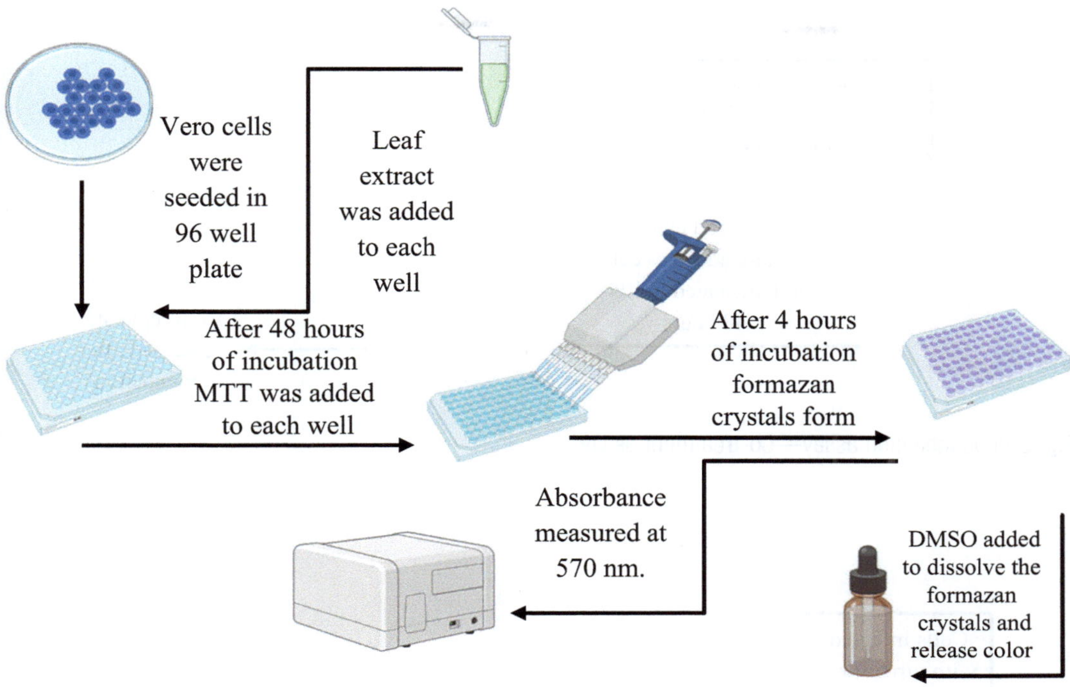

Fig. 3 Cytotoxic/MTT assay

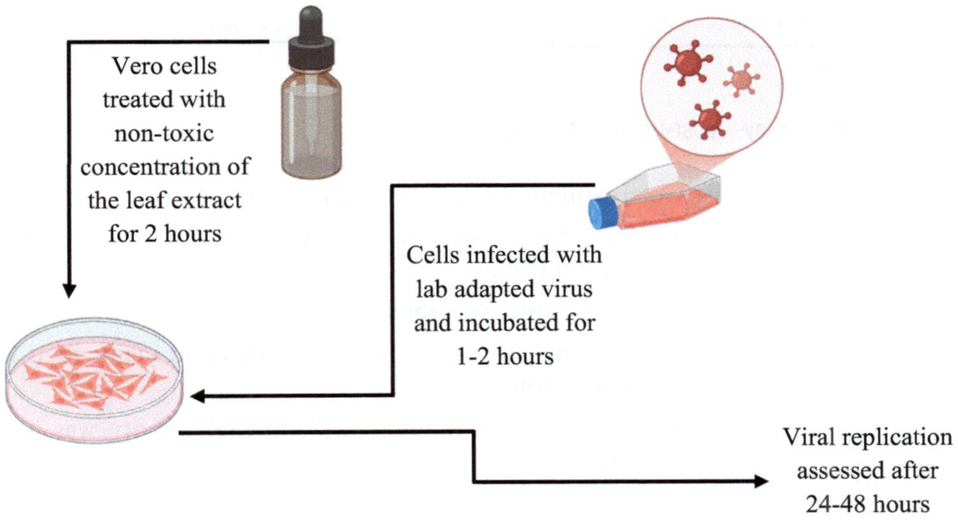

Fig. 4 Virus inhibition assay—Pretreatment assay

- Cells were then washed and infected with the virus at MOI (multiplicity of infection) = 0.1–1.
- Incubate for 1–2 h to allow virus adsorption.
- Viral replication was assessed after 24–48 h (Fig. 4).

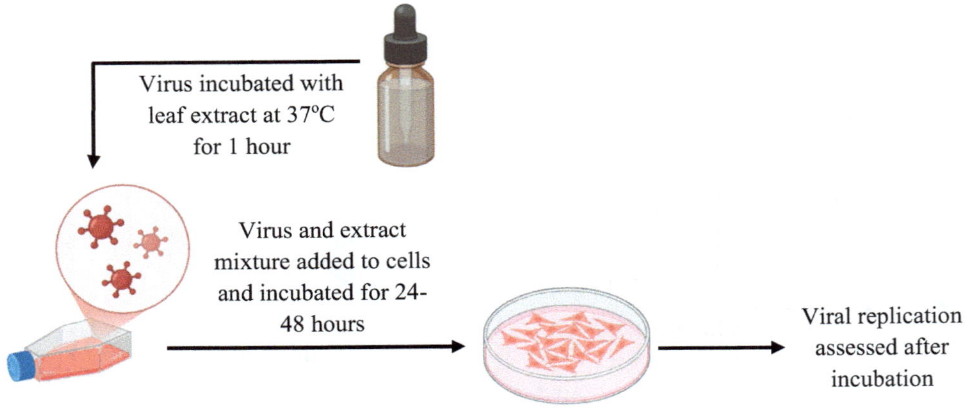

Fig. 5 Virus inhibition assay—Co-treatment assay

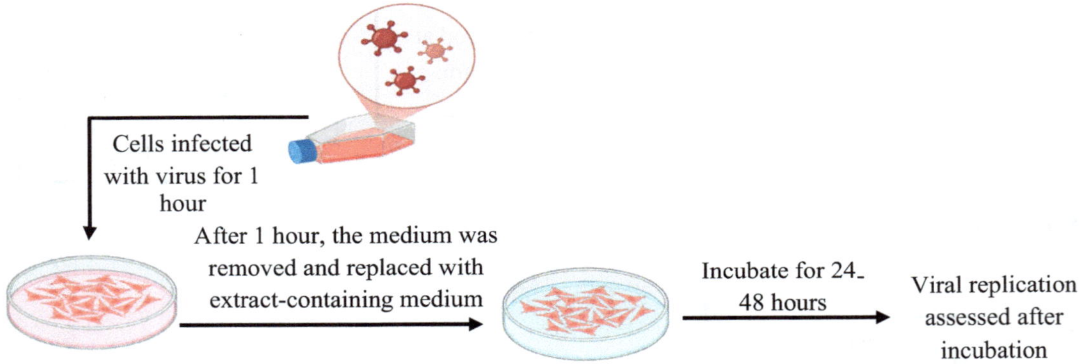

Fig. 6 Virus inhibition assay—post-treatment assay

(ii) Co-treatment Assay (Direct Inactivation).

- The virus was incubated with extract at 37 °C for 1 h before infecting cells.
- The mixture was then added to the cells.
- After adsorption, cells were washed and maintained with a fresh medium.
- Incubation continued, and antiviral activity was assessed (Fig. 5).

(iii) Post-treatment Assay.

- Cells were infected with the virus and allowed for adsorption.
- After 1 h, the medium was removed and replaced with extract-containing medium.
- Incubate for 24–48 h (Fig. 6).

3.6 Quantification of Antiviral Activity [4, 5]

- *Plaque Reduction Assay:* Plaques formed by the virus were counted; a reduction in number compared to control indicates antiviral activity.
- *Cytopathic Effect (CPE) Inhibition:* Microscopically observe a reduction in virus-induced cell damage.
- *qRT-PCR or ELISA:* Measure viral genome or protein expression levels.

3.7 Data Analysis [5–7]

- IC_{50} (concentration inhibiting 50% of viral activity) was calculated.
- Selectivity Index (SI) = CC_{50}/IC_{50}—a higher SI indicates better therapeutic potential.

References

1. Abdulabbas HS (2023) Antiviral effects of medicinal plants: Minireview. Res Rev. https://doi.org/10.52845/jmrhs/2023-6-2-4
2. Singh L, Dalapati P, Shah P, Mishra SS (2023) Future directions in antiviral potentials of medicinal plants. In: Promising Antiviral Herbal and Medicinal Plants, pp 295–306. https://doi.org/10.1201/9781003329169-19
3. Gamil Zeedan GS, Abdalhamed AM (2021) Antiviral effects of plant extracts used in the treatment of important animal viral diseases. World's Veterinary J 11(4):521–533. https://doi.org/10.54203/scil.2021.wvj67
4. Zeedan, G SG, M. Abdalhamed A (2023) Antiviral plant extracts: A treasure for treating viral diseases. Antiviral Strategies in the Treatment of Human and Animal Viral Infections. https://doi.org/10.5772/intechopen.111732
5. Atampugbire G, Adomako EEA, Quaye O (2024) Medicinal plants as effective antiviral agents and their potential benefits. Nat Prod Commun 19(9). https://doi.org/10.1177/1934578x241282923
6. Satapathy P, Tartte V, Babu YTR (2025) Medicinal plants as prospective sources of antiviral agents: a comprehensive review. Deep Science Publishing, pp 19–34. https://doi.org/10.70593/978-93-49307-18-6_3
7. Thomas E, Stewart LE, Darley BA, Pham AM, Esteban I, Panda SS (2021) Plant-based natural products and extracts: potential source to develop new antiviral drug candidates. Molecules 26(20):6197. https://doi.org/10.3390/molecules26206197

Chapter 18

Determination of Anti-allergic Effect from Plant Leaf

Siddhi A. Darji, Atharva Patil, Aakansha Yadav, and Devjani I. Banerjee

Abstract

The rising prevalence of allergic diseases, driven by environmental and genetic factors, underscores the need for alternative therapeutic strategies with fewer side effects. Plant-derived compounds offer promising potential in allergy management due to their natural origin and bioactivity. This chapter outlines a methodological framework for the extraction, characterization, and preliminary evaluation of anti-allergic compounds from plant leaves. Fresh leaves are collected, shade-dried, powdered, and subjected to ethanol-based crude extraction. To enhance yield and purity, fractionation is carried out through liquid-liquid partitioning followed by soxhlet extraction. The obtained phytochemicals are analyzed using techniques such as thin-layer chromatography (TLC), high-performance liquid chromatography (HPLC), and gas chromatography–mass spectrometry (GC–MS) to identify major bioactive constituents like flavonoids, alkaloids, and phenolics. Preliminary screening methods, including skin prick and patch tests, are discussed for assessing anti-allergic activity. The protocol described serves as a foundational guide for researchers aiming to explore plant-based approaches in allergy research and drug development.

Key words Anti-allergic activity, Plant leaves, Phytochemicals, Soxhlet extraction, HPLC, TLC, GC-MS, Natural antihistamines, Liquid-liquid partitioning, Skin patch test

1 Introduction

The frequency and occurrence of allergic infections have vividly increased over period mainly in developed nations. The allergic disease has risen due to various factors such as dietary choices, environmental contamination, genetic susceptibility, and lifestyle changes while having a significant negative impact on public health, resulting in poor quality of life and financial expenses leading to extreme situations. Although, the increasing number of allergies are widely observed, there is a gap in identifying the mechanism that conducts these conditions. Filling this gap is crucial for understanding, treating, and preventing this allergic reaction. Further we need to understand how the immune system reacts to this and its progression to treat it [1, 2].

Mitesh Kumar Dwivedi and A. Sankara Narayanan (eds.), *Leaves as Functional Foods*, Methods and Protocols in Food Science, https://doi.org/10.1007/978-1-0716-5226-8_18,

Chronic allergic conditions often involve Type I hypersensitivity, where immunoglobulin E (IgE) plays a central role. IgE binds to allergens and activates mast cells and basophils, leading to the release of pro-inflammatory mediators such as histamine, cytokines. This cascade causes the typical symptoms of allergies: swelling, mucosal inflammation, itching, and airway constriction [3].

Conventional medications like antihistamines, monoclonal antibodies, antileukotrienes, and corticosteroids are often prescribed to manage these symptoms. However, growing interest in natural, plant-derived anti-allergic medicines has fueled research into alternative therapies [4]. Medicinal plants have been integral to traditional healing systems such as Ayurveda, Traditional Chinese Medicine (TCM), Unani, and Kampo for centuries. Modern pharmacognosy and phytochemistry have provided a scientific foundation for these traditional practices by isolating, separating, and characterizing bioactive components responsible for therapeutic effects [1–3].

Plant-derived anti-allergic phytochemicals have attracted attention due to their immunomodulatory, antihistaminic, and anti-inflammatory properties. Particularly, secondary metabolites such as flavonoids, alkaloids, terpenoids, glycosides, phenolic acids, and tannins are known to:

- Reduce cytokine production.
- Block IgE binding.
- Prevent histamine release.
- Modulate key signaling pathways including NF-κB, MAPK, and JAK-STAT [2, 5].

Extracting these active compounds from plants represents a viable and sustainable strategy for the development of new anti-allergic medications. Among various plant parts, leaves offer a highly promising source of anti-allergic compounds for several reasons:

1. *High Content of Bioactive Compounds:* Leaves are rich in volatile oils, phenolic acids, and flavonoids, all known for potent anti-inflammatory and anti-allergic activities.
2. *Sustainable Harvesting:* Collecting leaves is less intrusive compared to harvesting roots or bark, making it more environmentally friendly and sustainable.
3. *Ease of Processing:* Leaves are easier to dehydrate, grind, and extract compared to other plant parts.
4. *Traditional Usage:* Leaf infusions and brews have been commonly used in traditional treatments for managing skin and respiratory allergies.

Leaves, being the most metabolically active organs, are not only crucial for photosynthesis but also for the synthesis of secondary metabolites. Their regenerative ability further supports their use without sacrificing the entire plant, making them an ideal focus for sustainable medicinal extraction [1].

2 Materials

2.1 Plant Material

- Fresh plant leaves.

2.2 Chemicals and Solvents

- Ethanol or methanol (analytical grade).
- Hexane, chloroform, ethyl acetate, n-butanol.
- Solvent mixtures for chromatography (*see* **Note 1**).
- Molecular biology grade water.

2.3 Instruments and Equipment

- Mechanical grinder.
- Conical flasks.
- Mechanical shaker.
- Whatman filter paper.
- Rotary evaporator.
- Separating funnel.
- High-performance liquid chromatography (HPLC) system.
- Column.
- Sample vials.
- Syringe filters.
- Pre-coated aluminum thin layer chromatography (TLC) plates.
- TLC chamber.
- GC-MS system.
- Capillary tubes.
- Spray bottles.
- Beakers.
- Hot air oven.
- Round bottom flask.
- Thimble.
- Condenser.
- Heating mantle.
- Soxhlet chamber.
- Application chamber for patch test.

3 Methods

3.1 Plant Material Preparation [1]

1. Collect fresh leaves of the selected plant species.
2. Wash thoroughly with distilled water to remove any dirt on them.
3. Air-dry the leaves at room temperature (25–28 °C) for 7–10 days (*see* **Note 2**).
4. Grind the dried leaves into a fine powder using a mechanical grinder.
5. The powdered plant material can be stored for further use.

3.2 Crude Extraction [2]

1. On a weighing scale, weigh approximately 300 g of the (stored) leaf powder.
2. Soak the leaf powder in 1000 mL of 70% ethanol in a conical flask (*see* **Note 3**).
3. Place the flask on a mechanical shaker at room temperature for 72 h to extract the phytochemicals (*see* **Note 4**).
4. Filter the extract using Whatman filter paper (No. 3) to remove any solid residues.
5. Using a rotary evaporator, concentrate the filtrate under reduced pressure at a temperature not exceeding 40 °C (*see* **Note 5**).

3.3 Liquid–Liquid Partitioning for Fractionation [6]

1. Rehydrate the semi-solid crude extract in a small volume of distilled water (*see* **Note 6**).
2. Sequentially partition the aqueous extract with solvents of increasing polarity—hexane, ethyl acetate, chloroform, and n-butanol—using a separating funnel.
3. For each solvent, add an equal volume to the aqueous solution, shake gently, and allow the layers to separate.
4. Collect the appropriate layer for each solvent (*see* **Note 7**).
5. Repeat the partitioning process 2–3 times with each solvent for complete extraction.
6. Concentrate each solvent layer using a rotary evaporator under reduced pressure (<40 °C) (*see* **Note 8**).

3.4 Soxhlet Extraction [7]

1. Dry and grind the collected plant leaves to a uniform powder of appropriate particle size.
2. Load approximately 20–30 g of powdered plant material into a thimble.
3. Insert the thimble into the soxhlet extractor chamber, and assemble the apparatus by connecting it to a round-bottom

flask containing 250 mL of either ethanol or methanol and a reflux condenser.

4. Heat the flask using a heating mantle to initiate solvent vaporization.
5. Allow the solvent vapors to rise, condense through the water-cooled condenser, and drip into the soxhlet chamber containing the plant powder.
6. As the solvent level in the chamber rises and reaches the siphon point, it carries the extracted compounds back into the boiling flask.
7. Repeat the cycle continuously for 6–8 h until the dry powder in thimble becomes colorless, indicating completion of extraction.
8. Concentrate the collected extract using a rotary evaporator under reduced pressure (temperature < 40 °C) to remove the solvent and obtain a semisolid crude extract (*see* **Note 5**).

   Soxhlet extraction is a continuous, automatic process for sorting and isolating anti-allergic compounds from plant leaves. Using a thimble, heat the solvent to evaporate it, which then liquidizes the extracts; the necessary complexes from the plant material are obtained (*see* **Note 9**). The entire process was presented in Fig. 1.

3.5 Chromatographic Analysis

Depending on the type of compounds expected, each dried fraction was tested with both qualitative and quantitative phytochemical analysis using appropriate chromatographic techniques like thin-layer chromatography (TLC), high-performance liquid chromatography (HPLC), or gas chromatography-mass spectrometry (GC-MS). These methods were employed to determine the bioactive components and assess the yield and purity of each component separately.

3.5.1 TLC [8]

1. Pour the solvent mixture into the TLC chamber to a depth of about 0.5–1 cm.
2. Place a piece of filter paper inside the chamber to saturate the atmosphere and ensure uniform solvent rise.
3. Cover the chamber with a lid and allow it to stand for 20–30 min for saturation.
4. Cut the TLC plate to the desired size (commonly 5 cm × 10 cm).
5. Using a pencil, draw a baseline about 1 cm from the bottom edge of the plate (*see* **Note 10**).
6. Mark equidistant points along the baseline for spotting multiple samples or standards.

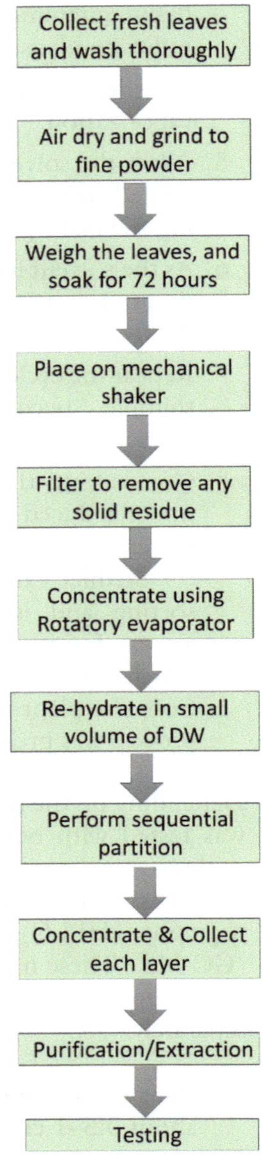

Fig. 1 Preparatory steps involved in the production of Anti allergic compounds from leaves

7. Dip a clean capillary tube into the sample solution.
8. Gently touch the tube to one of the marked spots on the baseline to apply a small, concentrated drop of the sample.
9. Let the spot air dry completely.
10. Repeat 2–3 times (if needed) to concentrate the sample at a single spot. Allow each application to dry before re-spotting.

11. Spot any reference compounds on separate marks if comparison is needed.
12. Carefully place the spotted TLC plate vertically into the developing chamber without touching the sides.
13. Make sure the baseline is above the solvent level in chamber.
14. Close the chamber and allow the solvent to ascend the plate through capillary action.
15. Once the solvent front reaches approximately 1–1.5 cm from the top edge of the plate, carefully remove it from the chamber.
16. Immediately draw a pencil line across the solvent front (before evaporation).
17. Allow the plate to dry at room temperature or place it in a hot air oven at 105 °C for 5 min (optional).
18. Observe the plate under UV light (254 nm or 365 nm) to detect fluorescent compounds.
19. Alternatively, spray the plate with a suitable visualizing/staining reagent (e.g., ninhydrin for amino acids, iodine vapors for alkaloids, sulfuric acid etc.).
20. Heat gently in an oven if required by the reagent.
21. Use a pencil to mark the center of each visible spot, then measure the distance from the baseline to each marked center.
22. Measure the distance from the baseline to the solvent front.
23. Calculate the Rf value using the formula (*see* **Note 11**).

3.5.2 HPLC [9]

1. Start the HPLC system, which includes the software, pump, and detector, and purge lines using solvent A, B, C, and D for 10 min (*see* **Note 12**).
2. Equilibrate the column with mobile phase for 15 min (*see* **Note 13**).
3. For at least fifteen minutes, acclimatize the column to the mobile phase pre-heat the column oven to 30 °C.
4. Prepare the mobile phase via mixing the solvent in required ratio.
5. Filter through a membrane followed by degas via ultrasonication.
6. Prepare the extract in mobile phase at desired concentration (ppm).
7. Filter through 0.45 μm syringe filter.
8. Set method parameters suitable for target compounds.
9. Inject samples using auto-samples.
10. Record and analyze the chromatograms.

11. Identify the peaks by retention time (typically 2–5 min).
12. Quantify using peak area or height.
13. Compare this with available standards.

3.5.3 GC-MS [10]

1. Use the GC-MS instrument equipped with an appropriate column such as HP-5 MS or DB-35 MS, with helium as the carrier gas.
2. Set the oven temperature program including initial temperature, ramp rates, and isothermal periods to ensure effective compound separation.
3. Inject the prepared extract into the GC-MS system using either a split or splitless inlet mode.
4. Record mass spectrum data using appropriate software such as Trace Finder or Turbo-Mass.
5. Identify compounds by comparing mass spectra and retention times with reference libraries like NIST or Wiley.
6. Determine the relative percentage of each identified compound in the extract.
7. Correlate identified compounds with reported anti-allergic properties using in vitro and in vivo studies and literature surveys.

3.6 Testing of Purified Anti-allergic Compound (See Note 14)

3.6.1 Skin Prick Test [11]

1. Screen volunteers following ethical guidelines and obtain informed consent.
2. Place a drop of purified compound on the volar side of the forearm.
3. Prick the skin lightly using a sterile lancet to introduce the compound into the epidermis.
4. Observe the site after 15–20 min for signs of wheal-and-flare reactions.
5. Compare results with positive control (histamine) and negative control (saline) to evaluate the anti-allergic response.

3.6.2 Patch Test [12]

Pre-test Instructions to Patients

1. Inform patients about the objective.
2. Identifying allergens causing allergic contact dermatitis.
3. Advise patients to avoid soaking the back, excessive sweating, and UV exposure 15 days before the test.
4. Describe likely symptoms: mild pain, redness, or itching are expected and normal.

3.6.3 Preparation of Allergens and Vehicles

1. Select allergens based on common exposures and sensitivities.
2. Choose suitable vehicles: use petroleum as the standard medium; others include hydrophilic gel, organic solvents (e.g., acetone, ethanol), or distilled water. For liquid substances, use filter paper to absorb and deliver the dose uniformly.

3.6.4 Chamber and Battery Setup

1. Use standardized chambers such as Finn Chamber (Epi-test, Finland), IQ Chambers (Sweden), or Van der Bend (Netherlands).
2. Maintain uniform allergen distribution and occlusion to reduce sensitization.
3. Use standard test batteries such as: Brazilian Standard Battery (30 petroleum-based compounds, except formaldehyde) and Cosmetics Battery for testing cosmetic-related allergens.
4. Arrange allergens in the battery based on chemical similarity and avoid placing cross-reactive substances adjacent to each other.

3.6.5 Application and Test Execution

1. Select the upper back as the application area for its flat, immobile surface. If unavailable, use the upper arms or thighs.
2. Prepare the skin by gently cleansing with ethanol to remove oils without irritation.
3. Apply chambers firmly on dry skin and secure with adhesive tape to prevent displacement, especially in humid environments.

3.6.6 First Reading and Chamber Removal (48 h)

1. Remove chambers in a well-lit room and mark the skin using indelible ink for follow-up.
2. Wait 15–20 min post-removal to allow transient redness to subside.
3. Record the reaction morphology (e.g., erythema, oedema, papules) and location in the patient's file.

3.6.7 Second Reading (72–96 h)

1. Re-evaluate the marked sites after 72–96 h to detect delayed allergic reactions.
2. Compare with 48-hour readings to distinguish irritant responses from true allergic sensitization.
3. Classify reactions based on ICDRG (International Contact Dermatitis Research Group) guidelines for standard interpretation.

4 Notes

1. All the chemicals and solvents should be analytical grade, which can be procured from certified suppliers.
2. Maintain the temperature to prevent degradation of heat-sensitive phytochemicals as well as take care of mold growth.
3. Cover the flask with aluminum foil to prevent evaporation.
4. During this time, the phytochemicals would dissolve in Ethanol.
5. Store the semi-solid crude extract at 4 °C for subsequent analysis.
6. It increases the chances of solubility.
7. Upper layer for hexane, ethyl acetate, and n-butanol, lower layer for chloroform.
8. Store the dried fractions for further analysis.
9. This method will help to obtain pure extract.
10. There sample will be spotted.
11. Rf = Distance traveled by solvent front/distance traveled by compound.
12. To eliminate air bubbles, purge every line using solvents A, B, C, and D for ten minutes.
13. Solvent combination (it may vary depending for your chemical), to ensure that the atmosphere in the jar is saturated with the solvent, the chamber should be left undisturbed for approximately half an hour.
14. Follow the ethical guidelines and obtain ethical approval before proceeding.

References

1. Murata K, Abe Y, Shinohara K, Futamura-Masuda M, Uwaya A, Isami F, Matsuda H (2014) Anti-allergic activity of the Morinda citrifolia extract and its constituents. Pharm Res 6(3):260
2. Guo MM, Xue WT, Liao LY, Ling X, Yu D, Lan XL, Zhou WQ (2022) Anti-allergic activity of natural plant products for the treatment of sensitive skin: a review. Pharmacol Res 3: 100117
3. Yong YK, Zakaria ZA, Kadir AA, Somchit MN, Ee Cheng Lian G, Ahmad Z (2013) Chemical constituents and antihistamine activity of Bixa orellana leaf extract. BMC Complement Altern Med 13:1–7
4. Abd Rani NZ, Kumolosasi E, Jasamai M, Jamal JA, Lam KW, Husain K (2019) In vitro anti-allergic activity of Moringa oleifera Lam. Extracts and their isolated compounds. BMC Complement Altern Med 19:1–16
5. Gong Y, Li HX, Guo RH, Widowati W, Kim YH, Yang SY, Kim YR (2021) Anti-allergic inflammatory components from the leaves of Piper crocatum Ruiz & Pav. Biol Pharm Bull 44(2):245–250
6. Wang J, Gu D, Wang M, Guo X, Li H, Dong Y, Guo H, Wang Y, Fan M, Yang Y (2017) Rational approach to solvent system selection for liquid–liquid extraction–assisted sample pre-treatment in counter–current chromatography. J Chromatogr B 1053:16–19

7. Kumar D, Bhat ZA, Singh P, Khatanglakar V, Bhujbal SS (2011) Antiasthmatic and antiallergic potential of methanolic extract of leaves of Ailanthus excelsa. Rev Bras Farmacogn 21: 139–145
8. Wagner H, Bladt S (1996) Plant drug analysis: a thin layer chromatography atlas. Springer Science & Business Media
9. Chandra P, Patel AK, Sachan N, Kabra A, Singh N, Alshammari A, Albekairi NA, Murthy HA (2025) Evaluation of antiallergic conjunctivitis effects of extract of Nelumbo nucifera rhizome and its isolated constituents against chemical-induced allergic conjunctivitis in experimental rats. J Chem 2025(1):6747119
10. Gomathi D, Kalaiselvi M, Ravikumar G, Devaki K, Uma C (2015) GC-MS analysis of bioactive compounds from the whole plant ethanolic extract of Evolvulus alsinoides (L.) L. J Food Sci Technol 52:1212–1217
11. Heinzerling L, Mari A, Bergmann KC, Bresciani M, Burbach G, Darsow U, Durham S, Fokkens W, Gjomarkaj M, Haahtela T, Bom A, Wöhrl S, Maibach H, Lockey R (2013) The skin prick test–European standards. Clin Transl Allergy 3:1–10
12. Lazzarini R, Duarte I, Ferreira AL (2013) Patch tests. An Bras Dermatol 88:879–888

Chapter 19

Determination of Anti-hypolipidemic Potential of Plant Leaf

Jeenatara Begum, Tamalika Chakraborty, Piyali Basak, and Rajarshi Das

Abstract

Hyperlipidemia is a high-risk factor for cardiovascular diseases. So, there is an increase in natural lipid-lowering agents. This study investigated the anti-hypolipidemic potential of selected plant leaf extracts using in vitro and in vivo assays. Phytochemical screening confirmed the presence of bioactive compounds responsible for the lipid-lowering effect, such as flavonoids, alkaloids, and saponins. The extract gave the result on hyperlipidemia animal models, significantly lowering total cholesterol, triglycerides, and low-density lipoprotein (LDL) levels while increasing high-density lipoprotein (HDL) levels. The enzymatic assays and histopathological studies add support to the therapeutic efficacy of the extract. This work suggests that the plant leaf extract has potential anti-hypolipidemic activity, which will be followed for the future development of natural lipid-lowering therapeutics.

Key words Hyperlipidemia, Cardiovascular diseases, Anti-hypolipidemic activity, Phytochemical screening, LDL, HDL, Histopathological studies

1 Introduction

Hyperlipidemia denotes an increase in the lipid and lipoprotein portion of the blood, which predisposes an individual to cardiovascular diseases, including atherosclerosis, stroke, and coronary artery disease. Hygienic practices that include sedentary lifestyles and poor eating habits combined with hereditary disposition make hyperlipidemia one of the increasing threats to one's life; hence, it necessitates effective lipid-lowering interventions (*see* **Note 1**). However, the majority use synthetic hypolipidemic agents, e.g., statins and fibrates, which lead to adverse effects on long-term use, including hepatotoxicity, myopathy, and gastrointestinal disturbances. Therefore, there has been increased interest in natural, plant-derived, side-effect-poor, effective liquid forms [1] (*see* **Note 2**).

We have been using medicinal plants in the traditional medicines for centuries, based on the claims of these plants to possess pharmacological efficacy, including lipid lowering [2].

Mitesh Kumar Dwivedi and A. Sankara Narayanan (eds.), *Leaves as Functional Foods*, Methods and Protocols in Food Science,
https://doi.org/10.1007/978-1-0716-5226-8_19,

'Bioactive compounds such as flavonoids, alkaloids, tannins, and saponins available in plant extracts have not only shown considerable hypolipidemia potential through effecting favorable changes in lipid metabolism, oxidative stress, and divergent improvement in enzymes associated with lipid regulation but also form an important part of the active ingredients present in these traditional formulations.' However, scientific validation of the data that increasingly gets recognition from the general public on plant-based remedies remains highly essential [1, 2] (*see* **Note 3**).

The study intends to investigate the possible anti-hypolipidemic attributes of a selected plant leaf extract through phytochemical analysis, in vitro enzyme inhibition assays, and in vivo lipid profile assessment in hyperlipidemic animal models. The research intends to provide support and evidence for the development of plant hypolipidemic agents that are safer and more sustainable alternatives to conventional lipid-lowering therapies by investigating their biochemical and physiological effects [2] (*see* **Note 4**).

2 Materials

I.
Plant Material Collection (*see* **Note 5**)

- Fresh leaves of the selected plant.
- Collection tools (scissors, gloves, sterile bags).

II.
Chemicals and Reagents (*see* **Note 6**).

- High-fat diet (HFD) for inducing hyperlipidemia.
- Standard hypolipidemic drug (e.g., Atorvastatin) as a positive control.
- Solvents for extraction (ethanol/methanol).
- Assay kits for lipid profile analysis: (*see* **Note 7**).
  - Total Cholesterol (TC).
  - Triglycerides (TG).
  - Low-Density Lipoprotein (LDL).
  - High-Density Lipoprotein (HDL).
- Normal saline for dilution.
- Hemotoxylin and Eosin (H&E) stains for histopathology.

III.
Experimental Animals (*see* **Note 8**).

- Male Wister rats (180–220 g).
- Standard laboratory diet and water ad libitum.

IV. *Equipment* (*see* **Notes 9** and **10**).

- UV-Visible spectrophotometer for biochemical analysis.
- Centrifuge for plasma separation.
- Biochemical analyzer for lipid profile measurement.
- Soxhlet apparatus for plant extraction.
- Analytical balance for weighing chemicals.
- Histopathology microscope for tissue analysis.
- Micro-pipettes and test tubes for sample handling.
- Glassware (beakers, flasks, Petri dishes).

3 Methods

I. *Preparation of Plant Extract* [3] (*see* **Note 11**)

Fresh plant leaves are collected and washed with distilled water to remove impurities and then dried at room temperature. The dried leaves are ground into a powder. This leaf powder is extracted by the Soxhlet apparatus using ethanol or methanol as an extraction solvent. This extract is filtered and concentrated using a rotary evaporator under reduced pressure. It is stored at 4 °C for further use.

II.
Induction of Hyperlipidemia [3, 4] (*see* **Note 12**)

1. 180–220 g of male Wister rats are maintained in standard laboratory conditions and kept under constant access to food and water.
2. They are being fed with a high-fat diet (HFD) for 4 weeks to induce hyperlipidemia.
3. Eventually, before the treatment begins, a baseline lipid profile will be drawn.

III.
Experimental Design [6] (*see* **Note 13**)

The rats have been divided into five groups (n = 6 per group): (Fig. 1).

1. Group I (Normal Control): Normal diet, no treatment.
2. Group II (Hyperlipidemic Control): HFD, no treatment.
3. Group III (Standard Drug Control): HFD + Standard hypolipidemic drug.
4. Group IV (Low Dose Plant Extract): HFD + Plant extract (200 mg/kg).

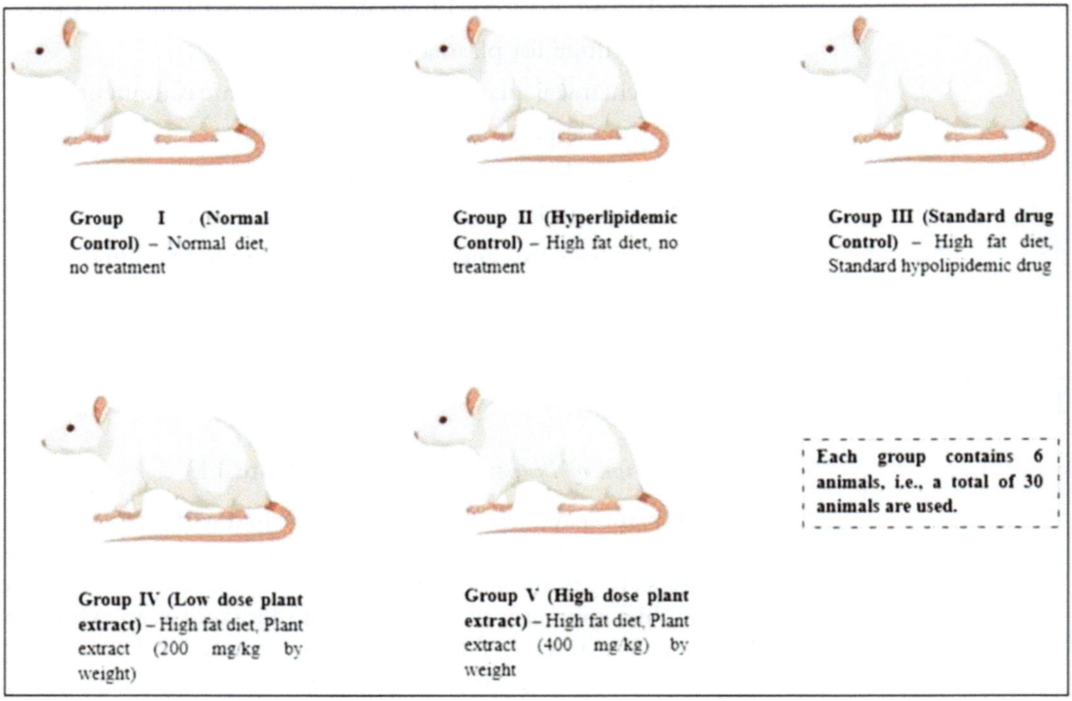

Fig. 1 Grouping of rats for the experiment

5. Group V (High Dose Plant Extract): HFD + Plant extract (400 mg/kg).

IV. *Treatment and Sample Collection* [5, 6] (*see* **Notes 14** and **15**)

1. Oral administration of plant extract and standard medicine would be used for four weeks.
2. Blood sample collection through "retro-orbital plexus" would take place under mild anesthesia at the end of the experiment.
3. The serum is separated through centrifugation at 3000 rpm for 10 min and stored at −20 °C for analysis.

V. *Biochemical Analysis* [5, 6] (*see* **Note 16**).

1. Complete serum cholesterol, serum triglycerides, low-density lipoprotein levels, and high-density lipoprotein are estimated using various biochemical assay kits.
2. The atherogenic index and the percent lipid reduction are calculated (Fig. 2).

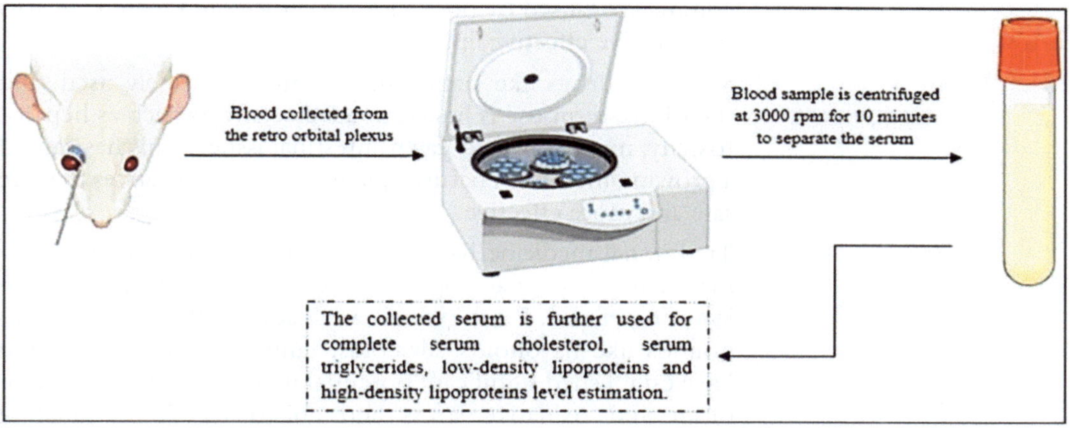

Fig. 2 Biochemical tests on the serum extracted from the blood from retro orbital plexus

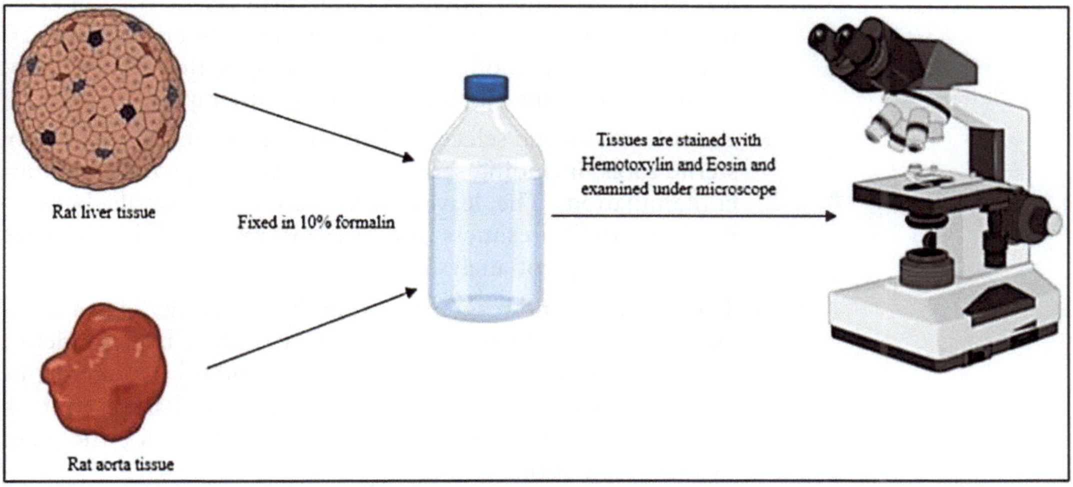

Fig. 3 Histopathological examination of liver and aorta tissues

VI. *Histopathological Examination* [5, 6] (*see* **Notes 17** and **18**)

1. Tissues of the liver and aorta were excised and fixed in 10% formalin solution and processed for histopathology.
2. Microscopic analysis of tissue sections previously stained by Hematoxylin and Eosin (H & E) will be performed for lipid accumulation and structural changes. (Fig. 3).

4 Notes

1. Hyperlipidemia is characterized by lipid and lipoprotein levels in the blood that contribute to cardiovascular diseases such as atherosclerosis, stroke, and coronary artery disease. It is

commonly caused by poor lifestyle choices and genetic factors that involve effective lipid-lowering treatments.

2. Synthetic drugs like statins and fibrates are widely used, and their long-term use is linked to adverse effects such as hepatotoxicity, myopathy, and gastrointestinal issues. This has caused a growing interest in natural, plant-derived alternatives that are safe and more effective.
3. Traditional medicine has long used medicinal plants for their maintained pharmacological benefits, which include lipid-lowering effects. Plant extracts are loaded with bioactive compounds like flavonoids, alkaloids, tannins, and saponins and have established significant potential in improving lipid metabolism and oxidative stress and enhancing lipid-regulating enzymes.
4. Investigate the anti-hyperlipidemic potential of a selected plant leaf extract through phytochemical analysis, in vitro enzyme inhibition assays, and in vivo lipid profile evaluations in hyperlipidemic animal models, therefore supporting the development of safe, plant-based hypolipidemic therapies.
5. Fresh leaves of the selected plant were collected by the use of sterile tools, including scissors and gloves, to provide minimal contamination. The leaves were placed into sterile bags to maintain their freshness and prevent microbial interference during subsequent analyses.
6. The chemicals and reagents used in this study included a high-fat diet (HFD), which was formulated to induce hyperlipidemia in experimental models, along with a standard hypolipidemic drug such as Atorvastatin, which acted as a positive control. Solvents like ethanol or methanol were utilized for the plant extract preparation.
7. Different assay kits were analyzing the lipid profile, including total cholesterol (TC), triglycerides (TG), low-density lipoprotein (LDL), and high-density lipoprotein (HDL) for the biochemical estimations. Normal saline was used for dilutions, and Hematoxylin and Eosin (H & E) stains were applied for histopathological examinations.
8. Male Wister rats weighing between 180 and 220 g were used as the experimental animals for the study. They were maintained under standard laboratory conditions and provided with a regular diet and water ad libitum throughout the experimental period to verify an optimal health and stability in physiological responses.
9. The equipment utilized in the study that's included the UV-Visible spectrophotometer is used for conducting biochemical analyses, and the centrifuge is used for plasma

separation. A biochemical analyzer was employed to measure lipid profiles, while a Soxhlet apparatus was used for plant extraction.

10. An analytical balance secured exact weighing of chemicals and histopathological examinations, which were performed by using a microscope. Additionally, micropipettes and test tubes are promoted in accurate sample handling and supported by standard laboratory glassware such as beakers, flasks, and Petri dishes.
11. The preparation of the plant extract is washing the freshly collected leaves of the selected plant using distilled water, which is followed by shade drying to preserve their phytochemical character. The leaves were dried and mechanically ground into a fine powder, which was then extracted with Soxhlet extraction using the solvents such as ethanol or methanol. Then the extract was filtered and concentrated under reduced pressure using a rotary evaporator. Finally, the concentrated extract was stored at 4 °C until further use.
12. Male Wister rats weighing between 180 and 220 g that were situated under standard laboratory conditions with continuous access to food and water. To induce hyperlipidemia, the animals were fed a high-fat diet (HFD) over a period of four weeks. Before starting any treatment, baseline lipid profile measurements were obtained to assess the extent of hyperlipidemia induction.
13. The experimental design involved dividing the rats into 5 groups, with each group consisting of 6 animals. Group I served as the normal control and received a standard diet without any treatment. Group II consists of the hyperlipidemic control, which was fed a high-fat diet (HFD) without treatment to induce hyperlipidemia. Group III received the HFD along with a standard hypolipidemic drug as the positive control. Groups IV and V were administered the HFD along with plant extract at doses of 200 mg/kg and 400 mg/kg body weight, respectively, which represent the low- and high-dose treatment groups.
14. The treatment phase involved an oral administration of the plant extract and standard hypolipidemic drug for a duration of four weeks.
15. At the end of the experimental period, blood samples were collected from the retro-orbital plexus under mild anesthesia. The collected blood was centrifuged at 3000 rpm for 10 min to separate the serum, which was stored at −20 °C until further biochemical analysis.
16. Biochemical analysis involved the estimation of complete serum lipid parameters, including total cholesterol, serum

triglycerides, low-density lipoprotein (LDL), and high-density lipoprotein (HDL), by the use of specific biochemical assay kits. Additionally, the atherogenic index and the percentage reduction in lipid levels were calculated to evaluate the hypolipidemic efficacy of provided the treatments.

17. For histopathological examination, liver and aorta tissues were removed and fixed in a 10% formalin solution before being progressed for histological analysis.
18. Tissue sections were stained with Hematoxylin and Eosin (H & E) and a microscopic examination was done to evaluate the lipid accumulation and any structural changes in the tissues.

References

1. Shedage AT, Pinjari RM (2023) A review on antihyperlipidemic activity of medicinal plants. Int J For Anim Fish Res 7(5):33–38. https://doi.org/10.22161/ijfaf.7.5.5
2. Rayadurgam S, Manikandan K (2022) Overview of effective traditional medicinal plants having antihyperlipidemic activity. J Nat Remedies:310–317. https://doi.org/10.18311/jnr/2022/29791
3. Lawal H, Ado AM, Usman M, Garba US (2024) Antihyperlipidemic activity of Plectranthus neochilus leaf ethanolic extract on fat-fed male Wistar rats. Int J Sci Global Sustain 10(2):28–32. https://doi.org/10.57233/ijsgs.v10i2.639
4. Nwaji AR (2022) Antihyperlipidemic property of Murraya koenigii ethanolic leaf extract on lipid profile of rats fed with cholesterol-enriched diets. Niger J Physiol Sci 37(2):215–218. https://doi.org/10.54548/njps.v37i2.8
5. Atanu FO, Jegede RE, Apeh DO, Suleiman MS (2021) Anti-hyperglycemic and anti-hyperlipidemic effects of hydroethanol leaf and stem extracts of Vitex doniana in diabetic rats. Asian J Res Biochem:53–59. https://doi.org/10.9734/ajrb/2021/v8i330184
6. Sudirman S, Pujiastuti N, Janna M, Herpandi, Widiastuti I (2023) Antihyperlipidemic activity of purified polyphenol extracted from water lettuce (Pistia Stratiotes) leaf: an in vitro analysis. Trop J Nat Prod Res 7(6):3177–3181. https://doi.org/10.26538/tjnpr/v7i6.17

Chapter 20

Determination of Antihyperlipidemic Potential of Plant Leaf

Maya P. Radeva-Ilieva and Stanila S. Stoeva-Grigorova

Abstract

Hyperlipidemia, a type of dyslipidemia, involves elevated levels of LDL, VLDL, total cholesterol, and triglycerides. It is a major risk factor for atherosclerosis and cardiovascular disease. Therefore, animal models of hyperlipidemia are commonly used to test the lipid-lowering effect of different natural compounds. Among various methods described in the literature, high-fat diet feeding is preferred. Such diets typically include ingredients like cholesterol, lard, sucrose, or fructose.

Key words Hyperlipidemia, Dyslipidemia, High-fat diet, LDL, VLDL, Total cholesterol, Triglycerides, Animal models, Rats, Mice

1 Introduction

The term "dyslipidemia" refers to abnormally high or low plasma levels of lipids such as total cholesterol, low-density lipoprotein (LDL) cholesterol, very low-density lipoprotein (VLDL) cholesterol, high-density lipoprotein (HDL) cholesterol, and triglycerides [1]. In particular, hyperlipidemia is characterized by an increase in LDL cholesterol, VLDL cholesterol, total cholesterol, and/or triglycerides [2]. Additionally, a decrease in HDL cholesterol may also be observed [3]. Hyperlipidemia is the main cause of atherosclerosis development, thus leading to an increased risk for life-threatening cardiovascular incidents [4]. Moreover, a low plasma level of HDL cholesterol is thought to be a single risk factor for cardiovascular diseases [5].

Animal hyperlipidemia model is widely used for investigating the lipid-lowering effect of many natural compounds. There are different methods to induce hyperlipidemia in laboratory animals, such as high-fat diet feeding, high-fat emulsion gavage, corticosteroid administration, etc. However, high-fat diet-induced hyperlipidemia most closely resembles hyperlipidemia associated with a long-term cholesterol-rich diet in humans. Various compositions

Mitesh Kumar Dwivedi and A. Sankara Narayanan (eds.), *Leaves as Functional Foods*, Methods and Protocols in Food Science, https://doi.org/10.1007/978-1-0716-5226-8_20,

of high-fat diets are reported in the literature, but most of them contain basal feed as well as cholesterol, lard, and sucrose, or fructose [6].

2 Materials

2.1 Assessment of Antihyperlipidemic Potential in a High-Fat Fed ICR Mouse Model [7–9]

- Test compound.
- Animals—young male ICR mice (3-week-old) [7] or adult male ICR mice (6-week-old) [8]; 10 animals per group ($n = 10$) [3, 7].
- Animal cage and house facility.
- Normal diet.
- High Fat Diet (HFD)—rich in cholesterol and lard [3, 7, 8] (*see* **Note 1**).
- Standard diagnostic kits [7] or automatic blood biochemical analyzer [3] for estimation of lipid profile markers and liver enzymes.
- 10% formalin.
- Paraffin.
- Hematoxylin-eosin dyes.
- Centrifuge.
- Microscope.

2.2 Assessment of Antihyperlipidemic Potential in a High-Fat-Fed Mouse Model [2]

- Test compound.
- Atorvastatin.
- Carboxymethylcellulose (CMC).
- Animals—adult healthy male Swiss Albino mice; weight 25–30 g; 6 animals per group ($n = 6$).
- Animal cage and house facility.
- Normal diet pellets.
- High Fat Diet (HFD) pellets (*see* **Note 2**).
- Standard diagnostic kits for estimation of lipid profile markers and liver enzymes (serum total cholesterol, total triglyceride, LDL, VLDL, HDL, SGOT, SGPT, serum ALP, and total protein levels).
- Phosphate-buffered saline.
- 10% neutral buffered formalin.
- Paraffin.
- Hematoxylin-eosin dyes.
- Centrifuge.

- Microtome.
- Fluorescence microscope with an attached camera.

2.3 Assessment of Antihyperlipidemic Potential in a High-Fat-Fed Rat Model [10, 11]

- Test compound.
- Simvastatin [12].
- Animals—adult healthy male Sprague–Dawley rats; weight 300 ± 20 g; 8 animals per group ($n = 8$).
- Animal cage and house facility.
- Basal diet.
- High Fat Diet (HFD) (*see* **Note 3**).
- Standard diagnostic kits for estimation of lipid profile markers (serum total cholesterol, total triglyceride, LDL, HDL) and liver enzymes (aspartate aminotransferase (AST), alanine aminotransferase (ALT)) [12].
- Saline [12].
- 10% formalin.
- Paraffin.
- Hematoxylin-eosin dyes.
- Centrifuge.
- Microtome.
- Microscope with an attached camera.

3 Methods

3.1 Assessment of Antihyperlipidemic Potential in a High-Fat-Fed ICR Mouse Model [7]

ICR mice represent a well-established experimental model for investigating lipid metabolism [7]. It should be noted that the induction of hyperlipidemia with HFD usually takes a long time (several weeks to months) [9].

3.1.1 Animals' Growth Conditions

- Use young or adult male ICR mice (25–35 g) [3, 7, 8].
- Animals are housed in cages under standard lighting (12 h day/night cycle) and relative humidity (55 ± 5%) conditions in temperature-controlled rooms (23 ± 1 °C), with free access to pellet diets and water ad libitum [8].
- Acclimatization of the animals is carried out for a certain time interval (at least one week) prior to the experimental procedures [3, 7, 8].
- Body weight is measured at a certain time interval (once daily [3], every three days [8] or once weekly [7]).
- All experimental procedures are carried out according to the relevant national requirements.

3.1.2 Experimental Design

- Randomly assign the experimental animals to several groups, ensuring an equal number of animals in each group ($n = 10$):
  - Control group (normal diet).
  - Hyperlipidemia model group (HFD).
  - Treatment group (HFD + tested substance).
- Form more than one treatment group if different doses of the test substance or combinations of different test substances are to be administered [3, 8].
- The total duration of the experiment is 8 weeks [7].

3.1.3 Blood Collection

At the end of the experiment, collect blood samples from the retro-orbital sinus of mice following a 12-hour fasting period. Allow the blood to stand at room temperature for 30 min, then centrifuge at 1500 rpm for 15 min. Use the resulting serum for biochemical analysis of lipid and liver biomarkers.

3.1.4 Biochemical Studies

- Evaluate biomarkers for lipid profile (serum total cholesterol, total triglycerides, HDL), and liver enzymes (ALT and AST) using standard diagnostic kits.
- Calculate LDL-C values using the following equation:
  - $C_{LDL} = C_{plasma}—C_{HDL}—TG/5$ [13].
- Calculate VLDL-C values using the equation: *TG/5*.

3.1.5 Histopathological Studies

- After blood sample collection, sacrifice the animals. Remove the liver, pancreas, epididymal, and subcutaneous fat tissue samples, then fix them in 10% formalin and embed them in paraffin.
- Perform microtomy on the paraffin-embedded tissues to obtain ultrathin slides.
- Stain the prepared slides using hematoxylin-eosin dyes.
- Examine each slide under a microscope at magnifications of 20× and 40× .

3.1.6 Statistical Analysis

Conduct statistical analysis on the obtained results to prove statistical significance.

3.2 Assessment of Antihyperlipidemic Potential in a High-Fat-Fed Swiss Albino Mouse Model [2]

The inclusion of a positive control group allows the obtained lipid-lowering results of the test substance to be compared with the effect of a standard lipid-lowering drug such as atorvastatin or simvastatin [2, 14].

3.2.1 Animals' Growth Conditions

- Use adult healthy male Swiss Albino mice (25–30 g).
- House the animals in polypropylene cages under standard environmental conditions, including a 12-hour light/dark cycle, relative humidity of 50 ± 15%, and temperature maintained at 22 ± 1 °C. Provide the animals with free access to a standard pellet diet for mice and water ad libitum. Acclimate the animals for 10 days prior to the initiation of experimental procedures. Conduct all experimental procedures in accordance with the relevant national regulatory requirements.

3.2.2 Experimental Design

- Randomly assign the experimental animals to several groups, ensuring that each group contains at least six mice ($n = 6$):
  - Control group (normal diet +0.3% aqueous solution of CMC or another vehicle, p.o.).
  - Hyperlipidemia model group (HFD + 0.3% aqueous CMC or another vehicle, p.o.).
  - Positive control group (HFD + Atorvastatin, 10 mg/kg daily, p.o.).
  - Treatment group (HFD + tested substance dissolved or suspended in the vehicle, p.o.).
- Form more than one treatment group if different doses of the test substance or combinations of different test substances are to be administered.
- Set the total duration of the experiment to 90 days. Administer the high-fat diet (HFD) continuously throughout the entire period, while the test substance and the standard lipid-lowering drug, atorvastatin, are administered during the second 45-day interval of the study [2].

3.2.3 Blood Collection

At the end of the experiment, collect blood samples from the retro-orbital sinus of mice following a 24-hour fasting period. Centrifuge the blood samples at 2500 rpm for 15 min. Use the obtained serum for biochemical analysis of lipid and liver biomarkers.

3.2.4 Biochemical Studies

Evaluate biomarkers for lipid profile (serum total cholesterol, total triglycerides, LDL, VLDL, and HDL) and liver enzymes (SGOT, SGPT, serum ALP, and total protein levels) using standard diagnostic kits.

3.2.5 Histopathological Studies

- After blood sample collection, sacrifice the animals under anesthesia. Then, remove the liver for subsequent histopathological examination.

- Process the extracted liver tissues sequentially as follows:
  - Rinse the liver tissues in phosphate-buffered saline.
  - Fix the tissues in 10% neutral buffered formalin.
  - Dehydrate the tissues using standardized ascending grades of alcohol.
  - Embed the tissues in paraffin.
- Perform microtomy on paraffin-embedded tissues to obtain ultrathin slides of 5 μm thickness.
- Stain the obtained slides with hematoxylin-eosin dyes.
- Examine five slides per animal's liver.
- Screen each slide for signs of liver injury and hepatotoxicity.

3.2.6 Statistical Analysis

Conduct statistical analysis on the obtained results to prove statistical significance.

3.3 Assessment of Antihyperlipidemic Potential in a High-Fat-Fed Rat Model [10, 11]

The addition of bile acid salts such as sodium cholate to the rat diet leads to increased fatty acid absorption, while propylthiouracil administration is linked to reduced lipid oxidation and enhanced plasma levels of cholesterol [11].

3.3.1 Animals' Growth Conditions

- Use adult healthy male Sprague-Dawley rats (300 ± 20 g).
- House the animals in cages under standard environmental conditions, including a 12-hour light/dark cycle, relative humidity of 55 ± 5%, and temperature maintained at 23 ± 1 °C. Provide animals with free access to food and water ad libitum. Acclimate the animals for 7 days prior to commencing experimental procedures [11, 12].
- Measure body weight at regular intervals, specifically once weekly [12].
- Conduct all experimental procedures in compliance with the relevant national regulatory requirements.

3.3.2 Experimental Design

- Randomly assign the experimental animals to several groups, ensuring each group contains eight rats ($n = 8$). Control group (basal diet).
  - Hyperlipidemia model group (HFD).
  - Positive control group (HFD + Simvastatin, 4 mg/kg daily, p.o.) [12].
  - Treatment group (HFD + tested substance).
- Form more than one treatment group if different doses of the test substance or different test substances are to be administered.

- Set the total duration of the experiment to 9 weeks [10–12].
- Administer the high-fat diet (HFD) to the animals continuously throughout the experiment. Administer the tested substances during the final 5 weeks of the study. According to Zhao et al. (2019), administer simvastatin and the tested substances for 3 weeks [12].

3.3.3 Blood Collection

At the end of the experiment, collect blood samples from the retro-orbital sinus of rats under anesthesia using ketamine (30 mg/kg) and xylazine (0.5 mg/kg) administered intramuscularly. Centrifuge the blood samples at 3500 rpm for 15 min. Use the obtained serum for biochemical analysis of lipid and liver biomarkers.

3.3.4 Biochemical Studies

Determine plasma levels of lipids and liver enzymes using a spectrophotometer analyzer.

- *Histopathological Studies.*
- After blood sample collection, sacrifice the animals under anesthesia. Subsequently, remove liver, aorta, and myocardium tissue samples for further histopathological analysis.
- Fix the extracted tissues in 10% formalin and embed them in paraffin.
- Perform microtomy on the paraffin-embedded tissues to obtain ultrathin slides of 4–5 μm thickness.
- Stain the prepared slides with hematoxylin-eosin dyes.
- Screen each slide for signs of tissue injury.

3.3.5 Statistical Analysis

Conduct statistical analysis on the obtained results to prove statistical significance.

4 Notes/Observation (Tables, Images/Figures)

1. An example of normal and high-fat diet composition is presented on Table 1 [7].
2. Composition of normal and high-fat diet pellets is shown on Table 2 [2].
3. A high-fat diet composition for establishment of an experimental model of hyperlipidemia in rats is presented on Table 3 [11].
4. In addition to the high-fat diet, high-fat emulsion may be administered by oral gavage once daily at a dose of 1 mL/100 g b.w. to induce hyperlipidemia in rats. The emulsion consist of oil phase and aqueous phase as is shown on Table 4 [15].

Table 1
Composition of normal and high-fat diet

Ingredients	Content of normal diet (g)	Content of HFD (g)
Corn starch	315	106
Casein	117.5	117.5
L-Cystine	2.75	2.75
Sucrose	50	50
Maltodextrin 10	50	50
Cholesterol	0	1
Lard	16.5	110
Soybean oil	10	10
Cellulose	25	25
Mineral mix	17.5	17.5
Vitamin mix	5	5
Calcium carbonate	15	15
Choline bitartrate	1.25	1.25

Table 2
Composition of normal and high-fat diet pellets

Ingredients	Content of normal diet (%)	Content of HFD (%)
Protein	16.5	19
Fat	2	17.50
Fibers	8	3.5
Ashes	8	3.5
NaCl	1	0
Vitamins		
Minerals		

Preparation of the oil phase:

1. Add lard to a beaker and heat to 100 °C.
2. Add the cholesterol.
3. When cholesterol is melted add propylthiouracil tablets.
4. After the mixture is homogenized, add polysorbate 80.

Table 3
High-fat diet composition

Ingredient	Content (%)
Basal diet	75
Lard	9
Sugar	7.5
Yolk powder	5.5
Cholesterol	2.5
Sodium cholate	0.3
Propylthiouracil	0.2

Table 4
Content of high-fat emulsion

Oil phase		Aqueous phase	
Lard	25 g	Distilled water	30 mL
Cholesterol	10 g	Propylene glycol	20 mL
Propylthiouracil tabl.	1 g	Sodium deoxycholate	2 g
Polysorbate 80	25 mL		

Table 5
Experimental model of metabolic syndrome in rats

	Food	Water	Duration of the experiment
Experimental design	Lard (17%) and fructose (17%) enriched rat chow	Water containing 10% fructose	8 weeks

Preparation of the aqueous phase:

1. Add distilled water to a beaker and heat in a 60 °C in water bath.
2. Add propylene glycol and sodium deoxycholate.

When sodium deoxycholate is fully dissolved, aqueous phase is added into the oil phase. The mixture is stirred, and the finished emulsion is stored in a refrigerator (4 °C).

5. An easy reproducible and cheap way to induce metabolic syndrome, particularly hyperlipidemia, in rats is to enrich the standard rat chow with lard and fructose as well as to add fructose in the drinking water [16]. The method is presented in Table 5:

References

1. Pappan N, Awosika AO, Rehman A (2025) Dyslipidemia. [Updated 2024 Mar 4]. In: StatPearls [Internet]. StatPearls Publishing, Treasure Island. https://www.ncbi.nlm.nih.gov/books/NBK560891/
2. Firdous SM, Hazra S, Gopinath SCB, El-Desouky GE, Aboul-Soud MAM (2021) Antihyperlipidemic potential of diosmin in Swiss Albino mice with high-fat diet induced hyperlipidemia. Saudi J Biol Sci 28(1): 109–115. https://doi.org/10.1016/j.sjbs.2020.08.040
3. Liu P, Yu S, Liu J, Zhou Y, Cao R, Zhou Y, Shi L, Du J (2020) Effects of *Lactobacillus* on hyperlipidemia in high-fat diet-induced mouse model. Arch Med Sci 19(3):792–799. https://doi.org/10.5114/aoms.2020.98927
4. Blinc L, Mlinaric M, Battelino T, Groselj U (2020) High-sensitivity C-reactive protein and carotid intima media thickness as markers of subclinical inflammation and atherosclerosis in pediatric patients with hypercholesterolemia. Molecules 25(21):5118. https://doi.org/10.3390/molecules25215118
5. Cooney MT, Dudina A, De Bacquer D, Wilhelmsen L, Sans S, Menotti A, De Backer G, Jousilahti P, Keil U, Thomsen T, Whincup P, Graham IM, SCORE investigators. (2009) HDL cholesterol protects against cardiovascular disease in both genders, at all ages and at all levels of risk. Atherosclerosis 206(2): 611–616. https://doi.org/10.1016/j.atherosclerosis.2009.02.041
6. Chen X, Zhou Y, Yang J, Yang R, Xue S, Wang Q, Niu W (2025) Animal model screening for hyperlipidemic ICR mice. Int J Mol Sci 26(5):2142. https://doi.org/10.3390/ijms26052142
7. Edward OC, Jeong DY, Yang HJ, Han A, Cha YS (2024) *Doenjang* ameliorates diet-induced hyperlipidemia and hepatic oxidative damage by improving lipid metabolism, oxidative stress, and inflammation in ICR mice. Foods 13(10):1471. https://doi.org/10.3390/foods13101471
8. Li Y, Luo Y-C, Chen G, Xiao Z-L, Wang Q, Zhao L, Ji B-P (2011) Comparison of protective effects between oat β-glucan and phenol-rich extracts in hyperlipidemic ICR mice. J Food Drug Anal 19(1):Article 8. https://doi.org/10.38212/2224-6614.2190
9. Andreadou I, Schulz R, Badimon L, Adameová A, Kleinbongard P, Lecour S, Nikolaou PE, Falcão-Pires I, Vilahur G, Woudberg N, Heusch G, Ferdinandy P (2020) Hyperlipidaemia and cardioprotection: animal models for translational studies. Br J Pharmacol 177(23):5287–5311. https://doi.org/10.1111/bph.14931
10. Zălar DM, Pop C, Buzdugan E, Kiss B, Ştefan MG, Ghibu S, Crişan D, Buruiană-Simic A, Grozav A, Borda IM, Mogoşan CI (2022) Effects of colchicine in a rat model of diet-induced hyperlipidemia. Antioxidants (Basel) 11(2):230. https://doi.org/10.3390/antiox11020230
11. Zălar DM, Pop C, Buzdugan E, Kiss B, Ştefan MG, Ghibu S, Bâlteanu VA, Crişan D, Buruiană-Simic A, Grozav A, Mogoşan CI (2021) Pharmacological effects of methotrexate and infliximab in a rats model of diet-induced dyslipidemia and Beta-3 overexpression on endothelial cells. J Clin Med 10(14): 3143. https://doi.org/10.3390/jcm10143143
12. Zhao X, Zhu J, Wang L, Li Y, Zhao T, Chen X, Sun Y, Dai Y, Wei G, Altamirano A, Zhang T, Yan Z (2019) U. diffracta extract mitigates high fat diet and VD3-induced atherosclerosis and biochemical changes in the serum liver and aorta of rats. Biomed Pharmacother 120: 109446. https://doi.org/10.1016/j.biopha.2019.109446
13. Friedewald WT, Levy RI, Fredrickson DS (1972) Estimation of the concentration of low-density lipoprotein cholesterol in plasma, without use of the preparative ultracentrifuge. Clin Chem 18(6):499–502. https://doi.org/10.1093/clinchem/18.6.499
14. Tunsophon S, Chootip K (2016) Comparative effects of piperine and simvastatin in fat accumulation and antioxidative status in high fat-induced hyperlipidemic rats. Can J Physiol Pharmacol 94(12):1344–1348. https://doi.org/10.1139/cjpp-2016-0193
15. Kong X, Gao Y, Geng X, Xie G, Hao S, Li Y, Zhang Z (2018) Effect of lipid lowering tablet on blood lipid in hyperlipidemia model rats. Saudi J Biol Sci. 25(4):715–718. https://doi.org/10.1016/j.sjbs.2017.12.002
16. Gancheva S, Zhelyazkova-Savova M, Galunska B, Chervenkov T (2015) Experimental models of metabolic syndrome in rats. Scripta Scientifica Medica 47(2):14–21

Chapter 21

Determination of the Antihyperglycemic Potential of Plant Leaf

Chetna Rathi, Bhupender Yadav, Simran Rani, Priyanka Dahiya, Nitin Chawaria, H. Vajiha Banu, Amita Suneja Dang, and Pooja Suneja

Abstract

Diabetes mellitus is a chronic metabolic disorder characterized by persistent hyperglycemia. Plant leaves have long been used in traditional medicine to treat diabetes mellitus, prompting research into their ability to decrease blood sugar levels. Antihyperglycemic potential of plant leaf extracts is assessed by targeting carbohydrate breakdown enzymes (α-amylase and α-glucosidase). Plant leaves, which are high in phytochemicals such as alkaloids, flavonoids, tannins, and saponins, show promise regarding carbohydrate metabolism regulation, insulin sensitivity improvement, and postprandial glucose spike reduction.

Key words Diabetes mellitus, Antihyperglycemic, Carbohydrates, Traditionally, Postprandial

1 Introduction

Diabetes mellitus is a common metabolic condition that affects people of all ages. It is characterized by hyperglycemia caused by deficits in insulin function, secretion, or both [1]. High blood sugar, or hyperglycemia, is a major contributing factor to the emergence of severe complications from type 2 diabetes, including harm to the kidneys, nerves, and eyes. The maintenance of glucose homeostasis is a complex interplay of hormonal regulation, enzymatic activity, and cellular signaling pathways, all of which can be disrupted in metabolic disorders such as diabetes mellitus [2]. Conventional treatments for diabetes often involve insulin therapy or oral hypoglycemic agents, but these approaches can sometimes lead to adverse effects, including hypoglycemia [3, 4]. Synthetic medications that are already used for this purpose frequently have unpleasant side effects such as bloating, diarrhea, and discomfort in the abdomen. The use of medicinal plants for treating diabetes dates back centuries, with various cultures worldwide employing plant-based remedies to manage blood sugar levels [5]. These

Mitesh Kumar Dwivedi and A. Sankara Narayanan (eds.), *Leaves as Functional Foods*, Methods and Protocols in Food Science, https://doi.org/10.1007/978-1-0716-5226-8_21,

plants often contain diverse bioactive compounds, including glycosides, alkaloids, terpenoids, and flavonoids, which have been implicated in their antidiabetic effects [6]. α-amylase and α-glucosidase are two enzymes that are essential to this process because they convert complex carbohydrates into simple sugars that the circulation can absorb rapidly [7]. By reducing the activity of these enzymes, the blood sugar rise after meals can be controlled. Recent research supports its potential function in regulating high blood sugar by indicating that its extracts may also inhibit α-amylase and α-glucosidase [8]. Plant leaves are abundant in natural substances with various biological activities, such as anti-inflammatory, anticancer, antidiabetic, and antioxidant properties, such as gallic acid, trilobatin, resveratrol, and quercetin. Recent studies on plant leaves have shown that they can block α-glucosidase, control intestinal cell glucose uptake and transporter activity, and even reduce blood sugar and oxidative stress [9]. Plant leaf extract provides promising natural sources of bioactive chemicals that could be used as safer plant-based substitutes for synthetic medications to treat diabetes [10].

2 Materials

2.1 Chemicals and Reagents

- Fresh plant leaves.
- Distilled water.
- Ethanol.
- Acetone.
- Starch.
- α-Amylase enzyme (Himedia).
- α-Glucosidase enzyme (Himedia).
- Dinitrosalicylic acid.
- 10% Dimethyl sulfoxide (DMSO).
- p-Nitrophenyl-α-D-glucopyranoside (pNPG).
- Sodium carbonate (0.1 M).
- Sodium phosphate buffer (0.02 M, pH 6.9).

2.2 Equipment and Instruments

- Pipettes and micropipettes (10–1000 μL range).
- Measuring cylinders.
- Conical flasks.
- Beakers.
- Test tubes.
- Micropipette tips.
- Spectrophotometer.

- Lyophilizer.
- Rotary evaporator.
- Grinder.
- Water bath.
- Vortex.

3 Methodology

3.1 Collection and Preparation of Plant Extracts [11–13]

- Collect fresh leaves from plants.
- Wash the collected leaves thoroughly to remove dust and debris.
- Shade-dry the leaves at room temperature until completely dehydrated. Using a grinder, powder the dried leaves.
- Use powdered leaves for extraction using distilled water/ ethanol/ acetone.
- Steep each extract in closed containers for 24 h at room temperature.
- Filter the resulting solutions using filter paper. Concentrate each filtrate using a rotary evaporator. Lyophilize the concentrated extracts using a freeze dryer.
- Reconstitute the freeze-dried powder in 10% dimethyl sulfoxide (DMSO) and use for further assay (*see* **Note 1**).

3.2 α-Amylase Inhibitory Assay [14]

- In a test tube, containing 250 μL of plant extract, add 250 μL of 0.02 M sodium phosphate buffer (pH 6.7), containing the enzyme α-amylase.
- To enable enzyme interaction, preincubate the mixture for ten minutes at 25 °C (*see* **Note 2**).
- To initiate the reaction, add 250 μL of 1% starch solution in 0.02 M sodium phosphate buffer.
- Incubate again at 25 °C for 10 min to allow enzymatic hydrolysis of starch.
- Stop the reaction by adding 500 μL of dinitrosalicylic acid (DNS) reagent.
- Place the tubes in boiling water for 5 min to develop the color. Cool the tubes to room temperature, then dilute the contents with 5 mL of distilled water.
- Measure absorbance at 540 nm and prepare a control sample using distilled water instead of the extract (*see* **Note 3**).
- Calculate the inhibition percentage of α-amylase using:

  % Inhibition = [(Abs control—Abs extracts)/Abs control] × 100

- Determine the IC_{50} value (the concentration of extract causing 50% inhibition) graphically from inhibition vs. concentration plots.

3.3 α-Glucosidase Inhibitory Assay [12, 15]

- Prepare the p-nitrophenyl-α-D-glucopyranoside (pNPG) substrate solution in 20 mM phosphate buffer (pH 6.7).
- Pipette 100 μL of α-glucosidase into a test tube.
- Add 50 μL of plant extract.
- Pre-incubate the mixture at room temperature (25 °C) for 10 min. Initiate the reaction by adding 50 μL of 3.0 mM pNPG solution (*see* **Note 2**).
- At 37 °C, incubate the mixture for 20 min.
- To stop the reaction, add 2 mL of 0.1 M Na_2CO_3.
- Measure the absorbance at 405 nm to detect the amount of p-nitrophenol released, which indicates enzyme activity (*see* **Note 3**).
- Use distilled water as a control for 100% enzyme activity.
- Calculate the percentage inhibition using the formula:

  % Inhibition = [(Abs control—Abs extract)/Abs control] × 100.
- Determine the IC_{50} value (concentration of extract that inhibits 50% of enzyme activity) graphically from the inhibition curve.

4 Notes

1. Ensure that all extracts are dissolved in a suitable solvent (e.g., 10% DMSO) that does not interfere with enzyme activity.
2. Pre-incubation of the enzyme with the extract allows proper interaction and reflects inhibitory potential. However, incubation time and temperature should be optimized and consistent across samples.
3. Spectrophotometric measurements must be taken immediately after the reaction is stopped to avoid changes in absorbance due to degradation or precipitation.

Acknowledgments

The authors are thankful for the infrastructural support provided by the Department of Microbiology and the Centre for Medical Biotechnology at Maharshi Dayanand University (MDU), Rohtak, India. CR, BY, SR, and NW are grateful to the Council of Scientific & Industrial Research (CSIR) in New Delhi, India, for the CSIR fellowship. PD acknowledges the fellowship support from the

University Grants Commission (UGC), New Delhi, India. The authors are thankful to the Department of Science and Technology, Govt. of Haryana, India (Grant No. HSCSIT/R&D/2022/2968) for providing infrastructural and financial support to undertake this research work.

References

1. Rathee S, Mogla OP, Sardana S, Vats M, Rathee P (2010) Antidiabetic activity of Capparis decidua Forsk Edgew. J Pharm Res 3: 231–234
2. Ononamadu CJ, Ibrahim A (2021) Molecular docking and prediction of ADME/drug-likeness properties of potentially active antidiabetic compounds isolated from aqueous-methanol extracts of Gymnema sylvestre and Combretum micranthum. Biotechnologia 102:85
3. Wang Z, Wang J, Chan P (2013) Treating type 2 diabetes mellitus with traditional Chinese and Indian medicinal herbs. Evid Based Complement Alternat Med 2013:343594
4. Udayakumar R, Kasthurirengan S, Mariashibu TS, Rajesh M, Anbazhagan VR, Kim SC, Choi CW (2009) Hypoglycaemic and hypolipidaemic effects of Withania somnifera root and leaf extracts on alloxan-induced diabetic rats. Int J Mol Sci 10:2367–2382
5. Mayyas F, Omeish A (2025) Comparison of the cardioprotective effects of liraglutide, dapagliflozin and their combination in a rat model of diabetes induced by streptozotocin. Life Sci 375:123721
6. Patel DK, Prasad SK, Kumar R, Hemalatha S (2012) An overview on antidiabetic medicinal plants having insulin mimetic property. Asian Pac J Trop Biomed 2:320–330
7. Bhatnagar A, Mishra A (2022) α-Glucosidase inhibitors for diabetes/blood sugar regulation. In: Natural products as enzyme inhibitors: an industrial perspective. Springer Nature Singapore, Singapore, pp 269–283
8. Kazeem MI, Ogungbe SM, Saibu GM, Aboyade OM (2014) In vitro study on the hypoglycemic potential of Nicotiana tabacum leaf extracts. Bangla J Pharmacol 9:140–145
9. Goli AS, Sato VH, Sato H, Chewchinda S, Leanpolchareanchai J, Nontakham J, Maitree M (2023) Antihyperglycemic effects of Lysiphyllum strychnifolium leaf extract in vitro and in vivo. Pharm Biol 61:189–200
10. Tran N, Pham B, Le L (2020) Bioactive compounds in anti-diabetic plants: from herbal medicine to modern drug discovery. Biology 9:252
11. Kazeem MI, Ashafa AOT (2015) In-vitro antioxidant and antidiabetic potentials of Dianthus basuticus Burtt Davy whole plant extracts. J Herbal Med 5:158–164
12. Sofowara A (2006) Medical plants and traditional medicine in Africa. Rep. edition
13. McCue PP, Shetty K (2004) A role for amylase and peroxidase-linked polymerization in phenolic antioxidant mobilization in dark-germinated soybean and implications for health. Process Biochem 39:1785–1791
14. Ali H, Houghton PJ, Soumyanath A (2006) A-Amylase inhibitory activity of some Malaysian plants used to treat diabetes; with particular reference to Phyllanthus amarus. J Ethnopharmacol 107:449–455
15. Aslan M, Orhan N, Orhan DD, Ergun F (2010) Hypoglycemic activity and antioxidant potential of some medicinal plants traditionally used in Turkey for diabetes. J Ethnopharmacol 128:384–389

Chapter 22

Determination of Anti-ulcer Potential of Plant Leaf

Stanila S. Stoeva-Grigorova and Maria T. Petrova

Abstract

Peptic ulcer disease is a common condition, affecting between 5% and 10% of the population during their lifetime. It is characterized by tissue erosion along the gastrointestinal tract due to an imbalance between protective factors (such as the mucosa, prostaglandins, and bicarbonate) and damaging factors (such as hydrochloric acid, pepsin, and others). Conventional treatment includes the use of antisecretory medications, antibiotics for the eradication of *Helicobacter pylori*, mucosal cytoprotectors, prostaglandin analogs, and antacids. In addition to conventional therapies, there is evidence suggesting that the disease may also respond favorably to functional foods and phytoproducts (including alkaloids, terpenes, flavonoids, chalcones, xanthones, anthocyanins, and capsaicinoids). In vivo experimental research on novel therapeutics for gastric ulcers commonly employs chemical (e.g., NSAIDs, ethanol, serotonin, histamine, reserpine, diethyldithiocarbamate, methylene blue, iron, and ascorbic acid), surgical (e.g., pyloric ligation-induced gastric ulcer), or stress-induced ulcer models to induce gastric damage. Among these, experimental models based on factors that reflect human lifestyle, such as ulcers induced by ethanol or NSAIDs, are preferred for the evaluation of new therapeutic agents.

Key words Peptic ulcer, Gastric damage, Antiulcer activity, NSAIDs, Ethanol, Stress-induced ulcer, Pyloric ligation

1 Introduction

Peptic ulcer disease is a pathological condition marked by mucosal erosion within the gastrointestinal tract. While it most frequently involves the duodenum, it may also affect the stomach or esophagus, depending on the site of mucosal injury. The development of peptic ulcer disease results from an imbalance between protective factors—such as the mucosal barrier, prostaglandins, and bicarbonate—and injurious factors—including hydrochloric acid, pepsin, and other noxious agents—where the latter predominate [1, 2]. Clinically, peptic ulcer disease presents with symptoms such as epigastric pain, melena, dyspepsia, and abdominal bloating. If left untreated, it may lead to complications including iron-deficiency anemia, unintended weight loss, and severe, potentially

Mitesh Kumar Dwivedi and A. Sankara Narayanan (eds.), *Leaves as Functional Foods*, Methods and Protocols in Food Science, https://doi.org/10.1007/978-1-0716-5226-8_22,

life-threatening events such as gastrointestinal perforation. Globally, peptic ulcer disease is a prevalent condition, affecting approximately 5% to 10% of the population over the course of a lifetime [2, 3]. The primary risk factor for peptic ulcer disease is the presence of *Helicobacter pylori* infection. Additional significant risk factors include the use of specific medications such as nonsteroidal anti-inflammatory drugs (NSAIDs) and corticosteroids, consumption of irritant foods, irregular eating patterns, etc. Furthermore, psychological stress, smoking, certain neoplasms (e.g., adenocarcinoma, lymphoma), and genetic predisposition have also been identified as contributing factors to disease susceptibility [4, 5]. Contemporary pharmacotherapeutic guidelines for peptic ulcer disease encompass not only antisecretory agents—such as proton pump inhibitors and H_2-receptor antagonists—but also antibiotic regimens aimed at the eradication of *Helicobacter pylori*, as well as adjunctive therapies including antacids, mucosal cytoprotective agents, and prostaglandin analogues. In addition to conventional treatment modalities, emerging evidence suggests that the disease may also respond favorably to functional foods and phytotherapeutic agents. The majority of bioactive compounds demonstrating therapeutic potential include alkaloids, terpenes, flavones, isoflavones, flavonols, chalcones, flavanones, xanthones, flavan-3-ols, anthocyanins, and capsaicinoids. These compounds exhibit their effects in a dose-dependent manner, primarily through the modulation of pro-inflammatory cytokines—such as tumor necrosis factor-α, interleukin-1β, and interleukin-6—as well as by reducing prostaglandin E_2 levels [4, 6, 7].

To evaluate the antiulcer potential of novel pharmacological agents, functional foods, or phytotherapeutics, preclinical studies are typically conducted using experimental animal models, predominantly involving rats and mice [8]. Various established methods are employed to induce gastric mucosal injury in these models, including:

- Stress-induced gastric ulceration.
- Chemically induced gastric lesions using agents such as NSAIDs, ethanol, serotonin, histamine, reserpine, diethyldithiocarbamate, methylene blue, iron, and ascorbic acid.
- Surgically induced gastric ulceration [9].

Among these, the most commonly utilized models for evaluating the efficacy of potential antiulcer agents are those that simulate lifestyle-related risk factors in humans, particularly NSAID-induced, ethanol-induced, and stress-induced gastric ulcer models. These approaches offer relevant insights into the pathophysiological mechanisms underlying peptic ulceration and the therapeutic efficacy of investigational compounds [9].

2 Materials

2.1 Stress-Induced Ulcer Model (Water-Immersion-Stress-Induced Ulcer Model)

- Experimental animals: Male Sprague-Dawley rats, aged 6 weeks and weighing between 180 and 220 g, housed individually, with 6÷10 animals assigned to each experimental group.
- Standard pellet diet provided ad libitum.
- Individual animal cages.
- Water immersion tanks, suitable for rats.
- Vehicle.
- The functional food under investigation.
- Reference drug—Omeprazole.
- Ether.
- 10% formalin used for the preservation of dissected tissue samples.
- 0.9% saline.
- Oral gavage (appropriate for the selected experimental animals).

2.2 NSAID-Induced Gastric Ulcer (Indomethacin-Induced Gastric Ulcer)

- Experimental animals: Male Wistar rats, 6–7 weeks old, weighing 180 ± 20 g, with 6÷10 animals assigned to each experimental group.
- Standard pellet diet provided ad libitum.
- Animal cages within a controlled housing facility.
- Vehicle (*see* **Note 1**).
- Functional food under investigation.
- Ulcer-inducing NSAID—Indomethacin.
- Reference drug—Famotidine.
- Ether.
- 0.9% saline.
- 10% formalin for tissue fixation following dissection.
- Oral gavage (appropriate for the selected experimental animals).

2.3 Ethanol-Induced Gastric Ulcer

- Experimental animals: Male Wistar rats, 6–7 weeks old, weighing 180 ± 20 g, with 6÷10 animals assigned to each experimental group.
- Standard pellet diet provided ad libitum.
- Animal cages within a controlled housing facility.
- Vehicle.
- Functional food under investigation.
- Reference drug—Omeprazole.
- Ulcerogenic agent—Absolute ethanol (≧ 99.8%).

- 10% formalin solution for fixing dissected organs.
- Ether.
- 0.9% saline.
- Oral gavage (appropriate for the selected experimental animals).

2.4 Pyloric Ligation-Induced Gastric Ulcer (Shay Rat Model)

- Experimental animals: Male Wistar rats, 6–7 weeks old, weighing 180 ± 20 g, with 6÷10 animals assigned to each experimental group.
- Standard pellet diet provided ad libitum.
- Animal cages within a controlled housing facility.
- Functional food under investigation.
- Reference Drug—Omeprazole.
- Oral gavage (appropriate for the selected experimental animals).
- 10% formalin solution for fixing dissected organs.
- Ether.
- 0.1 N NaOH (for titration).
- Indicators—Methyl orange and phenolphthalein.

3 Methods

3.1 Stress-Induced Ulcer Model (Cold Water Immersion Restraint Stress) [10–16]

Stress-induced gastric ulcers result mainly from histamine release, causing elevated acid secretion, diminished mucus production, and impaired mucosal blood flow. Oxidative stress and reduced prostaglandin synthesis further exacerbate ulcer formation. This model is widely used to assess the gastroprotective potential of agents, especially those with cytoprotective and mucus-enhancing effects [12].

3.1.1 Preparation of Test Substance

Perform both qualitative and quantitative analyses of the functional food composition in advance. Calculate the dosage of the functional food based on the animals' weight, ensuring precise determination of the active ingredient(s) concentration for maximum accuracy. Prepare the functional food in a qualitative and quantitative composition aligned with the requirements of the planned experiment, and in a formulation that ensures optimal stability and, preferably, allows for oral administration (see oral gavaging procedure in Chapter *Determination of Antidiabetic Potential from Plant Leaf*).

3.1.2 Acclimatization of the Animals

Acclimate the animals for 7 days prior to the experiment under controlled conditions. Maintain them in a well-ventilated room with a 12-hour light/dark cycle. Keep the temperature at 23 ± 2 °C and the humidity at 50 ± 10%. Provide standard rodent chow and water ad libitum. Handle the animals and perform experimental techniques in accordance with national and international regulations for the protection and humane treatment of laboratory animals. Exclude animals that show signs of poor health, behavioral abnormalities, or deviations in body weight from the study. Deprive the animals of food 24–36 hours before the experiment while ensuring free access to water.

3.1.3 Randomization of the Animals

Weigh the experimental animals individually and assign them to experimental groups randomly, ensuring an equal number of subjects per group. Designate one group as the control, which typically receives the vehicle treatment (Group 1). Assign a second group as the disease model, which does not receive any subsequent treatment with test or reference antiulcer agents (Group 2). Include a third group that will receive a reference antisecretory agent (Omeprazole 20 mg/kg, *p.o.*) prior to ulcer induction (*see* **Note 4**). Include additional groups that receive escalating doses of the test functional food (e.g., high, medium, and low doses) to enable a more comprehensive analysis of the results (Groups 4, 5, and 6).

- Treatment Procedure.
- Pre-treat all animal groups with the respective substances for seven consecutive days (or longer, if required).
- Induce gastric ulcers by placing each animal individually in a restraint cage and immersing it vertically in a water tank (20 ± 1 °C), gradually up to the level of the xiphoid process.
- Maintain immersion for 7 h.
- Following ulcer induction, anesthetize the animals and collect blood samples for serum analysis.
- Sacrifice the animals under cervical dislocation.
- Ligate the stomach of each animal at both the pyloric and cardiac ends. Immerse the organ in cold 0.9% saline, rinse its external surface, and then open it along the greater curvature. Remove the gastric contents and gently rinse the mucosal surface with cold 0.9% saline.
- Measure the ulcerated area using planimetry (mm^2).
- Fix all stomachs in 10% formalin solution.

3.1.4 Assessment of End-Points [11, 15–17]

(a) Macroscopic Examination (Using Computer-Assisted Image Analysis):

Count and measure the ulcerative lesions and calculate their area (with the naked eye or using image analysis software). Consider five petechial lesions as equivalent to one 1 mm^2 ulcer. Evaluate the ulcerations according to the following scoring system:

0	No visible damage
1	Presence of blood in the lumen
2	Pinpoint erosions
3	One to five small erosions (< 2 mm)
4	More than five small erosions (< 2 mm)
5	One to three large erosions (> 2 mm)
6	More than three large erosions (> 2 mm)

Determine the ulcer index and percentage of protection using the following formulas (*see* **Notes 5** and **6**):

$$\text{Ulcer Index (UI)} = \frac{\text{Total ulcer score}}{\text{Number of animals}}$$

$$\text{Percent Protection (PP)} = \frac{\left[\left(\text{UI}_{\text{Vehicle group}} - \text{UI}_{\text{Treated group}}\right) \times 100\right]}{\text{UI}_{\text{Vehicle group}}}$$

(b) Microscopic Examination

Fix tissue samples in 10% formalin, embed in paraffin, section, mount on microscope slides, and stain with hematoxylin and eosin. Examine the slides under light microscopy. Evaluate the presence of deep erosions, hemorrhages, epithelial necrosis, and inflammatory cell infiltration according to the following grading scale:

0	None
1	Mild
2	Moderate
3	Severe

3.1.5 Biochemical and Molecular Biomarkers Assessment

- *Gastric Tissue Oxidative Stress Markers*: Malondialdehyde, Catalase, Superoxide dismutase, Myeloperoxidase, Glutathione peroxidase, 8-hydroxy-2′-deoxyguanosine, Heme oxygenase-1.
- *Gastric Tissue Inflammatory Markers*: Prostaglandin E_2, interleukin-6, interleukin-1β, tumor necrosis factor-α, nitric oxide, cyclooxygenase-1, cyclooxygenase-2.
- *Gastric Tissue Protein Expression*: Expression levels of nuclear and cytoplasmic proteins, including nuclear factor-kappa B and inhibitor of nuclear factor kappa-B kinase subunit beta.

3.1.6 Statistical Analysis

Conduct statistical analysis on the recorded results.

3.2 NSAID-Induced Gastric Ulcer (Indomethacin-Induced Gastric Ulcer) [11–13, 15–17]

NSAIDs induce ulcers by inhibiting the synthesis of prostaglandins through the blockade of the cyclooxygenase enzymes. Prostaglandins are essential for protection, as they promote bicarbonate and mucus secretion, support blood flow, and regulate the turnover and repair of mucus-producing cells. NSAID-induced prostaglandin inhibition impairs gastric mucosal defense by reducing blood flow, mucus/bicarbonate secretion, epithelial repair, and platelet aggregation. Additionally, NSAIDs trigger neutrophil activation, ROS release, and lipid peroxidation, further promoting mucosal injury (Fig. 1). Gastric acid worsens this damage by deepening lesions and hindering healing [18]. The NSAID-induced gastric ulcer model is used to assess the anti-ulcer, cytoprotective, and antisecretory effects of test substances [16].

3.2.1 Preparation of Test Substance

Prepare the functional food in a qualitative and quantitative composition aligned with the requirements of the planned experiment and in a formulation that ensures optimal stability and, preferably, allows for oral administration.

3.2.2 Acclimatization of the Animals

Acclimate animals for 7 days prior to the experiment under controlled conditions: 12-hour light/dark cycle, temperature at 23 ± 2 °C, humidity at 50 ± 10%. Provide standard chow and water ad libitum. Handle and conduct experiments per ethical guidelines. Exclude animals showing health issues or abnormal behavior. Fast animals 24 h before the experiment, ensuring water access.

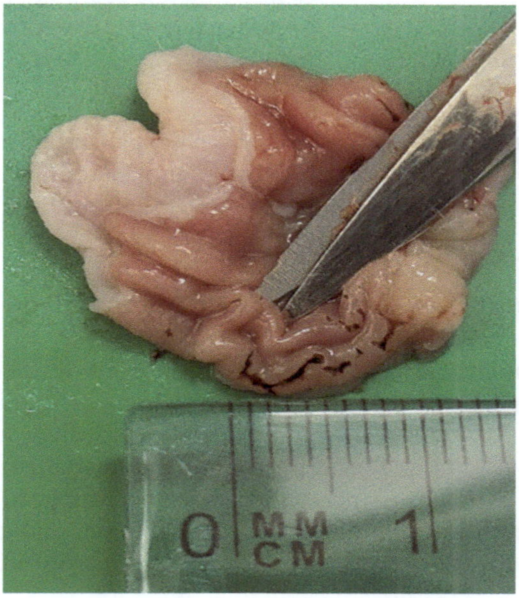

Fig. 1 A representative image of the damaged stomach in a rat model of Indomethacin-induced ulcer

3.2.3 Randomization of the Animals

Individually weigh the animals and randomly assign them to experimental groups of equal size. Designate one group as the control, receiving vehicle only (Group 1). Assign another group as the ulcer model, treated with a NSAID without subsequent therapeutic intervention (Group 2) (*see* **Notes 1**, **2** and **3**). Administer an antisecretory agent (Famotidine) to one of the ulcer-induced groups (Group 3) (*see* **Note 4**). Include additional ulcer-induced groups receiving graded doses of the test functional food (e.g., low, medium, and high) to enable comprehensive evaluation of its therapeutic potential (Groups 4, 5, and 6).

3.2.4 Treatment Procedure

- For eight consecutive days, the experimental groups are administered the respective substances orally, as follows:

  Group 1: Vehicle.

  Group 2: Vehicle.

  Group 3: Famotidine at a dose of 10 mg/kg body weight, suspended in 2 mL of vehicle.

  Groups 4, 5, and 6: Low, medium, and high doses of the test functional food, respectively.
- On Day 8, all animals in Groups 2÷6 receive a single oral dose of Indomethacin (60 mg/kg body weight), via oral gavage to induce gastric ulceration (*see* **Note 1**).
- Six hours after Indomethacin administration, sacrifice animals under cervical dislocation.
- Carefully excise the stomachs and open them along the greater curvature to collect the gastric contents.
- Gently isolate a representative portion of each stomach, weigh it, and fix it in 10% formalin solution for histopathological analysis.

3.2.5 Assessment of End-Points

Perform macroscopic and microscopic examination of gastrointestinal tissues, and determine relevant biochemical biomarkers, following the procedures outlined in Subheading 3.1 (*see* **Notes 5** and **6**).

3.2.6 Statistical Analysis

Conduct statistical analysis on the recorded results.

3.3 Ethanol-Induced Gastric Ulcer [11, 15, 16, 19]

Oral administration of absolute ethanol is a common model for inducing gastric injuries in animals. Ethanol causes gastric lesions through the actions of hydrochloric acid and pepsin, stimulating acid secretion, microcirculatory damage, and increased vascular permeability. This disrupts microcirculation, reduces mucus and prostaglandin secretion, and elevates oxidative stress. The model is useful for testing gastroprotective substances with cytoprotective and antioxidant properties but not for evaluating protection against acid secretion-dependent ulcers [12, 15].

3.3.1 Preparation of Test Substance

Prepare the functional food with a qualitative and quantitative composition that aligns with the requirements of the planned experiment and in a formulation that ensures optimal stability and, preferably, allows for oral administration.

3.3.2 Acclimatization of the Animals

Acclimate animals for 7 days prior to the experiment under controlled conditions: 12-hour light/dark cycle, temperature at 23 ± 2 °C, humidity at 50 ± 10%. Provide standard chow and water ad libitum. Handle and conduct experiments per ethical guidelines. Exclude animals showing health issues or abnormal behavior. Fast animals 24 h before the experiment, ensuring water access.

3.3.3 Randomization of the Animals

Individually weigh the animals and randomly assign them to experimental groups of equal size. Designate one group as the control, receiving vehicle only (Group 1). Assign another group as the ulcer model, treated with an absolute ethanol only, without any subsequent therapeutic intervention (Group 2). Administer the antisecretory agent to one of the ulcer-induced groups (Group 3) (*see* **Note 4**). Include additional ulcer-induced groups receiving graded doses of the test functional food (e.g., low, medium, and high) to enable comprehensive evaluation of its therapeutic potential (Groups 4, 5, and 6).

3.3.4 Treatment Procedure

- Pre-treat all animal groups with the respective substances for seven consecutive days (or longer, if required).
- On the final day, 30–60 min after the administration of the respective pharmaceutical or test substances, induce gastric injury in Groups 2÷6 by treating them with 1 mL of absolute ethanol.
- One hour after ulcer induction, under anesthesia, collect serum samples.
- Sacrifice all animals under ether anesthesia. Subsequently, remove the stomachs, wash them in cold saline, and incise them along the greater curvature. Then, measure the ulcerated area planimetrically (mm^2).
- Gently isolate a representative portion of each stomach, weigh it, and fix it in a 10% formalin solution for histopathological analysis.

3.3.5 Assessment of End-Points

Perform macroscopic and microscopic examination of gastrointestinal tissues, and determine relevant biochemical biomarkers, following the procedures outlined in Subheading 3.1 (*see* **Notes 5** and **6**).

3.3.6 Statistical Analysis

Conduct statistical analysis on the recorded results.

3.4 Pyloric Ligation-Induced Gastric Ulcer (Shay Rat Model) [11–13, 15, 16, 20–23]

3.4.1 Preparation of Test Substance

Prepare the functional food in a qualitative and quantitative composition aligned with the requirements of the planned experiment and in a formulation that ensures optimal stability and, preferably, allows for oral administration.

3.4.2 Acclimatization of the Animals

Acclimate animals for 7 days prior to the experiment under controlled conditions: 12-hour light/dark cycle, temperature at 23 ± 2 °C, humidity at 50 ± 10%. Provide standard chow and water ad libitum. Handle and conduct experiments per ethical guidelines. Exclude animals showing health issues or abnormal behavior. Fast animals 36–48 h before the experiment, ensuring water access.

3.4.3 Randomization of the Animals

Weigh the experimental animals individually and assign them to experimental groups randomly, ensuring an equal number of subjects per group. Designate one group as the control, which typically receives the vehicle treatment (Group 1). Assign a second group as the disease model, which does not receive any subsequent treatment with test or reference antiulcer agents (Group 2). Include a third group that will receive a reference antisecretory agent (Omeprazole 20 mg/kg, p.o.) prior to ulcer induction (*see* **Note 4**). Include additional groups that receive escalating doses of the test functional food (e.g., high, medium, and low doses) to enable a more comprehensive analysis of the results (Groups 4, 5, and 6).

3.4.4 Treatment Procedure

- Pre-treat all animal groups with the respective drugs or functional food 1 h prior to performing pylorus ligation.
- Perform pylorus ligation under ether anesthesia, ensuring precautions are taken to prevent damage to blood vessels and the pylorus.
- Carefully isolate the stomach during the ligation procedure.
- Inject the distinct experimental groups with either vehicle, antisecretory agent, or extract at low, medium, or high doses, via the duodenum.
- Upon completion of the procedure, reposition the stomach to its anatomical position and then suture the abdominal wall.
- To prevent coprophagy, the animals are individually housed in cages with elevated floors made of wide wire mesh.

- Restrict food and water intake for the subsequent 4 h post-surgery;
- At the conclusion of the fourth hour, under ether anesthesia, collect serum samples and seal the esophagus to prevent leakage of the stomach contents;
- The contents of the stomach are transferred to a centrifuge tube, and their acidity is quantified using 0.1 N NaOH via titration;
- The stomach is then incised along the greater curvature, washed with saline solution, fixed on a cork plate, and the number and severity of ulcers are evaluated;
- Euthanize the animals by administering an overdose of ether anesthesia;
- Gently isolate a representative portion of each stomach, weigh it, and fix it in a 10% formalin solution for histopathological analysis.

3.4.5 Assessment of End-Points

Perform macroscopic and microscopic examination of gastrointestinal tissues, and determine relevant biochemical biomarkers, following the procedures outlined in Subheading 3.1 (*see* **Notes 5**, **6** and 7).

3.4.6 Statistical Analysis

Conduct statistical analysis on the recorded results.

4 Notes

1. In the Indomethacin-induced ulcer model, the choice of oral dose may range from 40 to 100 mg/kg body weight. Similarly, ulcer formation can be assessed between 4- and 8-hours post-administration. Therefore, to optimize the experimental protocol, it is advisable to conduct a preliminary pilot study [9].
2. Acetylsalicylic acid (125–150 mg/kg body weight, *p.o.*) or Ibuprofen (400 mg/kg body weight, *p.o.*) may be employed as alternative ulcerogenic NSAIDs [15].
3. For the Indomethacin-induced ulcer protocol, a 0.1–5% Tween 80 solution may be used as a vehicle [16, 24, 25].
4. As reference antisecretory drugs proton pump inhibitors (e.g., Omeprazole) or H_2-receptor antagonists (e.g., Famotidine) may be used [15].
5. Alternatively, during macroscopic examination, gross mucosal damage and the percentage of inhibition may be assessed using [26, 27]:

$$\text{Ulcer Index (U.I.)} = \frac{\text{Ulcerated area}}{\text{Total stomach area}} \times 100$$

$$\text{Percent Ulcer Inhibition} = \frac{\left[\left(\text{U.I.}_{\text{Ulcer group}} - \text{UI}_{\text{Test group}}\right) \times 100\right]}{\text{UI}_{\text{Ulcer group}}}$$

6. The acidity of gastric contents can be determined in all experiments. To do so, collect the gastric contents from the stomach and centrifuge at 4000 rpm for 10 min to separate the gastric juice. Transfer the contents into a graduated centrifuge tube, then measure the volume of the gastric juice and its pH using a pH meter. The acidity of the gastric contents is determined by titrating with 0.1 N NaOH. For the purpose, add 1–2 drops of methyl orange reagent to the gastric juice. The presence of free hydrochloric acid will cause the solution to turn bright red. Using a burette, titrate the gastric juice with 0.1 N NaOH until the color changes to canary yellow. The volume of NaOH used (in mL) reflects the amount of free hydrochloric acid. Next, add 1–2 drops of phenolphthalein to the gastric juice. Continue titrating until a deep pink color appears. Cease titration when further addition of NaOH no longer intensifies the pink color. Record the volume of NaOH used (in mL) and use this value to calculate the total acidity of the gastric juice:

$$Total\ acidity\ (mEq/L) = 0.1\ N\ NaOH\ (mL) \times 10$$

   Compare the acidity results obtained from the different experimental groups [28].

7. In the Pyloric Ligation-Induced Gastric Ulcer rat model, lesions primarily develop in the rumen and the antrum of the stomach due to weaker mucosal protective mechanisms against gastric acid in these regions [16].

References

1. Parmar NS, Parmar S (1998) Anti-ulcer potential of flavonoids. Indian J Physiol Pharmacol 42(3):343–351
2. Lanas A, Chan FKL (2017) Peptic ulcer disease. Lancet 390(10094):613–624. https://doi.org/10.1016/S0140-6736(16)32404-7
3. Cleveland Clinic. Peptic ulcer disease [Internet]. 2024 [cited 2025 Apr 28]. Available from: https://my.clevelandclinic.org/health/diseases/10350-peptic-ulcer-disease
4. Kulshreshtha M, Srivastava G, Singh MP (2017) Pathophysiological status and nutritional therapy of peptic ulcer: an update. Environ Dis 2(3):76–86. https://doi.org/10.4103/ed.ed_11_17
5. Kavitt RT, Lipowska AM, Anyane-Yeboa A, Gralnek IM (2019) Diagnosis and treatment of peptic ulcer disease. Am J Med 132(4): 447–456. https://doi.org/10.1016/j.amjmed.2018.12.009
6. Farzaei MH, Abdollahi M, Rahimi R (2015) Role of dietary polyphenols in the management of peptic ulcer. World J Gastroenterol 21(21): 6499–6517. https://doi.org/10.3748/wjg.v21.i21.6499
7. Anchang MM, Okafor JN, Okoyeuzu FC (2024) The management of peptic ulcer

disease using conventional methods, bioactive compounds and dietotherapy: a review. Japanese journal of gastroenterology research 4(5). https://doi.org/10.52768/jjgastro/1202
8. Satapathy T, Sen K, Sahu S, Pradhan B, Gupta A, Khan MA, Kumar D, Satapathy A, Yadav N (2024) Experimental animal models for gastric ulcer/peptic ulcer: An overview. J Drug Deliv Ther 14(1):182–192. https://jddtonline.info/index.php/jddt/article/view/6258
9. Fulga S, Pelin AM, Ghiciuc CM, Lupuşoru EC (2019) Particularities of experimental models used to induce gastric ulcer. ARS Medica Tomitana. Ovidius University of Constanta 25(4):179–184. https://doi.org/10.2478/arsm-2019-0035
10. Takeuchi K, Furukawa O, Okabe S (1986) Induction of duodenal ulcers in rats under water-immersion stress conditions: influence of stress on gastric acid and duodenal alkaline secretion. Gastroenterology 91(3):554–563. https://doi.org/10.1016/0016-5085(86)90622-0
11. Adinortey MB, Ansah C, Galyuon I, Nyarko A (2013) In vivo models used for evaluation of potential antigastroduodenal ulcer agents. Ulcers 2013:796405. https://doi.org/10.1155/2013/796405
12. Thabrew MI, Arawwawala LDAM (2016) An overview of in vivo and in vitro models that can be used for evaluating anti-gastric ulcer potential of medicinal plants. Austin Biol 1(2):1007
13. Meena DK, Jayanthi M (2018) In-vivo models used for pre-clinical evaluation of anti-ulcer activity. Austin Pharmacol Pharm 3(2):1017
14. Lu S, Wu D, Sun G, Geng F, Shen Y, Tan J et al (2019) Gastroprotective effects of Kangfuxin against water-immersion and restraint stress-induced gastric ulcer in rats: roles of antioxidation, anti-inflammation, and pro-survival. Pharm Biol 57(1):770–777. https://doi.org/10.1080/13880209.2019.1682620
15. Arunachalam K, Sasidharan SP (2021) Animal experiments of gastric ulcer. In: Bioassays in experimental and preclinical pharmacology. Springer protocols handbooks. Humana, New York, pp 295–302. https://doi.org/10.1007/978-1-0716-1233-0_21
16. Maharani B (2022) Screening methods for the evaluation of antiulcer drugs. In: Lakshmanan M, Shewade DG, Raj GM (eds) Introduction to basics of pharmacology and toxicology. Springer, Singapore, pp 383–396. https://doi.org/10.1007/978-981-19-5343-9_28
17. Nabil M, El Raey MA, Abdo W, Abdelfattah MAO, El-Shazly AM, Sobeh M, Mahmoud MF (2021) Gastro-protective effects of *Albizia anthelmintica* leaf extract on indomethacin-induced gastric ulcer in Wistar rats: in silico and in vivo studies. Antioxidants (Basel) 10(2):176. https://doi.org/10.3390/antiox10020176
18. Beiranvand M (2022) A review of the most common *in vivo* models of stomach ulcers and natural and synthetic anti-ulcer compounds: a comparative systematic study. Phytomed Plus 2(2):100264. https://doi.org/10.1016/j.phyplu.2022.100264
19. Hui S, Fangyu W (2017) Protective effects of bilobalide against ethanol-induced gastric ulcer *in vivo/vitro*. Biomed Pharmacother 85:592–600. https://doi.org/10.1016/j.biopha.2016.11.068
20. Monteiro KM, Spindola HM, Possenti A, Tinti SV, Ruiz AL, Longato GB, Fiorito GF, Marchetti GM, Shiozawa L, Piloni BU, de Oliveira AC, Miyagawa LM, Carvalho JE (2013) Characterization of a refinement of the "pylorus ligation" model of rat gastric ulceration resulting in "no pain" and a more specific pharmacological response. J Pharmacol Toxicol Methods 67(2):121–128. https://doi.org/10.1016/j.vascn.2012.09.002
21. Mishra AP, Bajpai A, Chandra S (2019) A comprehensive review on the screening models for the pharmacological assessment of antiulcer drugs. Curr Clin Pharmacol 14(3):175–196. https://doi.org/10.2174/1574884714666190312143846
22. Danai P, Patel S, Pandey V, Singh P, Yadav G, Kumar A, Agarwal T (2021) Antiulcerogenic activity of *Anogeissus pendula* hydroalcoholic extract on pylorus ligated induced gastric ulcers in albino Wistar rats. Phytomed Plus 1(4):100127. https://doi.org/10.1016/j.phyplu.2021.100127
23. Shay H, Komarov SA, Fels SS, Meranze D, Gruenstein M, Siplet H (1945) A simple method for the uniform production of gastric ulceration in the rat. Gastroenterology 5:43–61
24. Alves NM, Nunes PHM, Mendes Garcez A, Lima de Freitas MC, Oliveira IS, da Silva FV, Fernandes HB, de Sousa DP, Oliveira RCM, Arcanjo DDR, Martins MDCCE (2023) Antioxidant mechanisms underlying the gastroprotective effect of menthofuran on experimentally induced gastric lesions in rodents. Evid Based Complement Alternat Med 2023:9192494. https://doi.org/10.1155/2023/9192494

25. Mahmoud MF, Abdo W, Nabil M, Drissi B, El-Shazly AM, Abdelfattah MAO, Sobeh M (2023) Apple (Malus domestica Borkh) leaves attenuate indomethacin-induced gastric ulcer in rats. Biomed Pharmacother 160:114331. https://doi.org/10.1016/j.biopha.2023.114331
26. Sabiu S, Garuba T, Sunmonu T et al (2015) Indomethacin-induced gastric ulceration in rats: protective roles of Spondias mombin and Ficus exasperata. Toxicol Rep 2:261–267. https://doi.org/10.1016/j.toxrep.2015.01.002
27. Abdel-Gawad DRI, Ibrahim MA, Moawad UK, Kamel S, El-Banna HA, El-Banna AH, Hassan WH, Abo El-Ela FI (2023) Effectiveness of natural biomaterials in the protection and healing of experimentally induced gastric mucosa ulcer in rats. Mol Biol Rep 50:9085–9098. https://doi.org/10.1007/s11033-023-08776-9
28. Amin RR, Aziz TA (2022) Gastroprotective effect of azilsartan through ameliorating oxidative stress, inflammation, and restoring hydroxyproline, and gastrin levels in ethanol-induced gastric ulcer. J Inflamm Res 15:2911–2923. https://doi.org/10.2147/JIR.S365090

Chapter 23

Determination of Antidepressant Potential of Plant Leaf

Miroslav Eftimov, Silvia Gancheva, and Stanila S. Stoeva-Grigorova

Abstract

Depression is a common neuropsychiatric disorder with a multifactorial etiology and complex pathophysiology. Treatment with conventional antidepressants is often unsatisfactory and frequently associated with various adverse drug reactions. A number of plant extracts and plant-based food supplements have already demonstrated beneficial effects on the emotional and biological symptoms of depression and are widely used by patients as alternative treatments. Plant-derived biologically active substances exert anti-inflammatory and antioxidant effects and may prevent and/or reverse the neurobiochemical alterations typically associated with depression-like behavior. Animal models of depression are essential for screening the antidepressant activity of plant extracts and elucidating their underlying mechanisms of action. Among these, the chronic unpredictable mild stress model is favored by many research groups due to its reliability and reproducibility. This model successfully mimics a major etiological factor of the disorder—chronic stress—and induces core symptoms of human depression, such as anhedonia and behavioral despair. These symptoms serve as behavioral endpoints for validating the experimental model and assessing the antidepressant potential of the test compound, including the functional food of interest.

Key words Depression, Animal models, Rats, Mice, Chronic unpredictable mild stress, Sucrose preference test, Forced swim test, Tail suspension test

1 Introduction

Depression (also referred to as depressive disorder) is a common affective disorder characterized by disturbances in mood. Its clinical manifestations vary widely, ranging from mild symptoms to severe (psychotic) depression, which may be accompanied by hallucinations and delusions. The global prevalence of depression remains consistently high, affecting approximately 3.4% of the total population and 5% of adults. The lifetime prevalence is estimated at 17%, indicating that approximately one in six individuals will experience at least one depressive episode during their lifetime [1–3]. The COVID-19 pandemic has had a profound impact on the incidence

Mitesh Kumar Dwivedi and A. Sankara Narayanan (eds.), *Leaves as Functional Foods*, Methods and Protocols in Food Science,
https://doi.org/10.1007/978-1-0716-5226-8_23,

of mood disorders. In the first year of the pandemic alone, the global prevalence of depression increased by an estimated 27.6% [4]. The medical, social, and economic significance of depression is substantial, not only due to its high prevalence, but also because it is a leading cause of disability and suicide worldwide—particularly among young adults.

The clinical signs of depression encompass a range of emotional symptoms, including persistent low mood (sadness), negative thoughts, apathy, pessimism, low self-esteem, loss of motivation, and anhedonia (the inability to experience pleasure). In most patients, these are accompanied by biological symptoms, primarily psychomotor retardation, disturbances in sleep and appetite, and decreased libido [2]. The etiology of depression is multifactorial, involving both genetic predisposition and environmental factors, most notably stressful life events [5, 6]. The pathophysiology of depression is complex and involves impaired monoaminergic neurotransmission alongside a reduction in neurotrophic factors—particularly brain-derived neurotrophic factor (BDNF)—that regulate synaptic plasticity and neurogenesis in key brain regions, including the prefrontal cortex and ventral hippocampus [5–7]. Additional contributing factors to the disorder include overactivation of the hypothalamic-pituitary-adrenal (HPA) axis, neuroinflammatory processes, disruption of the oxidant–antioxidant balance and mitochondrial function, alterations in gut microbiota composition, and other pathophysiological mechanisms [6].

There are no objective biomarkers currently available for the diagnosis of depression in humans; the condition is diagnosed solely through clinical interviews based on symptomatology. This lack of definitive diagnostic criteria presents significant challenges in modeling depression in experimental animals. Animal models are inherently limited in their ability to replicate core features of human depression, such as low self-esteem, negative thought patterns, feelings of worthlessness and guilt, and suicidal ideation. Consequently, no existing model fully recapitulates the human condition. Nevertheless, certain observable behavioral correlates of depression can be detected in animals and are employed as endpoints to validate the model. In most experimental studies, the primary behavioral endpoints assessed are anhedonia and behavioral despair, which serve as proxies for core symptoms of the human disorder [8]. Anhedonia is typically assessed using the sucrose preference test, which measures the reduced ability of experimental animals to experience pleasure or reward. Despair-like behavior is most commonly evaluated through the forced swim test and the tail suspension test. In both of these tests, animals are placed in an inescapable, stressful situation. The duration of active movement reflects the animal's motivation and perceived ability to cope with the adverse circumstances. Prolonged immobility is considered indicative of despair and is used as a behavioral marker of depression-like symptoms [8, 9]. It should be noted that in recent years the validity of immobility time as a measure of despair has been questioned. According to some researchers, immobility of the animals is the result of cognitive functioning that underlie behavioral adaptation [8–10].

In humans, initial episodes of depression are usually induced by stressful life events. Therefore, the animal models are most often based on the application of stressors. In some models, such as the early mother separation, the stressors are applied during the development. Most researchers prefer to use models that rely on the application of stressors in adulthood. The stressors are uncontrollable and/or unpredictable, mild, or intense, and applied for different periods of time. Examples of such models are the chronic unpredictable mild stress (CUMS), learned helplessness, and social defeat [5, 8, 9]. Among them, CUMS is generally preferred because it is reliable and reproducible and is based on the application of mild stressors, which are more ethically acceptable than the intense stressors in the other models. Some experimental models aim to induce directly the neurobiological alterations that are observed in depression, e.g., lipopolysaccharide injection induces neuroinflammation, corticosteroid administration mimics HPA axis overactivation, and olfactory bulbectomy mimics neural degeneration within regions of the corticolimbic networks. There are also genetic models using knockout mice, e.g., knocking out serotonin transporters or receptors, which appear to be suitable for modeling of treatment-resistant depression [8].

As depression is a major problem of current society, many synthetic antidepressants have been designed and introduced to clinical practice. However, a large proportion of patients do not respond satisfactorily to the conventional therapy. The beneficial effects of most synthetic antidepressants develop slowly, and the therapy is often accompanied by various adverse drug reactions. Therefore, efforts are directed to the search for new therapeutic approaches. The nature is an inexhaustible source of biologically active substances, and their potential beneficial effects in depression should be actively studied. Plants are rich in flavonoids, terpenes, saponins, and alkaloids that are able to produce a number of favorable neurobiochemical changes by suppressing neuroinflammatory pathways, improving the antioxidant defense system, normalizing brain monoamine levels, increasing BDNF hippocampal levels, and preventing HPA axis overactivation [11–13]. A number of plant extracts has been registered as OTC drugs or dietary supplements as an alternative to classical antidepressant treatment [12].

2 Materials

2.1 Assessment of Antidepressant Potential in a CUMS Rat Model [14–16]

- Animals—Sprague-Dawley or Wistar rats; 150–200 g, 8–10 weeks old; 10 animals per treatment group (*see* **Note 1**).
- Test compound (functional food of interest).
- General lab materials:
  - Animal housing cages (for each animal) with standard bedding.
  - Standard pellet food and water plus and water dispensers.
  - Standard lab cleaning and disinfection supplies.
  - Weighing scale (for animal weight tracking and sucrose test measurement).
- Mild stressors and required items:
  - Cage tilting—tiltable cages or wedges.
  - Wet bedding—water ~25 °C, measuring cup, spray bottle.
  - Food deprivation—cage modifications to block food.
  - Water deprivation—removable water dispenser.
  - Empty cage—cages without bedding.
  - Light/dark cycle reversal—controlled light system.
  - Strobe light—flashing light/strobe device.
  - Noise stress—sound generator or radio (white noise).
  - Restraint stress—restraint tubes or plastic film.
  - Soiled cage—use bedding from another animal.
- Behavioral test equipment (for depressive-like behavior assessment):
  - Sucrose preference test setup—two bottles, one with water, and one with 1% sucrose solution.
  - Forced swim test setup—transparent cylindrical tank (55–60 cm in diameter) with water, stopwatch.
- Documentation and Schedules.
  - Daily stressor schedule (for the whole week).
  - Behavioral testing schedule.
  - Weight and health monitor sheets.

Ethical approval documents—protocol for ethical work with laboratory animals.

2.2 Assessment of Antidepressant Potential in a CUMS Mouse Model [17–19]

- Animals—C57BL/6 or BALB/c mice; 15–20 g, 4 weeks old; 15 animals per treatment group.
- Test compound (functional food of interest).
- General lab materials:
  - Animal housing cages with standard bedding.
  - Standard pellet food and water plus water dispensers.
  - Standard lab cleaning and disinfection supplies.
  - Weighing scale (for animal weight tracking and sucrose test measurement).
  - Handling gloves and tools for minimal stress handling.
- Mild stressors and required items:
  - Cage tilting—tiltable cages or angle support.
  - Wet bedding—water ~25 °C, measuring cup, spray bottle.
  - Empty cage—cages without bedding.
  - Food deprivation—empty food hoper.
  - Water deprivation—blocked water access.
  - Light/dark cycle reversal—controlled light system.
  - Isolation—single housing.
  - Restraint stress (1–2 h)—mouse restraint tubes.
  - Soiled cage—bedding from another animal's cage.
- Behavioral test equipment (for depressive-like behavior assessment):
  - Sucrose preference test setup—two bottles, one with water and one with 1% sucrose solution.
  - Forced swim test setup—transparent cylindrical container (15–20 cm in diameter) with water, stopwatch.
  - Tail suspension test—suspension bar and adhesive tape, stopwatch.
- Documentation and Schedules:
  - Daily stressor schedule log (for the whole week).
  - Behavioral testing schedule.
  - Weight and health monitor sheets.

Ethical approval documents—protocol for ethical work with laboratory animals.

3 Methods

3.1 Assessment of Antidepressant Potential in a CUMS Rat Model [14–16]

3.1.1 Acclimatization of the Animals

Begin with a 7-day habituation phase to allow the rats to adapt to the new environment prior to the start of the experiments. Place the rats in a well-ventilated room. Ensure a 12-h light/dark cycle is maintained to simulate natural environmental conditions. Set the room temperature to 23 ± 2 °C and maintain relative humidity at 50 ± 10%. Provide ad libitum access to food and water. Handle the rats for a brief period (a few minutes) each day to facilitate their habituation to human interaction. Assess the animals daily for any signs of poor health, behavioral abnormalities, or deviations in body weight. Exclude any animals displaying these symptoms from the study. Ensure all experimental techniques and procedures are in strict compliance with national and international regulations for the humane treatment and protection of laboratory animals.

3.1.2 Distribution of Animals into Groups

Weigh all experimental animals and mark them appropriately for identification. Divide the animals into groups, ensuring an equal number of rats in each group. [Due to variability in the results of behavioral studies, it is recommended to have a minimum of 10 animals per group.] Assign one group as the control group. This group will receive the vehicle of the active substance, without the active compound. Assign another group as the disease model group. These animals will be subjected to CUMS and treated with the vehicle of the active substance. The remaining groups will be exposed to CUMS but will receive the test compound (functional food of interest) in increasing doses, such as low, medium, and high. If applicable, include a positive control group, which is subjected to CUMS and treated with a drug known to have proven antidepressant effects. This serves as a benchmark for comparison.

3.1.3 Experimental Procedure

- Design the stressor protocol to last between 3 and 8 weeks. Use each stressor once or a maximum of twice per week. Change the schedule of the stressors each week to prevent habituation. Ensure the use of at least seven different stressors. Refer to Table 1 for a sample weekly schedule, but adjust it to maintain variability.
- On Day 1, after a week of acclimatization, separate the rats into individual cages for the CUMS group. Social isolation increases stress, so housing the rats individually will enhance the stress response. House the control rats in a separate room with a 12-h light/dark cycle, in cages containing 4–6 rats each, ensuring a stable social environment.
- To apply the CUMS protocol, follow the weekly schedule for random exposure to mild stressors, as detailed below [20]:

Table 1
Exemplary schedule of stressors and sucrose preference test in rats subjected to CUMS

Day	Stressor	Duration
Monday	Empty cage Soiled sawdust	8:00–12:00 14:00–20:00
Tuesday	Flashing light Collecting in pairs Overnight illumination	11:00–17:00 17:00–9:00 (next day) 19:00–7:00 (next day)
Wednesday	Cell tilting	9:00–15:00
Thursday	Food deprivation Water deprivation Sucrose preference test	0:00–20:00 (next day) 0:00–20:00 (next day) 20:00–21:00
Friday	White noise Collecting in pairs Wet sawdust Overnight illumination	9:00–12:00 13:00–17:00 17:00–8:00 (next day) 19:00–7:00 (next day)
Saturday	Flashing light Water deprivation	10:00–14:00 17:00–10:00 (next day)
Sunday	Placing empty water bottle White noise	9:00–10:00 10:00–14:00

- *Cage Tilting*: Tilt the animal's cage at a 45° angle for a duration of 6–12 h.
- *Wet Bedding*: Spray the bedding with water (24 ± 1 °C) until it is moderately damp. After 12–15 h, dry the animals and transfer them to a cage with fresh, dry sawdust.
- *Food Deprivation*: Block access to food pellets for a period of 12–24 h.
- *Water Deprivation*: Remove the water dispenser for 12–24 h. One hour before the end of the deprivation period, add an empty water bottle.
- *Empty Cage*: Transfer the animals from their home cage to an empty cage (without sawdust) for 4 h.
- *Light/Dark Cycle Reversal*: Reverse the light/dark cycle by turning the light on during the night and off during the day, maintaining a 12-h light/dark pattern.
- *Strobe Light:* Turn off the room light and use a strobe light for 6 h using a flashing light device.
- *Noise Stress*: Expose the animals to white noise for 3–6 h using a sound generator.
- *Restraint Stress:* Place each rat in a restrainer for 4 h.

- *Collecting in Pairs*: Place two rats in one cage for a period of 4–20 h.
- *Soiled Cage:* Replace the bedding with soiled bedding from other animals or add the scent of a predator (e.g., cat) for 6–12 h.

Ensure the stressors are applied randomly, as per the weekly schedule, to avoid predictability and habituation, which are critical for maintaining the efficacy of the protocol.

Treatment with the test compound may begin on Day 1 or at a later timepoint, depending on the experimental design. For precise dosing, use intraperitoneal injection or oral gavage as the preferred methods of administration. Both techniques ensure accurate delivery of the compound. Control rats should receive the same treatment regimen as the experimental group, but instead of the test compound, they should be administered the substance vehicle (the carrier solution or medium without the active compound).

3.1.4 Assessment of End-Points

3.1.4.1 Sucrose Preference Test

- Acclimation phase (during acclimatization week and 32 h before each test phase): Animals are first given access to *two bottles* for 12 h (one with plain water and one with 1% sucrose solution) to ensure familiarity and rule out neophobia; bottles are typically switched between sides to prevent side bias (*see* **Note 2**);
- Test phase (starts from week 3): For 20 h rats are food and water deprived, followed by simultaneous individual access to the same amount of 1% sucrose solution and plain water for 1 h [21];
- *Intake is measured* (in mL or grams) for each bottle;
- The sucrose preference is calculated using the formula:

$$\text{Sucrose Preference}(\%) = \left(\frac{\text{Sucrose intake}}{\text{Sucrose} + \text{Water intake}}\right) \times 100$$

- Interpretation: High sucrose preference (e.g., >65–70%): Suggests normal hedonic behavior. Low sucrose preference: Suggests anhedonia and possibly a depression-like behavior.

3.1.4.2 Forced Swim Test [22]

- Each rat is placed in a cylindrical container filled with water (typically 23–25 °C) deep enough (around 15–30 cm) so the animal *cannot touch the bottom* or *escape*.
- The rat activity is observed for test period of 5 min. The time of immobility (sec.) is recorded—the animal floats with minimal movements to keep its head above the water.

Interpretation: Increased immobility time is interpreted as depression-like behavior. Decreased immobility after experimental substance treatment is interpreted as antidepressant-like effect (*see* **Notes 3**, **4**, **5**, and **6**).

3.1.4.3 Statistical Analysis

Conduct statistical analysis on the recorded results.

3.2 Assessment of Antidepressant Potential in a CUMS Mouse Model [17–19]

3.2.1 Acclimatization of the Animals

Begin with a 7-day habituation phase to allow the animals to adapt to the new environment prior to the start of the experiments. Place the mice in a well-ventilated room. Ensure a 12-h light/dark cycle is maintained to simulate natural environmental conditions. Set the room temperature to 23 ± 2 °C and maintain relative humidity at 50 ± 10%. Provide ad libitum access to food and water. Handle the rodents for a brief period (a few minutes) each day to facilitate their habituation to human interaction. Assess the animals daily for any signs of poor health, behavioral abnormalities, or deviations in body weight. Exclude any animals displaying these symptoms from the study. Ensure all experimental techniques and procedures are in strict compliance with national and international regulations for the humane treatment and protection of laboratory animals.

3.2.2 Distribution of Animals into Groups

Weigh all experimental animals and mark them appropriately for identification. Divide the animals into groups, ensuring an equal number of animals in each group. [Due to variability in the results of behavioral studies, it is recommended to have a minimum of ten animals per group.] Assign one group as the control group. This group will receive the vehicle of the active substance, without the active compound. Assign another group as the disease model group. These animals will be subjected to CUMS and treated with the vehicle containing active substance. The remaining groups will be exposed to CUMS, but will receive the test compound in increasing doses, such as low, medium, and high. If applicable, include a positive control group, which is subjected to CUMS and treated with a drug known to have proven antidepressant effects. This serves as a benchmark for comparison.

3.2.3 Experimental Procedure

- The typical duration of the CUMS protocol ranges from 3 to 8 weeks. Design a CUMS schedule within this timeframe, incorporating a minimum of seven distinct mild stressors. Each stressor should be applied no more than once or twice per week. To prevent behavioral habituation, the weekly schedule must be randomized and altered from week to week.
- Following a 7-day acclimatization period, mice allocated to CUMS groups should be housed in groups of 4–6 per cage. Control animals should be maintained under standard laboratory conditions in a separate room to eliminate exposure to stress-related cues.
- According to the randomized weekly schedule, administer one of the following mild stressors per day to CUMS group animals [18, 19]:

- *Cage Tilting*: Tilt the home cage at a 45° angle for 4 h.
- *Wet Bedding*: Moisten bedding with water (24 ± 1 °C) until moderately damp. After 4 h, dry the animals and return them to a cage with clean, dry sawdust.
- *Empty Cage Exposure*: Transfer mice into a cage devoid of bedding material for 4 h.
- *Food Deprivation*: Restrict access to food for 6–10 h.
- *Water Deprivation*: Remove water dispensers for 6–10 h. Introduce an empty water bottle 1 h before the deprivation period ends.
- *Light/Dark Cycle Reversal*: Invert the circadian light cycle by illuminating the room at night and darkening it during daytime.
- *Social Isolation*: House animals individually for a 24-h period.
- *Restraint Stress*: Place each mouse in a restrainer device for 15–30 min.
- *Soiled Cage Exposure*: Replace the bedding with that previously used by another mouse or contaminated with rat sawdust for a duration of 1–2 h.

Administration of the test compound (e.g., functional food of interest) may commence on Day 1 or at a later stage within the CUMS schedule. The compound should be delivered via intraperitoneal injection or oral gavage to ensure accurate dosing. Control animals should receive an equivalent volume of the vehicle solution, delivered using the same administration route.

3.2.4 Assessment of End-Points

3.2.4.1 Sucrose Preference Test

- Acclimation phase (during acclimatization week and 18 h before each test phase): Animals are first given access to *two bottles* for 12 h (one with plain water and one with 1% sucrose solution) to ensure familiarity and rule out neophobia; bottles are typically switched between sides to prevent side bias (*see* **Note 2**).
- Test phase (starts from week 3): For 6 h mice are food and water deprived, followed by simultaneous individual access to the same amount of 1% sucrose solution and plain water for 12 h [21].
- *Intake is measured* (in mL or grams) for each bottle.
- The sucrose preference is calculated using the formula:

$$\text{Sucrose Preference}(\%) = \left(\frac{\text{Sucrose intake}}{\text{Sucrose} + \text{Water intake}}\right) \times 100$$

- Interpretation: High sucrose preference (e.g. > 65–70%): Suggests normal hedonic behavior. Low sucrose preference: Suggests anhedonia and possibly a depression-like behavior.

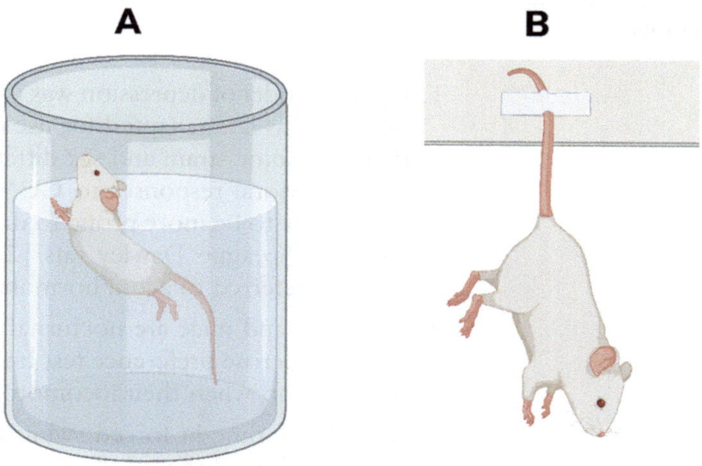

Fig. 1 Assessment of antidepressant potential in a CUMS mouse model: (**a**) Forced swim test; (**b**) Tail suspension test

3.2.4.2 Forced Swim Test [22]

- Each mouse is placed in a cylindrical container filled with water (typically 23–25 °C) deep enough so the animal *cannot touch the bottom* or *escape* (Fig. 1a).
- The mouse activity is observed for a test period of 5 min. The time of immobility (sec.) is recorded—the animal floats with minimal movements to keep its head above the water.
- Interpretation: Increased immobility time is interpreted as depression-like behavior. Decreased immobility after experimental substance treatment is interpreted as an antidepressant-like effect (*see* **Notes 3**, **4**, **5**, and **6**).

3.2.4.3 Tail Suspension Test [23]

- The mouse is suspended by the tail, using adhesive tape and fixed horizontal bar or a suspension apparatus, to hang about 30–50 cm above a surface (Fig. 1b).
- The mouse behavior is observed for 6 min. The immobility time is recorded (sec.)—when the animal is hanging passively, making only slight movements to keep balance.

Interpretation: Longer immobility time indicate increased behavioral despair—a depression-like behavior. Shorter immobility time after experimental substance treatment suggests a reversal of despair, interpreted as an antidepressant-like effect.

3.2.4.4 Statistical Analysis

Conduct statistical analysis on the recorded results.

4 Notes

1. CUMS model of depression was initially designed for rats [14–16] and was later adapted for use on mice [17–19]. However, there are some strain and sex differences in the neurochemical and behavioral responses to CUMS [24], e.g., Wistar Kyoto rats appear to be more prone to depression-like behavior compared to Sprague Dawley rats [25]. Male rats and mice are usually preferred to avoid hormonal cycle effects.
2. Since rats and mice are nocturnal animals, it is recommended that the sucrose preference test be performed at the beginning of the night, when their locomotor activity is at its peak [16].
3. Immobility time in forced swim test or tail suspension test is increased by substances that reduce the spontaneous motor activity of experimental animals. Therefore, to avoid false positive results for depression-like behavior, it is necessary to first perform an open field test for assessment of locomotor activity [9, 22].
4. In the forced swim test, a pre-test (training) session aiming to reduce novelty is optional. The training session is performed 24 h before the test session and usually lasts for 5 min. According to the original test designed by Porsolt, the duration of the test session is 15 min [22]. However, due to ethical considerations, tests that cause extreme physical exhaustion of experimental animals should be avoided.
5. In addition to anhedonia and despair, CUMS usually induces anxiety-like behavior in rodents. It is therefore appropriate to perform tests for its assessment, such as the social interaction test and the elevated plus maze test [19].
6. Depression-like behavior may be observed in experimental models that aim to induce other disorders, e.g., models of obesity, metabolic syndrome, and other disorders of energy metabolism [26, 27]; models of menopause, such as ovariectomy-induced osteoporosis [28, 29]; and models of hypothyroidism [30, 31].

Acknowledgments

The current work is supported by the Science Fund of Medical University of Varna, Bulgaria (project №24006).

References

1. World Health Organization (WHO) (2023) Depressive disorder (depression) [Internet]. 2023 [cited 2025 Apr 15]. Available from: https://www.who.int/news-room/fact-sheets/detail/depression
2. Ritter JM, Flower RJ, Henderson G, Loke YK, MacEwan D, Rang HP (2020) Antidepressant drugs. In: Rang & Dale's pharmacology, 9th edn. Elsevier
3. DeBattista C (2024) Antidepressant agents. In: Vanderah TW (ed) Katzung's basic & clinical pharmacology. 16th edn. McGraw-Hill, New York. 2024 [cited 2025 Apr 15]. Available from: https://accesspharmacy.mhmedical.com/content.aspx?bookid=3382&sectionid=281751777
4. COVID-19 Mental Disorders Collaborators (2021) Global prevalence and burden of depressive and anxiety disorders in 204 countries and territories in 2020 due to the COVID-19 pandemic. Lancet 398(10312):1700–1712. https://doi.org/10.1016/S0140-6736(21)02143-7
5. Song J, Kim YK (2021) Animal models for the study of depressive disorder. CNS Neurosci Ther 27(6):633–642. https://doi.org/10.1111/cns.13622
6. Tian H, Hu Z, Xu J, Wang C (2022) The molecular pathophysiology of depression and the new therapeutics. MedComm 3:e156. https://doi.org/10.1002/mco2.156
7. Ménard C, Hodes GE, Russo SJ (2016) Pathogenesis of depression: insights from human and rodent studies. Neuroscience 3(321):138–162. https://doi.org/10.1016/j.neuroscience.2015.05.053
8. Planchez B, Surget A, Belzung C (2019) Animal models of major depression: drawbacks and challenges. J Neural Transm (Vienna) 126(11):1383–1408. https://doi.org/10.1007/s00702-019-02084-y
9. Hao Y, Ge H, Sun M, Gao Y (2019) Selecting an appropriate animal model of depression. Int J Mol Sci 20(19):4827. https://doi.org/10.3390/ijms20194827
10. Molendijk ML, de Kloet ER (2015) Immobility in the forced swim test is adaptive and does not reflect depression. Psychoneuroendocrinology 62:389–391. https://doi.org/10.1016/j.psyneuen.2015.08.028
11. Farahani MS, Bahramsoltani R, Farzaei MH, Abdollahi M, Rahimi R (2015) Plant-derived natural medicines for the management of depression: an overview of mechanisms of action. Rev Neurosci 26(3):305–321. https://doi.org/10.1515/revneuro-2014-0058
12. Dobrek L, Głowacka K (2023) Depression and its phytopharmacotherapy-a narrative review. Int J Mol Sci 24(5):4772. https://doi.org/10.3390/ijms24054772
13. Urbanska N, Ashaolu TJ, Mattova S, Simko P, Kiskova T (2025) The potential of selected plants and their biologically active molecules in the treatment of depression and anxiety Disorders. Int J Mol Sci 26(5):2368. https://doi.org/10.3390/ijms26052368
14. Katz RJ, Roth KA, Carroll BJ (1981) Acute and chronic stress effects on open field activity in the rat: implications for a model of depression. Neurosci Biobehav Rev 5(2):247–251. https://doi.org/10.1016/0149-7634(81)90005-1
15. Willner P (1984) The validity of animal models of depression. Psychopharmacology 83(1):1–16. https://doi.org/10.1007/BF00427414
16. Papp M (2012) Models of affective illness: chronic mild stress in the rat. In: Enna SJ, Williams M, Barret JF, Ferkany JW, Kenakin T, Porsolt RD (eds) Current protocols in pharmacology. Wiley, Hoboken. https://doi.org/10.1002/0471141755.ph0509s57
17. Surget A, Belzung C (2008) Unpredictable chronic mild stress in mice. In: Kalueff AV, LaPorte JL (eds) Experimental animal models in neurobehavioral research. Nova Science, New York, pp 79–112. ISBN: 9781606920220
18. Nollet M, Le Guisquet AM, Belzung C (2013) Models of depression: unpredictable chronic mild stress in mice. Curr Protoc Pharmacol 61:5.65.1–5.65.17. https://doi.org/10.1002/0471141755.ph0565s61
19. Burstein O, Doron R (2018) The unpredictable chronic mild stress protocol for inducing anhedonia in mice. J Vis Exp 140:58184. https://doi.org/10.3791/58184
20. Frisbee JC, Brooks SD, Stanley SC, d'Audiffret AC (2015) An unpredictable chronic mild stress protocol for instigating depressive symptoms, behavioral changes and negative health outcomes in rodents. J Vis Exp 106:53109. https://doi.org/10.3791/53109
21. Muscat R, Willner P (1992) Suppression of sucrose drinking by chronic mild unpredictable stress: a methodological analysis. Neurosci Biobehav Rev 16:507–517. https://doi.org/10.1016/S0149-7634(05)80192-7

22. Porsolt RD, Bertin A, Jalfre M (1977) Behavioral despair in mice: a primary screening test for antidepressants. Arch Int Pharmacodyn Ther 229:327–336
23. Steru L, Chermat R, Thierry B, Simon P (1985) The tail suspension test: a new method for screening antidepressants in mice. Psychopharmacology 85(3):367–370. https://doi.org/10.1007/BF00428203
24. Carere C, Maestripieri D (2013) Animal personalities: behavior, physiology, and evolution. University of Chicago Press, Chicago. https://doi.org/10.7208/chicago/9780226922065.001.0001
25. O'Mahony CM, Clarke G, Gibney S, Dinan TG, Cryan JF (2011) Strain differences in the neurochemical response to chronic restraint stress in the rat: relevance to depression. Pharmacol Biochem Behav 97:690–699. https://doi.org/10.1016/j.pbb.2010.11.012
26. Gancheva S, Galunska B, Zhelyazkova-Savova M (2017) Diets rich in saturated fat and fructose induce anxiety and depression-like behaviours in the rat: is there a role for lipid peroxidation? Int J Exp Pathol 98(5):296–306. https://doi.org/10.1111/iep.12254
27. Vagena E, Ryu JK, Baeza-Raja B, Walsh NM, Syme C, Day JP, Houslay MD, Baillie GS (2019) A high-fat diet promotes depression-like behavior in mice by suppressing hypothalamic PKA signaling. Transl Psychiatry 9(1):141. https://doi.org/10.1038/s41398-019-0470-1
28. Georgieva A, Eftimov M, Todorova M, Kuzmanova V, Kuzmanov A, Kuzmanov K, Vlaskovska M, Valcheva-Kuzmanova S (2021) Effects of ovariectomy-induced estrogen deficit on rat behaviour, lipid metabolism, inflammation, bone mineral density, and turnover. Folia Med (Plovdiv) 63(3):385–391. https://doi.org/10.3897/folmed.63.e55267
29. Grigoryan GA (2022) Ovariectomy as a model of anxiety-depressive disorders. Neurochem J 16:1–13. https://doi.org/10.1134/S1819712422010068
30. Özgür E, Gürbüz Özgür B, Aksu H, Cesur G (2016) The effect of congenital and postnatal hypothyroidism on depression-like behaviors in juvenile rats. J Clin Res Pediatr Endocrinol 8(4):439–444. https://doi.org/10.4274/jcrpe.3498
31. Jin Z, Ling J, Yu J, He M, Ni P, Zhang F, Wang Y (2021) Serotonin 2A receptor function and depression-like behavior in rats model of hypothyroidism. Exp Brain Res 239(8):2435–2444. https://doi.org/10.1007/s00221-021-06129-1

Chapter 24

Determination of Anxiolytic Potential of Plant Leaf

Stanila S. Stoeva-Grigorova

Abstract

Anxiety is a prevalent emotional state, necessitating the development of effective therapeutic strategies. Recent research suggests that certain bioactive food compounds may exhibit anxiolytic properties while simultaneously offering a favorable safety profile compared to traditional pharmacological agents. However, validating their efficacy requires rigorously standardized behavioral testing. Animal models remain indispensable for assessing anxiolytic activity, as they allow the investigation of complex behavioral responses. Among the most commonly used paradigms are the elevated plus maze, light/dark box, open field test, and social interaction test, all of which are described in the present chapter.

Key words Anxiety, Anxiolytic activity, In vivo modeling, Rats, Mice, Elevated plus maze, Light/dark box, Open field test, Social interaction test

1 Introduction

Anxiety is an emotional state characterized by feelings of unease, discomfort, and fear, typically elicited by a perceived or unknown threat, affecting approximately one in eight individuals worldwide [1, 2]. Recent studies have increasingly demonstrated that certain food products—such as those containing curcumin, cannabidiol, rosmarinic acid, caffeic acid, ginsenosides, chrysin, fermented compounds, and food protein-derived anxiolytic peptides—may play a role in the prevention and/or treatment of anxiety [2–5]. These compounds are often regarded as having a more favorable safety profile compared to pharmaceutical agents. However, akin to traditional pharmacological interventions, the validation of the anxiolytic properties of functional foods requires the application of rigorously validated testing procedures. It is generally preferable to utilize animal-based models, as they allow for the representation of complex behavioral responses, thereby providing a more accurate assessment of the efficacy of these substances.

Anxiety tests in laboratory animals have been utilized for decades to investigate the neurobiological characteristics from a

Mitesh Kumar Dwivedi and A. Sankara Narayanan (eds.), *Leaves as Functional Foods*, Methods and Protocols in Food Science, https://doi.org/10.1007/978-1-0716-5226-8_24,

pathophysiological perspective, as well as for the screening of novel compounds with anxiolytic activity [6]. Rats and mice are the most frequently employed species in preclinical anxiety studies [7]. Popular methods for investigating anxiety include:

- *The Geller-Seifter test*—A method based on operant behavior, involving food deprivation and electric shock. A variation of this is the simplified "Vogel" conflict test, which requires little to no prior training.
- *The social interaction test*—Changes in the social behavior of rodents are measured in an environment with anxiogenic stimuli.
- *The light/dark box test*—The test is based on the conflict between the rodents' natural drive to explore their environment and their fear of bright light.
- *The elevated plus maze*—The test is based on the conflict between the rodents' exploratory behavior and their fear of open and/or elevated spaces.
- *The open field test*—This test is a commonly employed behavioral paradigm in neuroscience and pharmacology to evaluate anxiety-related responses, locomotor activity, and exploratory behavior in rodents. It capitalizes on the inherent tension between a rodent's instinct to explore novel environments and its innate avoidance of open, brightly illuminated areas, which can evoke anxiety-like reactions.
- *The defensive burying test*—This test is based on the species-specific behavioral response of rodents when exposed to nociceptive or predator-related stimuli. In a laboratory setting, this response can be triggered by an electrified shock sensor placed on one of the walls of an acrylic chamber. Rats that have received a shock from the stationary sensor typically exhibit stereotyped behavior, scattering bedding material across the chamber floor toward and over the sensor, while avoiding further contact with it. A variation of this model is the so-called "marble burying test" in mice.
- *The thirsty rat conflict test*—This test is a behavioral assay used to evaluate anxiety-like responses and the anxiolytic potential of xenobiotics in rodents. It is based on the approach-avoidance conflict, where a rat must choose between obtaining water to satisfy its thirst and avoiding an aversive stimulus, such as an electric shock [6, 8, 9].

To ensure that the observed level of anxiety is determined by the test conditions themselves rather than external factors, the following variables must be taken into account: the species, breed, sex, and age of the animals; housing conditions (individual or group housing, light intensity); the method and timing of animal transfer between the breeding area and the experimental room; diet; etc. [10–12].

In accordance with the Three Rs principle, this chapter presents some of the most commonly used yet least stressful methods for assessing anxiolytic activity in experimental animals [13]. The procedures presented do not involve food or water deprivation, nor the use of electroshock. They are of relatively short duration and moderate labor intensity, although they require access to specialized experimental setups, which should preferably be obtained from a licensed manufacturer.

2 Materials

2.1 Social Interaction Test [6, 8, 14, 15]

- Adult male rats of the Hooded Lister, Sprague-Dawley, or Wistar strains (150–200 g, 8–10 weeks old; at least 10–15 animals per group).
- Standard Pellet diet (if the functional food is not intended to fully meet the nutritional needs of the animals).
- Test substance.
- Vehicle solution (for the control group).
- A quiet testing room, isolated from external disturbances, with the capability to control lighting from dim (30 lux) to bright (250–300 lux).
- Animal cages (sufficient number for individual housing of the animals).
- Video camera, mounted above the arena (*see* **Note 1**).
- Arena for conducting the social interaction test (The use of the open field test arena is a practical solution).
- 70% ethanol.

2.2 Light-Dark Box Test [6, 16]

- Adult male mice (suitable strains include Swiss albino, C57BL/6 J, NIH/s strain, and BKW); weight ~25–30 g, with at least 10–15 animals per group.
- Standard Pellet diet (if the functional food is not intended to fully meet the nutritional needs of the animals).
- Test substance.
- Vehicle solution (for the control group).
- A quiet testing room, isolated from external disturbances.
- Light-dark box (preferably from a certified manufacturer).
- 70% ethanol.

2.3 Elevated Plus Maze Test [6, 15, 17]

- Animals—Sprague-Dawley or Wistar male rats; 150–200 g, 8–10 weeks old; at least 10–15 per treatment group.
- Standard pellet diet (if the functional food is not intended to fully meet the nutritional needs of the animals).
- Test substances.
- Vehicle solution (for the control group).
- A quiet testing room, isolated from external disturbances.
- Equipment for the elevated plus maze test (preferably with a video camera and an automated system for recording behavioral responses) (*see* **Note 1**).
- 70% ethanol.

2.4 Open Field Test [6, 9, 15, 18]

- Wistar or Sprague-Dawley rats (between 8 and 12 weeks old, 150–200 g), at least 10–15 animals per group.
- Standard pellet diet (if the functional food is not intended to fully meet the nutritional needs of the animals).
- Test substance.
- Vehicle solution (for the control group).
- A quiet testing room, isolated from external disturbances.
- An open field arena (preferably from a certified manufacturer, equipped with the capability for automated recording of behavioral responses).
- 70% ethanol.

3 Methods

3.1 Social Interaction Test [6, 8, 15, 19–23]

The social interaction test is the first animal model of anxiety that attempts to utilize ethologically relevant sources of anxiety and measure natural forms of behavior as dependent variables (e.g., the time spent by pairs of male rats in social interaction, such as olfactory exploration, following, or grooming the partner). Since the behavior of one rat influences that of the other, it is important that the pair be treated as a unit, and only one value for the pair should be used.

3.1.1 Preparation of Test Substance

To ensure high scientific validity and humane treatment in the experiments, carefully consider both the preparation of the test substance and its method of administration. It is recommended to perform both qualitative and quantitative analyses of the composition of the functional foods in advance. If the nutritional intake is to be provided solely by the test substance, it is crucial to account for its nutritional value. To achieve maximum accuracy, the dosage of the functional food should be calculated based on the animals' weight and needs, with precise determination of the active ingredient concentration. Depending on the research objectives, the functional food may be administered to the animals over varying time periods (e.g., daily for a specified number of days prior to testing) to assess its effects on behavior under stress conditions.

3.1.2 Acclimatization of the Animals

Before the experiment, the animals undergo a 7-day acclimatization period under controlled conditions. They are housed in a well-ventilated room with a 12-h light/dark cycle. The temperature is maintained at 23 ± 2 °C, and humidity is kept at 50 ± 10%. Standard rodent chow and water are provided ad libitum. Handling and experimental procedures follow national and international regulations for the protection and humane treatment of laboratory animals. Animals exhibiting signs of poor health, behavioral abnormalities, or deviations in body weight are excluded from the study.

Following the acclimatization period, the animals must be prepared for the actual behavioral experiment. To this end, house the rats individually for 5 days prior to the experiment, as this is expected to increase the levels of social interaction (*see* **Note 2**). The weight of the animals should be documented daily. The light should be turned on at least 1 h prior to the commencement of testing. Since the intensity of light can be used as an anxiogenic stimulus, the animals should be kept under low illumination (≤50 lux). Otherwise, continuous exposure to this stressor may result in damage to the photoreceptors.

3.1.3 Randomization of the Animals

The experimental animals are first weighed individually, then randomly assigned to different experimental groups, ensuring that each group contains an equal number of subjects. One group serves as the control group, typically receiving the vehicle treatment. To facilitate a more detailed analysis, additional groups are included, each receiving progressively higher doses of the test compounds (e.g., high, medium, and low doses). The results from these groups are then compared to those of a separate experimental group treated with a reference drug known for its anxiolytic effects (e.g., benzodiazepines, barbiturates). The former drug increases social interaction time across all test conditions, while the latter does so in all conditions except for the weak light and familiar arena scenario.

3.1.4 Treatment Procedure

- Standardize the testing time to minimize diurnal variations, with the optimal period being between 08:00 and 13:00 h.
- Move the rat cages to a dimly lit area for acclimatization (near the testing room) at least 1 h prior to the test. [Such experiments are conducted in licensed vivaria, which are typically equipped with such anterooms.]

- While in the waiting area, both rats in the pair should receive identical treatment with the test substance. Given that the experiment involves the study of functional foods, the oral route of administration is preferred. The methods for precise dosing via an orogastric tube are detailed in the chapter (Determination of Antidiabetic Potential from Plant Leaf).
- Due to the high sensitivity of the experimental animals, position the experimental setup in a quiet laboratory room with a minimal number of objects and features (such as external noises and movements) that could potentially influence behavior.
- Place the male rat pairs in an experimental arena. Conduct the procedure with a relatively unfamiliar partner (housed individually in an adjacent cage) to promote social behavior. Partner pairs should be selected such that their weights do not differ by more than 10 g.
- Based on the fact that bright light and an unfamiliar environment are stressors for the experimental rodents, the following combinations of testing conditions are possible:
  - Dim light + familiar arena.
  - Dim light + unfamiliar arena.
  - Bright light + familiar arena.
  - Bright light + unfamiliar arena.
- It is permissible to use one or more of the four possible testing conditions depending on the objective of the experiment. [An animal may not be tested more than once in an unfamiliar testing environment. It may first be tested in the unfamiliar environment, followed by testing in the familiar one. In this case, before the second test, the rat should be placed in the arena for a brief 5–10 min acclimation session over two consecutive days.]
- The duration of the test is not strictly fixed, but studies in the literature most commonly report experiments lasting 5–10 min.
- Record all experimental sessions via video, while monitoring them on a screen in an adjacent room. Commercially available specialized experimental setups from licensed manufacturers are now used, allowing for the recording of social interactions with a customized keyboard and walls integrated with photocells, which automatically register locomotor activity.

3.1.5 Assessment of End-Points

To assess social interaction, record the time spent in active social interaction. This includes the determination of:

- Aggressive behavior (kicking, boxing, fighting, submission).
- Non-aggressive behavior (sniffing, following, grooming the partner).

Test substances with anxiolytic effects should significantly increase the time spent in social interaction (most notably when the arena is unfamiliar and/or brightly lit). As a general rule, if the results are doubled, it likely indicates that the treatment has induced a significant biological effect.

3.1.6 Statistical Analysis

Conduct statistical analysis on the recorded results (*see* **Note 5**).

3.2 Light-Dark Box Test [6, 16, 24–30]

In its simplest form, the experimental setup consists of a box, typically constructed from PVC or plexiglass to facilitate easy cleaning. For experiments involving rats, an appropriate setup includes compartments with dimensions of 42 × 20 × 40(h)cm (Fig. 1). It comprises two compartments with distinct lighting conditions. The brightly lit compartment serves as a stress-inducing stimulus for the animal, characterized by illumination levels of 400–1000 lux, openness, and exposure. In contrast, the dark compartment functions as a secure refuge, providing a low-light environment. A small opening connects the two compartments, allowing the rodent to move freely between them.

3.2.1 Preparation of Test Substance

Ensuring high scientific validity and humane treatment in experiments requires careful attention to both the preparation of the test substance and the method of its administration.

3.2.2 Acclimatization of the Animals

Before the experiment, the animals undergo a 7-day acclimatization period under controlled conditions. They are housed in a well-ventilated room with a 12-h light/dark cycle. The temperature is maintained at 23 ± 2 °C, and humidity is kept at 50 ± 10%. Standard rodent chow and water are provided ad libitum. Handling and

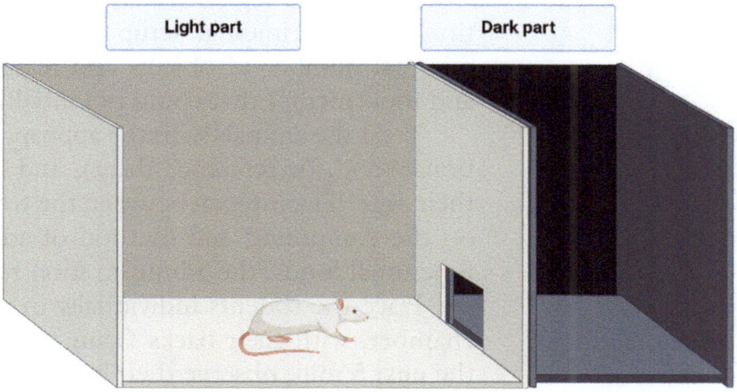

Fig. 1 Experimental setup for the light/dark box test in rodents

experimental procedures follow national and international regulations for the protection and humane treatment of laboratory animals. Animals exhibiting signs of poor health, behavioral abnormalities, or deviations in body weight are excluded from the study.

3.2.3 Randomization of the Animals

The experimental animals are first weighed individually, then randomly assigned to different experimental groups, ensuring that each group contains an equal number of subjects. One group serves as the control, typically receiving the vehicle treatment. To facilitate a more detailed analysis, additional groups are included, each receiving progressively higher doses of the test compounds (e.g., high, medium, and low doses). The results from these groups are then compared to those of a separate experimental group treated with a reference drug known for its anxiolytic effects.

3.2.4 Treatment Procedure

Standardize the testing time to minimize diurnal variations, with the optimal period being between 08:00 and 13:00 h.

Following the acclimatization period, the animals must be prepared for the actual behavioral study. To this end, house the rats individually for 5 days prior to the experiment, as this is expected to increase the levels of social interaction.

The weight of the animals should be documented daily.

The light should be turned on at least 1 h prior to the commencement of testing. Since the intensity of light can be used as an anxiogenic stimulus, the animals should be kept under low illumination ($\leq$50 lux). Otherwise, continuous exposure to this stressor may result in damage to the photoreceptors.

During the final days of acclimatization, the animals should be placed in the testing arena for brief periods to allow them to adapt to the test conditions (e.g., variations in light and dark within the light/dark box).

Move the rat cages to a dimly lit area for acclimatization (near the testing room) at least 1 h prior to the test.

Due to the high sensitivity of the experimental animals, position the experimental setup in a quiet laboratory room with a minimal number of objects and features (such as external noises and movements) that could potentially influence behavior.

Treat the animals with the appropriate substance (vehicle, functional food, or reference drug), and immediately return them to their cage. The interval between the treatment and the test depends on the compound and method of administration (preferably, for functional foods, the administration route should be oral).

Place the rodents individually in the center of the illuminated chamber, with their backs facing the intermediate opening. Over the next 5 min, observe their behavior according to the assessment endpoints outlined below.

Clean the arena with 70% ethanol immediately before testing the behavior of each experimental animal.

3.2.5 Assessment of End Points

Reduction of anxiety in the Light-Dark Box Test is indicated by an increased number of transitions between the dark and illuminated compartments, as well as enhanced exploratory behavior, characterized by increased time spent and a higher number of crossed lines in the illuminated zone. In this regard, the following parameters should be quantitatively characterized:

- The number of transitions from the illuminated to the dark zone.
- The time spent in each of the two zones (with time spent in the illuminated zone considered most relevant for assessing anxiolytic activity).
- The latency period before the first exit from the illuminated zone.

3.2.6 Statistical Analysis

Conduct statistical analysis on the recorded results (*see* **Note 5**).

3.3 Elevated Plus Maze Test [6, 8, 15, 17, 31–35]

The elevated plus maze test is an unconditioned spontaneous behavioral conflict model. The basis of the test lies in the conflict between the innate exploratory behavior of rodents, placed in a novel environment, and the avoidance of its aversive characteristics. It is the most widely used animal model for anxiety, as the effects of test substances on appetite or pain sensitivity are not expected to interfere with the experimental results. Additionally, the test serves to assess locomotor activity. The classic experimental setup is shaped like a plus sign. Two of the opposite arms are open, while the other two are enclosed by walls, with all four arms separated by a central zone. The apparatus is additionally elevated from the ground, allowing not only elements of unfamiliarity and openness but also incorporating height (Fig. 2).

3.3.1 Preparation of Test Substance

Ensuring high scientific validity and humane treatment in experiments requires careful attention to both the preparation of the test substance and the method of its administration.

3.3.2 Acclimatization of the Animals

Before the experiment, the animals undergo a 7-day acclimatization period under controlled conditions. They are housed in a well-ventilated room with a 12-h light/dark cycle. The temperature is maintained at 23 ± 2 °C, and humidity is kept at 50 ± 10%. Standard rodent chow and water are provided ad libitum. Handling and experimental procedures follow national and international regulations for the protection and humane treatment of laboratory animals. Animals exhibiting signs of poor health, behavioral abnormalities, or deviations in body weight are excluded from the study.

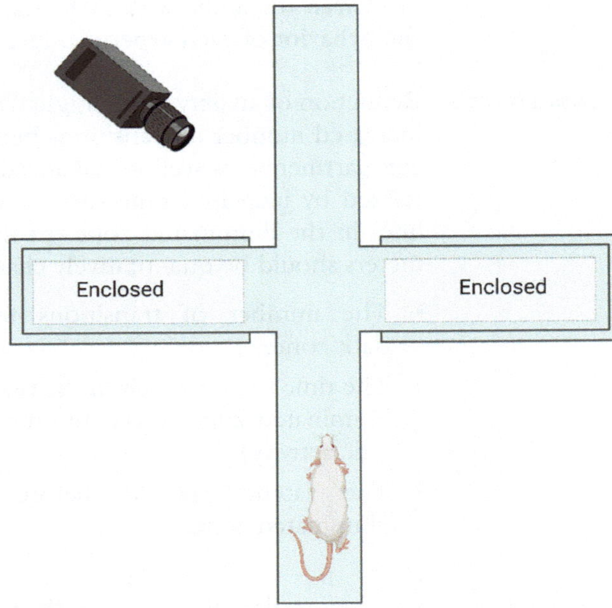

Fig. 2 Elevated plus maze test

3.3.3 Randomization of the Animals

The experimental animals are first weighed individually, then randomly assigned to different experimental groups, ensuring that each group contains an equal number of subjects (*see* **Note 3**). One group serves as the control, typically receiving the vehicle treatment. To facilitate a more detailed analysis, additional groups are included, each receiving progressively higher doses of the test compounds (e.g., high, medium, and low doses). The results from these groups are then compared to those of a separate experimental group treated with a reference drug known for its anxiolytic effects.

3.3.4 Treatment Procedure

- Standardize the testing time to minimize diurnal variations, with the optimal period being between 08:00 and 13:00 h.
- Move the rat cages to a dimly lit area for acclimatization (near the testing room) at least 1 h prior to the test.
- Due to the high sensitivity of the experimental animals, position the experimental setup in a quiet laboratory room with a minimal number of objects and features (such as external noises and movements) that could potentially influence behavior.
- Place the maze on the floor of the test room (commercially available models typically have the required height and do not require the experimenters to adjust a stand or table). The behavior of the animals can be assessed either through live observation or via video footage by a trained observer.
- Place the animal on the central platform, facing the open arm.

- Observe the entries (with all four paws) and exits (with both front paws) of the animal into the arms of the experimental setup over the next 5 min (*see* **Note 4**).
- Clean the arena with 70% ethanol immediately before testing the behavior of each experimental animal.

3.3.5 Assessment of End Points

The anxiolytic effect is indicated by an increase in activity in the open arms of the maze (i.e., an increase in the percentage of entries into the open arms, as well as the percentage of time spent in them). In this regard, the following parameters should be quantitatively characterized:

- Number of entries into the open arms.
- Number of entries into the closed arms.
- Time spent in the open arms.
- Time spent in the closed arms.
- Time spent in the central square of the experimental setup.

3.3.6 Statistical Analysis

Conduct statistical analysis on the recorded results (*see* **Note 5**).

3.4 Open Field Test [6, 9, 15, 18, 24, 29, 33, 36–39]

The experimental setup includes an arena (commonly referred to as the open field). Its dimensions may vary depending on the species of the animals being studied and the specific objectives of the experiment. Typically, the arena is either circular (for rats, diameter ~100 cm) or square (for rats, ~100 × 100 cm), with the walls needing to be sufficiently high to prevent the animals from jumping over them (>40 cm for rats). Although variations in size are observed in practice, all arenas are divided in the same manner—the surface is partitioned into a central and a peripheral zone (Fig. 3).

3.4.1 Preparation of Test Substance

Ensuring high scientific validity and humane treatment in experiments requires careful attention to both the preparation of the test substance and the method of its administration.

3.4.2 Acclimatization of the Animals

Before the experiment, the animals undergo a 7-day acclimatization period under controlled conditions. They are housed in a well-ventilated room with a 12-h light/dark cycle. The temperature is maintained at 23 ± 2 °C, and humidity is kept at 50 ± 10%. Standard rodent chow and water are provided ad libitum. Handling and experimental procedures follow national and international regulations for the protection and humane treatment of laboratory animals. Animals exhibiting signs of poor health, behavioral abnormalities, or deviations in body weight are excluded from the study.

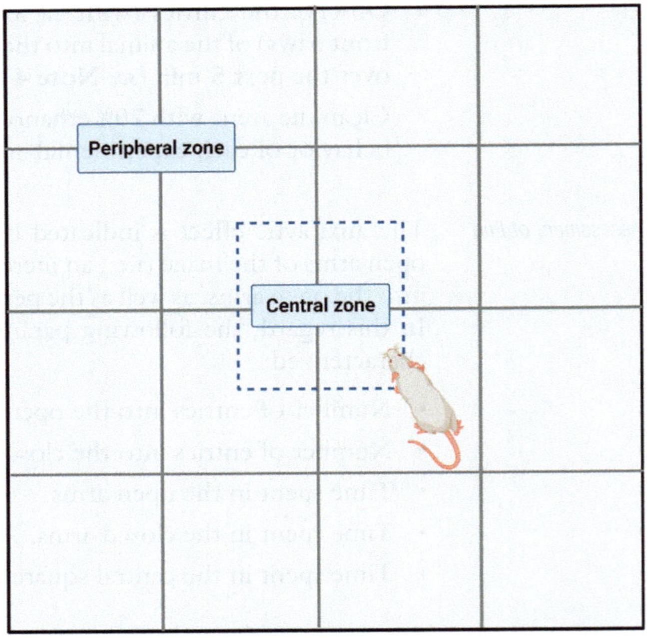

Fig. 3 Square-shaped open field test arena

3.4.3 Randomization of the Animals

The experimental animals are first weighed individually, then randomly assigned to different experimental groups, ensuring that each group contains an equal number of subjects. One group serves as the control, typically receiving the vehicle treatment. To facilitate a more detailed analysis, additional groups are included, each receiving progressively higher doses of the test compounds (e.g., high, medium, and low doses). The results from these groups are then compared to those of a separate experimental group treated with a reference drug known for its anxiolytic effects.

3.4.4 Treatment Procedure

- Standardize the testing time to minimize diurnal variations, with the optimal period being between 08:00 and 13:00 h.
- Move the animal cages to a dimly lit area for acclimatization (near the testing room) at least 1 h prior to the test.
- Due to the high sensitivity of the experimental animals, position the experimental setup in a quiet laboratory room with a minimal number of objects and features (such as external noises and movements) that could potentially influence behavior.
- The experimental procedure commences with the placement of a single animal from the group into the central zone of the open field.
- Monitor the rodent's behavior for a duration of 5 min, recording changes in location and the time spent in either the central or

peripheral zones. This approach allows for the assessment of both anxiety levels and locomotor activity.

- Proceed with the experiment by sequentially testing all remaining animals in the initially designated groups.
- Clean the arena with 70% ethanol immediately before testing the behavior of each experimental animal.

3.4.5 Assessment of End Points

During the 5-min observation period in the open field, the following criteria are assessed:

- Distance traveled.
- Number of crossed lines.
- Time spent in the central zone.
- Time spent in the peripheral zone.
- Time spent immobile.

3.4.6 Statistical Analysis

Conduct statistical analysis on the recorded results (*see* **Note 5**).

4 Notes

1. If a video camera is used in behavioral models, it should be positioned vertically above the test arena, with the video monitor located in an adjacent room to avoid disturbing the animals by the observer.
2. In the social interaction test, highly active pairs may be observed whose confrontations are followed by deliberate avoidance of the partner for a period of time. Such animals should be excluded from the study.
3. In the elevated plus maze test, animals should be subjected to a single trial, unless the aim is to study a distinct type of anxiety triggered by repeated exposure to the maze.
4. Throughout the experiment, animals may be observed falling from the elevated maze, particularly if they have been administered high doses of substances with sedative and/or muscle relaxant properties.
5. If both the observer and the person conducting the statistical analysis are unaware of the substances administered to the different animal groups, this will significantly reduce subjectivity in the interpretation of the experimental results.

References

1. Zhang SX, Li LZ (2025) War anxiety: a review. Curr Psychiatry Rep 27:140–146. https://doi.org/10.1007/s11920-024-01583-4
2. Suryavanshi A, Kumar V, Shukla D, Asati V, Mahapatra DK, Bharti SK, Ahire ED (2024) Chapter 27 – Nutraceuticals and anxiety. In: Keservani RK, Kesharwani RK, Emerald M, Sharma AK (eds) Nutraceutical fruits and foods for neurodegenerative disorders. Academic, pp 537–559. https://doi.org/10.1016/B978-0-443-18951-7.00029-3
3. Hilimire MR, DeVylder JE, Forestell CA (2015) Fermented foods, neuroticism, and social anxiety: an interaction model. Psychiatry Res 228(2):203–208. https://doi.org/10.1016/j.psychres.2015.04.023
4. Hafeez Z, Benoit S, Cakir-Kiefer C, Darya A, Micloa L (2021) Food protein-derived anxiolytic peptides: their potential role in anxiety management. Food Funct 12:1415–1431. https://doi.org/10.1039/D0FO02432E
5. Kumar P, Rai S, Joshi A (2024) Chapter 6 – Role of functional foods and nutraceuticals for depression, post-traumatic stress disorder, and suicidal behaviors. In: Keservani RK, Kesharwani RK, Emerald M, Sharma AK (eds) Nutraceutical fruits and foods for neurodegenerative disorders. Academic, pp 101–122. https://doi.org/10.1016/B978-0-443-18951-7.00006-2
6. File SE, Lippa AS, Beer B, Lippa MT (2004) Animal tests of anxiety. Curr Protoc Pharmacol 27:5.38.1–5.38.21. https://doi.org/10.1002/0471141755.ph0538s27
7. Cryan JF, Sweeney FF (2011) The age of anxiety: role of animal models of anxiolytic action in drug discovery. Br J Pharmacol 164(4):1129–1161. https://doi.org/10.1111/j.1476-5381.2011.01362.x
8. Treit D, Engin E, McEown K (2010) Animal models of anxiety and anxiolytic drug action. Curr Top Behav Neurosci 2:121–160. https://doi.org/10.1007/7854_2009_17
9. Kraeuter AK, Guest PC, Sarnyai Z (2019) The open field test for measuring locomotor activity and anxiety-like behavior. Methods Mol Biol 1916:99–103. https://doi.org/10.1007/978-1-4939-8994-2_9
10. Prasad A, Prasad C (1996) Short-term consumption of a diet rich in fat decreases anxiety response in adult male rats. Physiol Behav 60(3):1039–1042. https://doi.org/10.1016/0031-9384(96)00135-7
11. Hartley DE, Edwards JE, Spiller CE et al (2003) The soya isoflavone content of rat diet can increase anxiety and stress hormone release in the male rat. Psychopharmacology 167(1):46–53. https://doi.org/10.1007/s00213-002-1369-7
12. Souza CG, Moreira JD, Siqueira IR et al (2007) Highly palatable diet consumption increases protein oxidation in rat frontal cortex and anxiety-like behaviour. Life Sci 81(3):198–203. https://doi.org/10.1016/j.lfs.2007.05.001
13. Holley T, Bowe G, Campia I et al (2016) Accelerating progress in the replacement, reduction and refinement of animal testing through better knowledge sharing, EUR 28234 EN. Publications Office of the European Union. https://doi.org/10.2788/934083
14. Berton O, Ramos A, Chaouloff F et al (1997) Behavioral reactivity to social and nonsocial stimulations: a multivariate analysis of six inbred rat strains. Behav Genet 27(3):155–166. https://doi.org/10.1023/A:1025641509809
15. Abtulov M, Eftimov M, Marinova S, Salbashian M, Zhelyazkova-Savova M, Valcheva-Kuzmanova S (2024) Behavioral effects of Aronia melanocarpa fruit juice in experimental animals with diet-induced metabolic syndrome. Trakia J Sci 3:209–218. https://doi.org/10.15547/tjs.2024.03.003
16. Hascoët M, Bourin M (2009) The mouse light–dark box test. In: Gould T (ed) Mood and anxiety related phenotypes in mice, Neuromethods, vol 42. Humana Press, Totowa, pp 205–213. https://doi.org/10.1007/978-1-60761-303-9_11
17. Moneva-Marinova K, Rafailova E, Reyzov M, Todorova M, Eftimov M, Gancheva S, Zhelyazkova-Savova M, Valcheva-Kuzmanova S (2024) Behavioral effects of chronic Japanese quince fruit juice administration to rats with diet-induced metabolic syndrome. Acta Medica Bulgarica 51(2):34–38. https://doi.org/10.2478/amb-2024-0017
18. Arunachalam K, Sasidharan SP (2021) Animal experiments of central nervous system (CNS). In: Bioassays in experimental and preclinical pharmacology, Springer protocols handbooks. Humana, pp 347–357. https://doi.org/10.1007/978-1-0716-1233-0_18
19. File SE, Seth P (2003) A review of 25 years of the social interaction test. Eur J Pharmacol 463(1–3):35–53. https://doi.org/10.1016/s0014-2999(03)01273-1

20. Koob GF, Zimmer A (2012) Animal models of psychiatric disorders. In: Aminoff M, Boiler F, Swaab D (eds) Handbook of clinical neurology. Elsevier, New York, pp 137–166. https://doi.org/10.1016/B978-0-444-52002-9.00009-7
21. File SE, Hyde JR (1978) Can social interaction be used to measure anxiety? Br J Pharmacol 62(1):19–24. https://doi.org/10.1111/j.1476-5381.1978.tb07001.x
22. Kondrakiewicz K, Kostecki M, Szadzińska W, Knapska E (2019) Ecological validity of social interaction tests in rats and mice. Genes Brain Behav 18:e12525. https://doi.org/10.1111/gbb.12525
23. UCSF IACUC. Miscellaneous rodent procedures – oral gavage in mice and rats. Available at: https://iacuc.ucsf.edu/sites/g/files/tkssra751/f/wysiwyg/STD%20PROCEDURE%20-%20Misc%20Rodent%20Procedures%20-%20Oral%20Gavage%20In%20Mice%20and%20Rats.pdf
24. Kulesskaya N, Voikar V (2014) Assessment of mouse anxiety-like behavior in the light-dark box and open-field arena: role of equipment and procedure. Physiol Behav 133:30–38. https://doi.org/10.1016/j.physbeh.2014.05.006
25. Jun G (2024) Light-dark box test. In: The ECPH encyclopedia of psychology. Springer, Singapore, pp 1–2. https://doi.org/10.1007/978-981-99-6000-2_82-1
26. Bloch S, Belzung C (2023) The light–dark box test in the mouse. In: Harro J (ed) Psychiatric vulnerability, mood, and anxiety disorders, Neuromethods, vol 190. Humana, New York, pp 41–53. https://doi.org/10.1007/978-1-0716-2748-8_3
27. Doukkali Z, Taghzouti K, Bouidida EL, Nadjmouddine M, Cherrah Y, Alaoui K (2015) Evaluation of anxiolytic activity of methanolic extract of *Urtica urens* in a mice model. Behav Brain Funct 11:19. https://doi.org/10.1186/s12993-015-0063-y
28. Ugo Basile (2025) Light/Dark Box for Mice and Rats [Internet]. 2025 [cited 2025 Apr 5]. Available from: https://ugobasile.com/products/categories/mazes-tracking/light-dark-box-for-mice-and-rats
29. Rein B, Ma K, Yan Z (2020) A standardized social preference protocol for measuring social deficits in mouse models of autism. Nat Protoc 15(10):3464–3477. https://doi.org/10.1038/s41596-020-0382-9
30. Bourin M (2015) Animal models for screening anxiolytic-like drugs: a perspective. Dialogues Clin Neurosci 17(3):295–303. https://doi.org/10.31887/DCNS.2015.17.3/mbourin
31. Hogg S (1996) A review of the validity and variability of the elevated plus-maze as an animal model of anxiety. Pharmacol Biochem Behav 54(1):21–30. https://doi.org/10.1016/0091-3057(95)02126-4
32. Walf A, Frye C (2007) The use of the elevated plus maze as an assay of anxiety-related behavior in rodents. Nat Protoc 2:322–328. https://doi.org/10.1038/nprot.2007.44
33. Kraeuter AK, Guest PC, Sarnyai Z (2019) The elevated plus maze test for measuring anxiety-like behavior in rodents. In: Guest PC (ed) Pre-clinical models, Methods Mol Biol, vol 1916. Humana Press, New York, pp 69–74. https://doi.org/10.1007/978-1-4939-8994-2_4
34. Komada M, Takao K, Miyakawa T (2008) Elevated plus maze for mice. J Vis Exp 22:1088. https://doi.org/10.3791/1088
35. Vilmosh N, Delev D, Kostadinov I, Zlatanova H, Kotetarova M, Kandilarov I et al (2022) Anxiolytic effect of *Satureja montana* dry extract and its active compounds rosmarinic acid and carvacrol in acute stress experimental model. J Integr Neurosci 21(5):124. https://doi.org/10.31083/j.jin2105124
36. Prut L, Belzung C (2003) The open field as a paradigm to measure the effects of drugs on anxiety-like behaviors: a review. Eur J Pharmacol 463(1–3):3–33. https://doi.org/10.1016/S0014-2999(03)01272-X
37. Gould TD, Dao DT, Kovacsics CE (2009) The open field test. In: Gould TD (ed) Mood and anxiety related phenotypes in mice, Neuromethods, vol 42. Humana Press, Totowa, pp 1–20. https://doi.org/10.1007/978-1-60761-303-9_1
38. Perals D, Griffin AS, Bartomeus I, Sol D (2017) Revisiting the open-field test: what does it really tell us about animal personality? Anim Behav 123:69–79. https://doi.org/10.1016/j.anbehav.2016.10.006
39. Võikar V, Stanford SC (2023) The open field test. In: Harro J (ed) Psychiatric vulnerability, mood, and anxiety disorders, Neuromethods, vol 190. Humana, New York, pp 27–40. https://doi.org/10.1007/978-1-0716-2748-8_2

Chapter 25

Determination of Neuroprotective Effect of *Bacopa monnieri* and *Sterculia setigera* Leaf Extracts

Jigisha Roy Panda and Tamalika Chakraborty

Abstract

This protocol focuses on the extraction and determination of neuroprotective capacity (neuromodulators) of plant parts from *Bacopa monnieri* and *Sterculia setigera*. Both plants are traditionally recognized for their neuroprotective properties. The protocol includes extraction, in vitro and in vivo assessment of neuroprotective effects, behavioral and biochemical analyses, and molecular assays to evaluate neuroprotection against induced neurotoxicity. The methodology outlines systematic identification and characterization of neuroprotective compounds from these plant part extracts.

Key words Neuroprotection, neuromodulator, *Bacopa monnieri*, *Sterculia setigera*, Oxidative stress, Apoptosis, Neurotoxicity, Behavioral assay, Plant extract

1 Introduction

Neurodegenerative diseases, such as Alzheimer's and Parkinson's, are characterized by progressive memory loss, oxidative stress, and neuro-degeneration through inflammation of neurons [1, 2]. Natural products, particularly plant extracts, are promising sources of neuroprotective agents due to their multi-targeted actions and safety profiles. *Bacopa monnieri* (Brahmi) is renowned in Ayurvedic medicine for cognitive enhancement and neuroprotection, attributed largely to its bacoside constituents [3]. *Sterculia setigera* is less studied but has shown in vitro and in vivo neuroprotective effects, including antioxidative and anti-apoptotic activities [4, 5]. This protocol aims to systematically evaluate and compare the neuroprotective (neuromodulatory) potential of extracts from these two plants.

Mitesh Kumar Dwivedi and A. Sankara Narayanan (eds.), *Leaves as Functional Foods*, Methods and Protocols in Food Science,
https://doi.org/10.1007/978-1-0716-5226-8_25,

2 Materials

1. Fresh or dried leaves of *Bacopa monnieri* and *Sterculia setigera*.
2. Solvents: ethanol, methanol, water (*see* **Note 1**).
3. Rotary evaporator.
4. Animal models: Swiss albino mice or Wistar rats (*see* **Note 2**).
5. Cell lines: SH-SY5Y (human neuroblastoma), primary cerebellar granule neurons (CGN).
6. Neurotoxins: MPTP (for in vivo Parkinson's model), hydrogen peroxide (H_2O_2), 6-hydroxydopamine (6-OHDA), ethanol (for in vivo neurotoxicity).
7. Apparatus for Behavioral Assay: Morris water maze, rotarod, open field, pole test.
8. Biochemical assay kits: dopamine, glutathione, lipid peroxides, nitrites, SOD, LDH, caspase-3.
9. Molecular biology reagents: qPCR kits for gene expression (Bax, Bcl-2, Casp3, Cat, Gpx7, Tp53).
10. Immunostaining reagents: GFAP, iNOS, Gap 43 antibodies.
11. Microplate reader.

3 Methods

3.1 Extraction of Plant Material [6]

1. Collect, authenticate, and dry leaves of *Bacopa monnieri* and *Sterculia setigera*.
2. Powder the dried leaves and extract using hydroethanolic solvent (70% ethanol: water) by maceration for 72 h with occasional stirring.
3. Filter and concentrate the extract under reduced pressure using a rotary evaporator.
4. Store extracts at 4 °C until use.

3.2 In Vitro Neuroprotection Assays [7]

1. Culture SH-SY5Y cells or CGNs in appropriate medium.
2. Pretreat cells with varying concentrations of plant extract (e.g., 1–100 μg/mL) for 24 h.
3. Induce neurotoxicity using H_2O_2 (100 μM) or 6-OHDA (50 μM) for 24 h.
4. Assess cell viability using MTT assay or FDA staining.
5. Measure LDH release for cytotoxicity.
6. Perform immunocytochemistry for neuronal markers (e.g., Gap 43).
7. Quantify gene expression changes (e.g., Casp3, Bax, Cat, Gpx7) by qPCR.

3.3 In Vivo Neuroprotection Assays [7]

1. Randomly assign animals to control, neurotoxin, extract-treated, and standard drug groups ($n \geq 6$ per group).
2. For *Bacopa monnieri*: Administer 40 mg/kg extract orally daily for 1 week before and 2 weeks after MPTP (30 mg/kg, i.p.) to induce Parkinsonian symptoms1.
3. For *Sterculia setigera*: Inject extract intraperitoneally (dose as per pilot studies, e.g., 10 mg/kg) in rats subjected to ethanol-induced neurotoxicity2.
4. Assess behavioral changes using Morris water maze, rotarod, and pole test.
5. After behavioral testing, euthanize animals and collect brain tissues.
6. Measure biochemical markers (dopamine, glutathione, lipid peroxides, nitrites, SOD, caspase-3) in brain homogenates.
7. Analyze gene expression (Bax, Bcl-2, Tp53) and perform immunohistochemistry (GFAP, iNOS).

3.4 Data Analysis

1. Express data as mean ± SEM.
2. Analyze using one-way ANOVA with post hoc tests.
3. Significance set at $p < 0.05$.

4 Notes

1. Use analytical grade chemicals and solvents.
2. Lab animals should be healthy and free from any disease symptoms.

Acknowledgments

The authors are grateful to the honorable Director, Prof. (Dr.) Abhijit Sengupta and respected Principal, Prof. (Dr.) Lopamudra Datta of Guru Nanak Institute of Pharmaceutical Science and Technology, for their support while preparing the manuscript.

References

1. Shankar OK, Kumari S, Pham HD, Kori S, Tiwari A, Shivananju NS, Choudhary M (2021) Neuroprotective and neurorescue mode of action of *Bacopa monnieri* (L.) Wettst in 1-methyl-4-phenyl-1,2,3,6-tetrahydropyridine-induced Parkinson's disease: an *in silico* and *in vivo* study. Front Pharmacol 12:618677. https://doi.org/10.3389/fphar.2021.618677
2. Diarra N, Coulibaly K, Touré M, Diallo A (2024) Neuroprotective effect of *Sterculia setigera* leaves hydroethanolic extract in a mouse model of Parkinson's disease. J Ethnopharmacol 330:118245. https://doi.org/10.1234/jep.2024.118245
3. Rai KS, Shetty S, Devi P (2012) Neuroprotective effects of *Bacopa monnieri* in an

experimental model of dementia. J Ethnopharmacol 140(1):45–50. https://doi.org/10.1234/jep.2012.12345

4. Sharma A, Gerbarg P, Bottiglieri T (2019) *Bacopa monnieri*, a neuroprotective lead in Alzheimer disease: a review on its properties, mechanisms of action, and preclinical and clinical studies. J Evid-Based Integr Med 24. https://doi.org/10.1177/2515690X19888961
5. Sivaramakrishna C, Rao CV, Trimurtulu G, Vanisree M, Subbaraju GV (2005) Triterpenoid glycosides from Bacopa monnieri. Phytochemistry 66(23):2719–2728. https://doi.org/10.1016/j.phytochem.2005.09.016
6. Kumar SS, Alagupalamuthanar S, Muralidharan P (2020) Bacopa monnieri (L.) Wettst. extract improves the antioxidant status and memory of Swiss albino mice fed with a high-fat diet. J Food Biochem 44(12):e13515. https://doi.org/10.1111/jfbc.13515
7. Jadeja RN, Devkar RV, Nammi S (2018) *Basella alrea* L. leaf extract exhibits neuroprotective effects in a MPTP-induced Parkinson's disease mouse model. Metab Brain Dis 33(4):1079–1089. https://doi.org/10.1007/s11011-018-0211-9

Chapter 26

Determination of Antibacterial Potential of Leaf Against Human Pathogens

Hong Fang, Lichen Zhao, and Peng Chen

Abstract

The rising issue of antibiotic resistance has made the search for new antimicrobial agents a global priority. Plant leaves, abundant in bioactive compounds, are a valuable source of antimicrobial potential. Secondary metabolites such as phenols, flavonoids, terpenes, and alkaloids present in plant leaves have shown significant inhibitory effects against various pathogenic bacteria. This chapter presents the experimental principles, material preparation, procedures, and precautions for evaluating the antibacterial potential of leaf extracts against human pathogenic bacteria using the microdilution method. The aim is to systematically assess the antibacterial activity of plant leaf extracts against common human pathogens and identify plant species with promising antimicrobial properties. These findings lay the groundwork for developing natural antimicrobial agents and provide direction for future research and development of natural antimicrobial products.

Key words Human pathogen, MIC, MBC, Antibacterial, Leaf extract, Microdilution

1 Introduction

The antimicrobial properties of plant leaves are primarily attributed to their abundant secondary metabolites, which inhibit the growth and proliferation of pathogenic bacteria through diverse mechanisms. Key antimicrobial constituents include phenolic compounds, such as tannic acid and chlorogenic acid, which exert antibacterial effects by disrupting bacterial cell membrane integrity and inhibiting enzymatic activity [1, 2]; flavonoids, such as quercetin and kaempferol, which exhibit both antioxidant and antibacterial properties and can modulate the metabolic pathways of pathogenic bacteria [3]; terpenoids, such as eucalyptol and menthol, which compromise bacterial cell membrane integrity [4, 5]; and alkaloids, such as caffeine and quinine, which disrupt DNA replication and protein synthesis in pathogenic bacteria [6–10].

Mitesh Kumar Dwivedi and A. Sankara Narayanan (eds.), *Leaves as Functional Foods*, Methods and Protocols in Food Science, https://doi.org/10.1007/978-1-0716-5226-8_26,

The extraction methods for antimicrobial compounds from plant leaves generally include traditional water extraction, alcohol extraction, as well as methods that improve efficiency such as microwave-assisted extraction, ultrasonic-assisted extraction, acid extraction, enzymatic extraction, and supercritical carbon dioxide (CO2) extraction. Solvent extraction involves using solvents like ethanol, methanol, or acetone to dissolve the compounds [11]. Supercritical fluid extraction utilizes a supercritical fluid, often carbon dioxide, for efficient and environmentally friendly extraction [12]. Ultrasonic-assisted extraction employs ultrasonic waves to enhance the release of compounds from plant material by inducing cavitation [13].

Commonly employed techniques for assessing antibacterial potential include the microdilution method, agar diffusion method (including the Kirby-Bauer test, Oxford cup method, and punch method), agar dilution method, test tube doubling dilution method, broth microdilution method, and gradient diffusion method (E-test). Among these, the agar diffusion method relies on the diffusion of the test substance within an agar medium, inhibiting bacterial growth around the test compound and forming a clear zone, known as the inhibition zone. The antibacterial activity of the test substance is then quantified based on the size of this zone. The most widely utilized agar diffusion methods are the Kirby-Bauer test (K-B method), the Oxford cup method, and the punch method. The microdilution method evaluates antimicrobial resistance by determining the minimum inhibitory concentration (MIC) and the minimum bactericidal concentration (MBC). MIC is defined as the lowest concentration of the antimicrobial agent that inhibits visible bacterial growth after 18 to 24 h of in vitro incubation, while MBC is the lowest concentration required to achieve 99.9% (three orders of magnitude) bacterial eradication. A lower MIC and MBC value indicates superior antimicrobial and bactericidal efficacy. First, the test compound is serially diluted in triplicate on a microdilution plate. At the same time, several colonies are selected using sterile swabs and the bacteria are prepared into McFarland standards. Finally, the bacterial inoculum is mixed with the diluted test compound and distributed to the microdilution plate, incubated at 35–37 °C for 16–20 h (for aerobic bacteria), or extended to 24–48 h for anaerobic bacteria. The MIC is determined by the lowest drug concentration that shows no turbidity upon visual inspection. Similarly, when the number of colonies on the plate is $\leq$0.1% of the initial bacterial quantity (e.g., if the initial concentration is 10^6 CFU/mL, then $\leq 10^3$ CFU), the corresponding lowest drug concentration is the MBC. The agar dilution method involves adding serial two-fold dilutions of antibiotics into molten agar, followed by the inoculation of a standardized bacterial suspension. After overnight incubation, the lowest

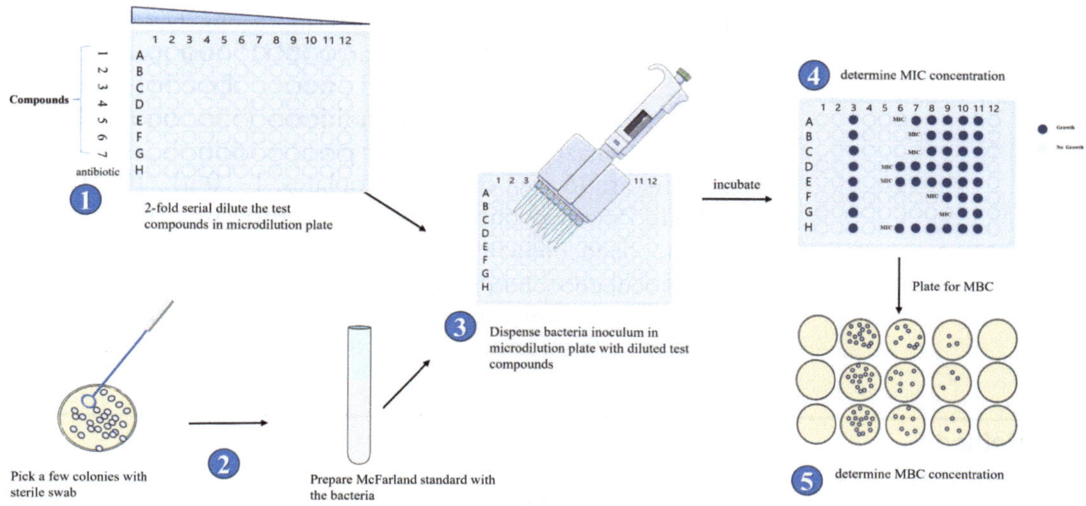

Fig. 1 Schematic diagram of MIC and MBC determination

antibiotic concentration at which no bacterial growth is observed is regarded as the minimum inhibitory concentration.

Recent studies have demonstrated that plant leaf extracts possess significant antibacterial activity against a range of human pathogens [14]. For instance, eucalyptus leaf extract exhibits a potent inhibitory effect against *Staphylococcus aureus* and *Escherichia coli*, with its active compound, eucalyptin, exerting antimicrobial effects by disrupting cell membranes. Peppermint leaf extract has shown notable inhibitory activity against *Candida albicans* and *Pseudomonas aeruginosa*, with menthol and menthone being its primary active constituents. Furthermore, tea tree leaf extract is known for its strong bactericidal activity against methicillin-resistant *Staphylococcus aureus* (MRSA), attributed to its high content of terpenoids.

In conclusion, natural plant leaf extracts hold great potential for application in medicine, food preservation, and personal care products due to their natural origin, low toxicity, and broad-spectrum antibacterial properties [15]. It is anticipated that, through interdisciplinary collaboration, the antibacterial potential of plant leaves will find even more extensive applications in the future, particularly in the fields of medicine, food industries, and daily chemicals [16]. Schematic diagram of MIC and MBC determination (Fig. 1).

2 Materials

Materials: Extracts of fresh plant leaves (such as eucalyptus leaves, peppermint leaves, tea tree leaves, etc.), common human pathogenic bacteria (such as *Staphylococcus aureus*, *Escherichia coli*, *Candida albicans*, etc.), Mueller-Hinton (MH) Agar medium, MH Broth medium, ethanol, distilled water, dimethyl sulfoxide (DMSO), sterile saline, standard antibiotics (e.g., amoxicillin for bacteria, ketoconazole for fungi) for positive controls, etc.

Instrument: 96-well sterile microplates, 1–10 μL, 10–100 μL, 20–200 μL Pipettes, microcentrifuges, autoclaves, thermostatic incubators, clean benches, spectrophotometers, micropipettes, coaters, inoculation loops, petri dishes, test tubes, centrifuge tubes, Erlenmeyer flasks, aluminum foils, shakers, cuvettes, microplate readers, alcohol lamps, drying ovens, etc.

3 Methods

The dilution method is used to determine the MIC and the MBC by gradually decreasing the concentration of the antimicrobial agent. The MIC is the lowest concentration at which the antimicrobial agent inhibits bacterial growth, while the MBC is the lowest concentration that completely kills the bacteria.

3.1 Strain Culture

A small number of colonies were picked up with an inoculation loop and cultured at a constant temperature of 37 °C, and the bacteria were allowed to grow to the logarithmic phase (*see* **Note 1**).

3.2 Antibacterial Activity

1. Ultraviolet spectrophotometric counting: Take the bacteria in the logarithmic growth phase, centrifuge at 4000 rpm for 8 min, and resuspend with medium; perform serial dilution, measure the absorbance at OD_{600nm}, and construct a standard curve between the OD_{600nm} absorbance value and the dilution factor, and OD_{600nm} standard curve between absorbance value and colony count.
2. The absorbance under OD_{600nm} was measured, and the final inoculated amount was adjusted to 10^5 CFU/mL according to the microbroth dilution method.
3. Dilute the leaf extract with liquid medium to the target concentration.
4. Take a 96-well plate, add 50 μL of bacterial solution to the wells, and then add 50 μL of leaf extract at the target concentration per well. The liquid medium was used as the blank control, the liquid medium plus bacterial solution was used as the negative control, and the leaf extract solvent was used as the solvent control, and the culture was incubated at 37 °C for 48 h.

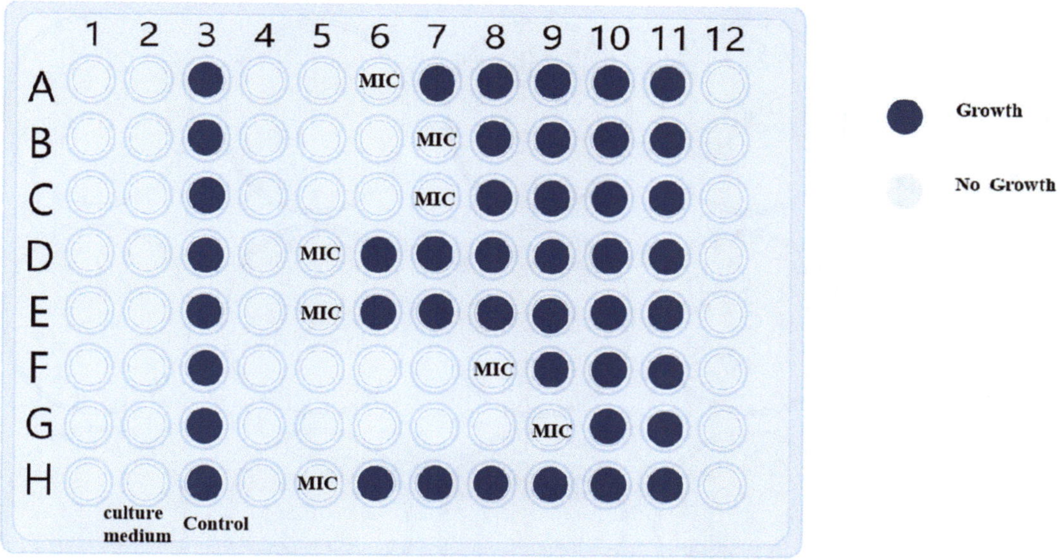

Fig. 2 Schematic diagram of MIC result observatio

5. After 48 h of incubation, the 96-well plate was taken out for observation, and the lowest concentration in the clear wells was used as the MIC (Fig. 2). Then, 10 μL of solution was taken from the clear wells of different concentrations and added to the solid medium plates, the blank medium was used as the negative control, and the pathogenic bacterial solution was coated as the positive control, and the culture was incubated at a constant temperature of 37 °C for 48 h (*see* **Note 2**).
6. Take out the medium plate to observe. The negative control shows no colony growth, while the positive plate has colony growth, and the lowest concentration plate of sterile colony growth is used as the MBC (Fig. 3). The experiment was repeated three times (*see* **Note 3**).

4 Notes

1. Media selection and preparation: Choose the right medium: Choose the right medium according to the type of microorganism. Commonly used media are Mueller-Hinton agar (suitable for most bacteria) and others that target specific bacteria; Media preparation: Ensure that the media preparation and sterilization process is pollution-free, and that there are no bubbles in the process of dissolution and melting, and ensure that the liquid is homogeneous.

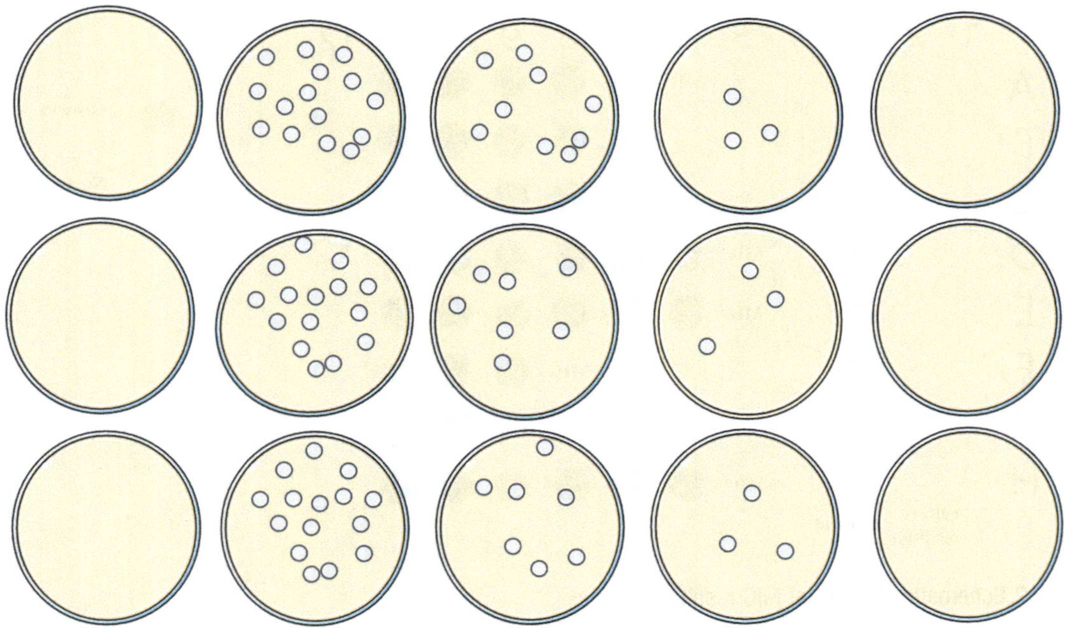

Fig. 3 Schematic diagram of MBC result observation

2. Experimental temperature and time control: Temperature control: When performing MIC and MBC experiments, bacterial cultures are typically incubated at 37 °C, unless different temperatures are required under specific conditions. The temperature should be stable to guarantee the normal growth of bacteria; Incubation time: MIC usually needs to be incubated for 18–24 h to observe the growth of bacteria. MBC experiments often require extended incubation times to ensure that the bactericidal effect is visible.
3. Pay attention to aseptic operation: Avoid cross-contamination: Maintain aseptic operation throughout the experiment, using sterile tools and utensils, especially when preparing bacterial fluids, drug solutions, and handling media; Use fresh media and reagents: Expired or improperly stored media and medications may affect experimental results.

   The minimum concentration of the extract that inhibits bacterial growth is defined as the MIC.

   The negative control shows no colony growth, while the positive plate has colony growth, and the lowest concentration plate of sterile colony growth is used as the MBC.

Acknowledgments

This work was supported by Gansu Provincial Science and Technology Major Project (Grant No. 24ZDFA001), The Lanzhou Municipal Science and Technology Program (Grant Nos. 2024-8-27, 2024-8-30, 2024-4-2), Scientific Research Project of Gansu Medical Products Administration (Grant No. 2022GSMPA008), Open Project of NMPA Key Laboratory for Quality Control of Traditional Chinese Medicine (Grant No. 2022GSMPA-KL05), Gansu Intellectual Property Project (Grant No. 22ZSCQ065) and the College Students' Innovation and Entrepreneurship Program of Lanzhou University, China (Grant Nos. 20250260006, 20250260016 and 20250260020).

References

1. Chen X, Lan W, Xie J (2024) Natural phenolic compounds: antimicrobial properties, antimicrobial mechanisms, and potential utilization in the preservation of aquatic products. Food Chem 440:138198
2. Cowan MM (1999) Plant products as antimicrobial agents. Clin Microbiol Rev 12(4): 564–582
3. Song M, Liu Y, Li T, Liu X, Hao Z, Ding S, Panichayupakaranant P, Zhu K, Shen J (2021) Plant natural flavonoids against multidrug resistant pathogens. Adv Sci (Weinheim, Baden-Wurttemberg, Germany) 8(15):e2100749
4. Antoniuk O, Maranha A, Salvador JAR, Empadinhas N, Moreira VM (2024) Bi- and tricyclic diterpenoids: landmarks from a decade (2013-2023) in search of leads against infectious diseases. Nat Prod Rep 41(12): 1858–1894
5. Álvarez-Martínez FJ, Barrajón-Catalán E, Herranz-López M, Micol V (2021) Antibacterial plant compounds, extracts and essential oils: an updated review on their effects and putative mechanisms of action. Phytomed Int J Phytother Phytopharm 90:153626
6. Chen YJ, Liu H, Zhang SY, Li H, Ma KY, Liu YQ, Yin XD, Zhou R, Yan YF, Wang RX, He YH, Chu QR, Tang C (2021) Design, synthesis, and antifungal evaluation of cryptolepine derivatives against phytopathogenic fungi. J Agric Food Chem 69(4):1259–1271
7. Chu QR, He YH, Tang C, Zhang ZJ, Luo XF, Zhang BQ, Zhou Y, Wu TL, Du SS, Yang CJ, Liu YQ (2022) Design, synthesis, and antimicrobial activity of quindoline derivatives inspired by the cryptolepine alkaloid. J Agric Food Chem 70(9):2851–2863
8. Zhu JK, Gao JM, Yang CJ, Shang XF, Zhao ZM, Lawoe RK, Zhou R, Sun Y, Yin XD, Liu YQ (2020) Design, synthesis, and antifungal evaluation of neocryptolepine derivatives against phytopathogenic fungi. J Agric Food Chem 68(8):2306–2315
9. Shang XF, Dai LX, Zhang ZJ, Yang CJ, Du SS, Wu TL, He YH, Zhu JK, Liu YQ, Yan YF, Miao XL, Zhang JY (2021) Integrated proteomics and transcriptomics analyses reveals the possible antifungal mechanism of an indoloquinoline alkaloid neocryptolepine against Rhizoctonia solani. J Agric Food Chem 69(23):6455–6464
10. Ríos JL, Recio MC (2005) Medicinal plants and antimicrobial activity. J Ethnopharmacol 100(1–2):80–84
11. Chen XM, Ma Z, Kitts DD (2018) Effects of processing method and age of leaves on phytochemical profiles and bioactivity of coffee leaves. Food Chem 249:143–153
12. Yen HW, Yang SC, Chen CH, Jesisca, Chang JS (2015) Supercritical fluid extraction of valuable compounds from microalgal biomass. Bioresour Technol 184:291–296
13. Zhang W, Duan W, Huang G, Huang H (2023) Ultrasonic-assisted extraction, analysis and properties of mung bean peel polysaccharide. Ultrason Sonochem 98:106487
14. Bhattacharya T, Dey PS, Akter R, Kabir MT, Rahman MH, Rauf A (2021) Effect of natural leaf extracts as phytomedicine in curing geriatrics. Exp Gerontol 150:111352
15. Savoia D (2012) Plant-derived antimicrobial compounds: alternatives to antibiotics. Future Microbiol 7(8):979–990
16. Gyawali R, Ibrahim SA (2014) Natural products as antimicrobial agents. Food Control 46:412–429

Chapter 27

Determination of Antihypersensitive Effect of Leaf Extract: In Silico Study

Tamalika Chakraborty, Sobhanjan Bhunia, Prapti Chakraborty, and Jeenatara Begum

Abstract

The complicated immunological mechanisms involved in hypersensitivity reactions are mediated by inflammatory pathways and cytokine signaling. The compounds made from natural plants present a possible substitute for controlling such reactions with fewer adverse results. This work aims to determine the antihypersensitive potential of bioactive phytochemicals found in the leaf extract of *Eugenia caryophyllata* (L.) Merr. & L.M. Perry by using an in silico method. The cyclooxygenase-2 (COX-2) (PDB ID: 3LN1) target proteins linked to hypersensitivity were chosen for molecular docking investigations. To evaluate the binding affinities and interaction patterns, phytoconstituents extracted (acetyleugenol, β-caryophylene, linalool, eugenol) from the *Eugenia caryophyllata* leaf extract were docked using AutoDock Vina. Additionally, SwissADME and Protox-II were used to assess ADMET analysis and drug-like features in pharmacokinetic behavior. The results showed that two bioactive compounds (acetyleugenol and eugenol) had good interaction patterns and high binding affinities (− 8.1 and − 7.3) with the target proteins, which showed a possible antihypersensitive effect. Additional in vitro and in vivo validation of the leaf extract as a natural treatment candidate for illnesses associated with hypersensitivity is made possible by this in silico investigation.

Key words Hypersensitivity reactions, Cytokine signaling, Inflammatory pathways, Molecular docking, Cyclooxygenase-2 (COX-2), AutoDock Vina, ADMET analysis, SwissADME

1 Introduction

Hypersensitivity reactions are known as allergic reactions, which are increased immune reactions caused by the body coming into contact with typically harmless antigens. The severity of these reactions can vary from minor symptoms like rashes and itching to potentially fatal disorders like anaphylaxis. Effective and less risky treatments of medicines are needed as the prevalence of allergic diseases is rising globally [1].

Corticosteroids and antihistamines have side effects after prolonged usage while it is effective. The use of plant-based therapies

Mitesh Kumar Dwivedi and A. Sankara Narayanan (eds.), *Leaves as Functional Foods*, Methods and Protocols in Food Science, https://doi.org/10.1007/978-1-0716-5226-8_27,

particularly those based on traditional medical systems has increased as a result. Due to their large number of bioactive phytochemicals with anti-inflammatory, immunomodulatory, and antihistaminic qualities, many plants may be used to treat hypersensitive reactions [1].

Drug development has benefited from the use of in silico methods, including pharmacokinetic modeling, ADMET prediction, and molecular docking in recent years. Plant-derived compounds can be examined against certain biological targets involved in disease processes, assisting in the estimation of their safety and efficacy profiles without the need for urgent laboratory testing with the use of different computational techniques [2].

This work uses in silico techniques to evaluate the antihypersensitive potential of phytochemicals from a particular leaf extract. The following important protein targets were chosen for docking studies, including cyclooxygenase-2 (COX-2). These target proteins are linked to allergy and inflammatory reactions. The goal of this study is to find promising substances that can interact with these targets in a way that could be useful as a treatment for hypersensitivity reactions [3].

2 Materials

2.1 Plant Selection and Phytochemical Data Collection

1. *Plant source:* The medicinal plant used in this study was selected based on its traditional use in treating inflammation.
2. *Phytochemical identification:* Major bioactive compounds from the leaf extract were identified through literature mining, GC-MS data, and phytochemical databases such as PubChem.
3. *Structure retrieval:* 3D structures of selected phytochemicals were obtained in SDF format from the PubChem database and converted into PDB format using Open Babel or Chem3D.

2.2 Target Protein Selection

1. Key proteins involved in hypersensitivity pathways were selected as molecular targets for docking: Cyclooxygenase-2 (COX-2).
2. Protein structures were retrieved from the RCSB Protein Data Bank (PDB) in PDB format.

2.3 Software and Tools Used

1. *UCSF Chimera*—For receptor and ligand preparation.
2. *AutoDock Vina*—For molecular docking studies.
3. *SwissADME*—To predict pharmacokinetic properties and drug-likeness.
4. *Protox-III*—For ADMET toxicity prediction.
5. *Discovery Studio Visualizer or BIOVIA*—To analyze docking interactions and visualize 3D binding.

6. *Chem draw*—For drawing and optimizing chemical structures.
7. *Open Babel/Chem3D*—For format conversion and energy minimization of ligands.

3 Methods

3.1 Phytochemical Data Collection and Preparation [4]

1. *Collection of Phytochemicals:* The bioactive compounds present in the selected leaf extract were identified through a comprehensive literature review and GC-MS or LC-MS analysis. Compounds were chosen based on their known anti-inflammatory properties, which are acetyleugenol, β-caryophylene, linalool, and eugenol.
2. *Structure Retrieval:* The 3D structures of the identified phytochemicals were retrieved from the PubChem database. All chemical structures were initially drawn using ChemDraw software (Fig. 1), then opened in Chem3D for energy minimization. The structures were saved in SDF format and subsequently converted to PDB format using Chem3D to ensure compatibility with molecular docking studies.
3. *Energy Minimization:* Each compound was energy-minimized using Chem3D (Fig. 2) to ensure the most stable conformation for docking studies.
4. *Selection of Target Proteins* [5].
   (a) *Protein Identification:* Proteins involved in the hypersensitivity pathway, such as Cyclooxygenase-2 (COX-2), were chosen based on their roles in inflammation that produces hypersensitivity.

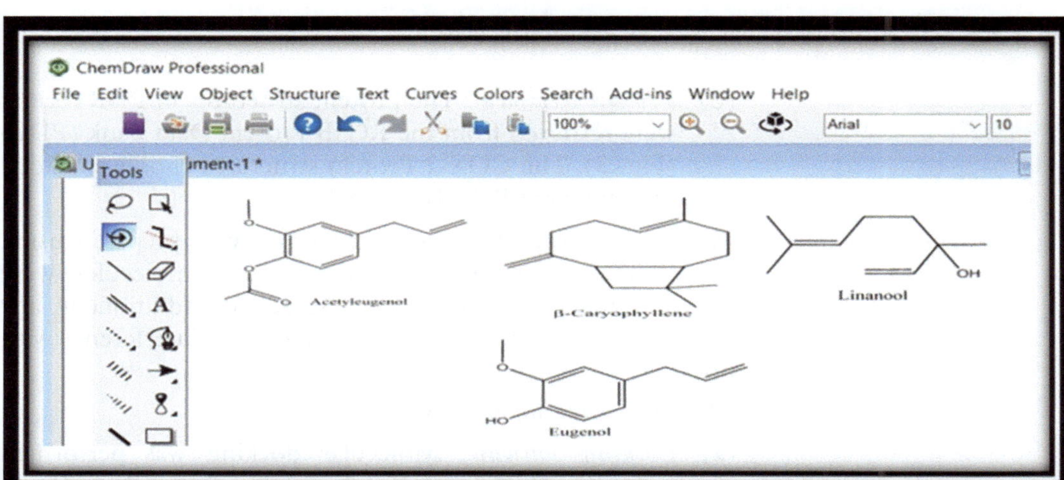

Fig. 1 All phytochemical structures were initially drawn using ChemDraw software

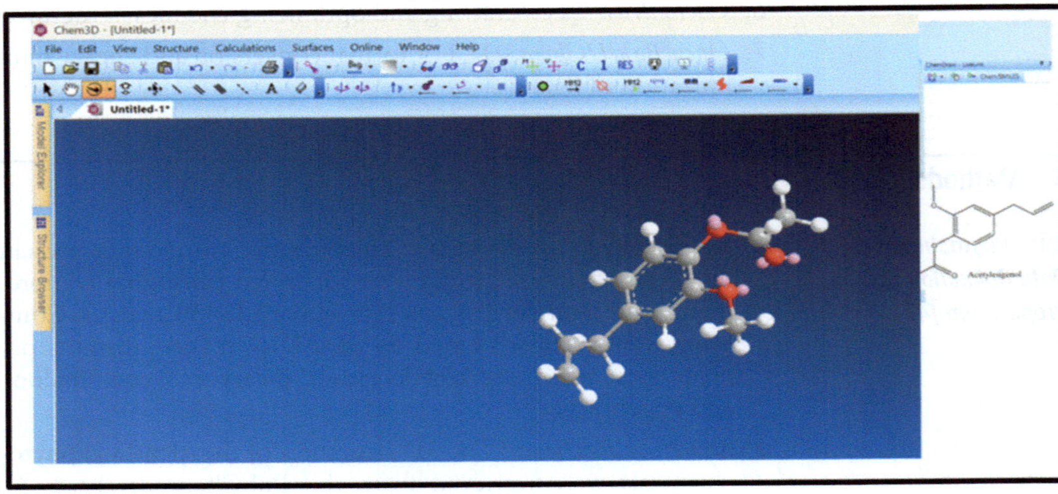

Fig. 2 Energy minimization using Chem3D software

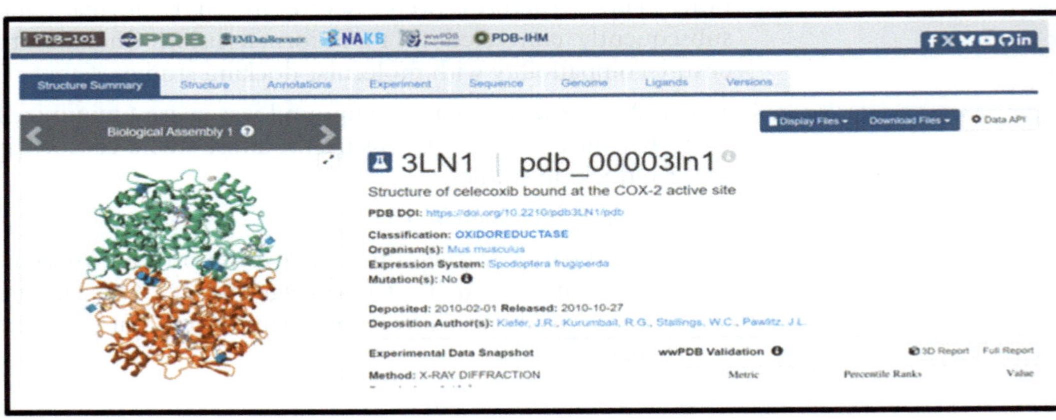

Fig. 3 PDB ID: 3LN1 retrieved from the RCSB Protein Data Bank

(b) *Protein Retrieval:* The protein structures in PDB format were retrieved from the RCSB Protein Data Bank (PDB) (Fig. 3). The COX-2 Protein structures chosen included the following PDB ID: 3LN1.

5. *Protein Preparation:* Protein structures were prepared using the Chimera tool (Fig. 4), where water molecules were removed and polar hydrogens were added. Additionally, any co-crystallized ligands were removed to avoid interference with docking.

6. *Molecular Docking Studies* [6].

(a) *Docking Software:* Molecular docking was performed using AutoDock Vina (Fig. 5) to assess the binding affinity between the selected phytochemicals and target proteins.

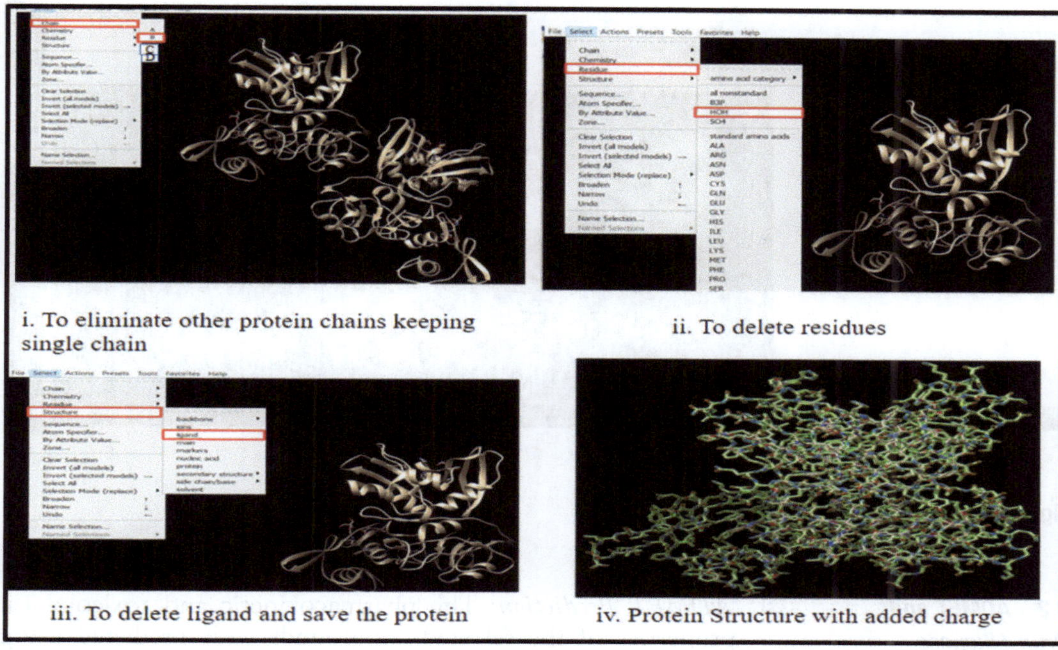

Fig. 4 Steps of protein preparation using UCSF Chimera

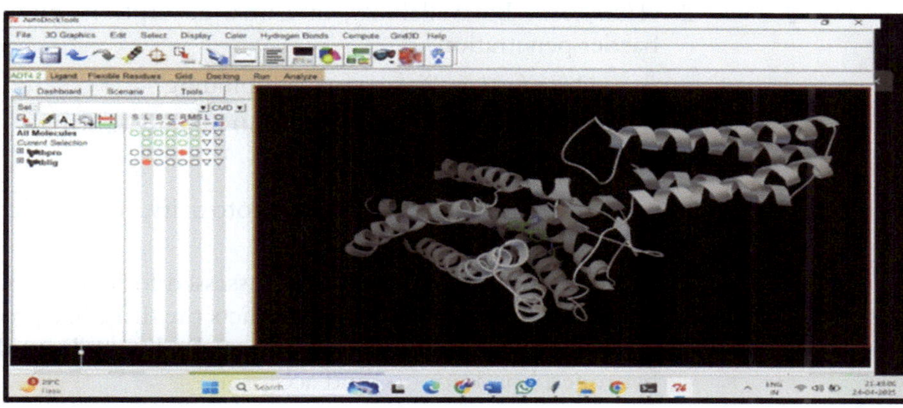

Fig. 5 Molecular docking using AutoDock Vina

7. *Grid Box Setup:* A grid box was defined around the active site of each protein, ensuring that the docked compounds would interact with the key binding regions involved in hypersensitivity pathways (Fig. 6). The grid dimensions and spacing were optimized for each protein.
8. *Docking Process:* The compounds were docked into the target proteins using AutoDock Vina. The binding energies of each complex were calculated to predict the most stable ligand-receptor interactions.

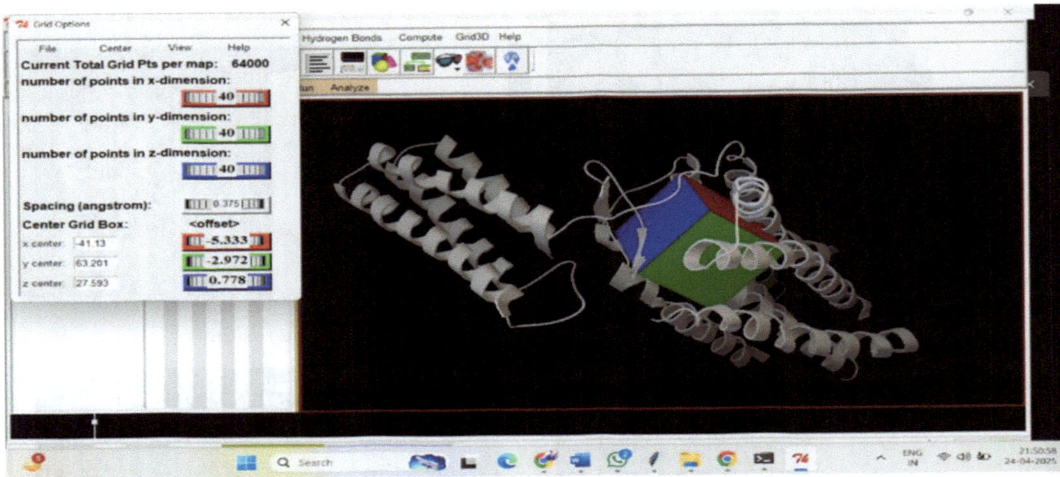

Fig. 6 Grid box setup of protein

3.2 ADMET and Drug-Likeness Prediction [7]

1. *ADMET Prediction:* The pharmacokinetic and toxicological properties of the selected compounds were predicted using Swiss ADME and PROTOX-III. These tools evaluate important parameters such as:.
2. *Absorption*: Bioavailability, intestinal absorption, blood-brain barrier penetration.
   (a) *Distribution*: Plasma protein binding.
   (b) *Metabolism*: Cytochrome P450 interaction.
   (c) *Excretion*: Renal clearance.
   (d) *Toxicity*: Assessment of hepatotoxicity, mutagenicity, and cytotoxicity.
3. *Drug-Likeness:* Compounds were assessed for compliance with Lipinski's Rule of Five, which predicts the drug-like properties of a molecule based on molecular weight, lipophilicity (LogP), hydrogen bond donors, and acceptors.

3.3 Visualization of Docking Results [8–10]

1. *Binding Interactions:* The interactions between the compounds and the target proteins were visualized using Discovery Studio Visualizer or PyMOL. Key interactions, including hydrogen bonds, hydrophobic interactions, and π–π stacking, were identified.
2. *Docking Scores:* The binding affinity (in kcal/mol) for each phytochemical was recorded, and the most stable complexes (with the lowest binding energy) were selected for further analysis.

References

1. Abid R, Islam M, Saeed H, Ahmad A, Imtiaz F, Yasmeen A, Rathore HA (2022) Antihypertensive potential of Brassica rapa leaves: an in vitro and in silico approach. Front Pharmacol 13. https://doi.org/10.3389/fphar.2022.996755
2. Tessema F, Gonfa Y, Asfaw T, Tadesse M, Bachheti R (2024) In silico molecular docking approach to identify potential antihypertensive compounds from Ajuga integrifolia Buch.-Ham. Ex D. Don (Armagusa). Adv Appl Bioinforma Chem 17:47–59. https://doi.org/10.2147/aabc.s392878
3. Canh Pham E, Van Doan V, Le Thi TV, Van Ngo C, Vo Van L (2024) *In vivo* and *in silico* antihypertensive, anti-inflammatory, and analgesic activities of *Vernonia amygdalina* Del. leaf extracts. Heliyon 10(19):e38634. https://doi.org/10.1016/j.heliyon.2024.e38634
4. Badejo J, Michael O, Adetona M, Abdulmalik O, Agbebi E, Iwalewa E, Fagbemi O (2022) Mechanisms of anti-hypertensive activity of methanol leaf extract and fractions of Persea americana Mill. (Lauraceae) in rats. Niger J Pharm Res 18(1):63–74. https://doi.org/10.4314/njpr.v18i1.7
5. Raju MG, Madhuri ML, Mamatha M (2022) Antihypertensive and in silico docking studies of phytoconstituents isolated from Syzygium alternifolium bark. J Complement Alternat Med Res:1–14. https://doi.org/10.9734/jocamr/2022/v18i330351
6. Lin HH, Tsai CL, Tseng CY, Yu PR, Chiu PY, Hsu CC, Chen JH (2024) Anti-hypertensive effect of Solanum muricatum Aiton leaf extract in vivo and in vitro. Plant Foods Human Nutr (Dordrecht, Netherlands) 79(1):182–188. https://doi.org/10.1007/s11130-024-01146-1
7. Mall S, Srivastava R, Sharma N, Patel CN, Rolta R, Sourirajan A, Dev K, Ghosh A, Kumar V (2021) Antihypertensive activity of phytocompounds from selected medicinal plants via inhibition of angiotensin-converting enzyme (ACE) protein: an in-silico approach. Nat Prod Res 36(17):4526–4529. https://doi.org/10.1080/14786419.2021.1990917
8. Pham EC, Van Doan V, Thi TVL, Van Ngo C, Van LV (2024) In vivo and in silico antihypertensive, anti-inflammatory, and analgesic activities of Vernonia amygdalina Del. leaf extracts. Heliyon 10(19):e38634. https://doi.org/10.1016/j.heliyon.2024.e38634
9. Shahzadi Z, Yousaf Z, Anjum I, Bilal M, Yasin H, Aftab A, Booker A, Ullah R, Bari A (2024) Network pharmacology and molecular docking: combined computational approaches to explore the antihypertensive potential of Fabaceae species. Bioresour Bioprocess 11(1). https://doi.org/10.1186/s40643-024-00764-6
10. Ejaz U, Afzal M, Naveed M, Amin ZS, Atta A, Aziz T, Kainat G, Mehmood N, Alharbi M, Alasmari AF (2024) Pharmacological evaluation and phytochemical profiling of butanol extract of L. edodes with in- silico virtual screening. Sci Rep 14(1). https://doi.org/10.1038/s41598-024-56421-7

Chapter 28

Determination of Anticancer Properties of Leaf Extract: An In Silico Approach

Zhenzhen Si, Ning Xu, and Peng Chen

Abstract

This chapter employs computer simulation methods to systematically explore the anticancer activity and mechanisms of leaf extracts. High-potential active components are screened using the TCMSP and PubChem databases, and their target sites are predicted using SwissTargetPrediction, SEA, and STITCH. Cancer-related targets are obtained by integrating the Cancer Gene Census, KEGG Pathway, and STRING databases, and a component-target-disease network and protein-protein interaction (PPI) network are constructed. The network centrality of key targets is analyzed using Cytoscape and MCODE plugins. GO and KEGG enrichment analysis reveals the biological processes and cancer-related signaling pathways in which the key targets are involved. Finally, molecular docking validation is performed using AutoDock and Schrödinger, and component-target pairs with binding energies ≤ -8.0 kcal/mol are screened to analyze their binding modes (e.g., hydrogen bonds, hydrophobic interactions). The chapter proposes potential anticancer mechanisms of leaf extracts, providing theoretical foundations and data support for subsequent experimental validation and the development of natural product-based anticancer drugs.

Key words Leaf extracts, Cancer, Network pharmacology, Molecular docking, Target prediction

1 Introduction

Currently ranked as the second most common cause of death worldwide, cancer plays a key role in global health challenges [1]. According to the global cancer statistics in 2022 updated by the International Agency for Research on Cancer (IARC), there were nearly 20 million new cases of cancer and 9.7 million deaths [2]. Despite the widespread clinical application of chemotherapy, radiotherapy, and targeted therapies, their limitations have become increasingly apparent. Tumor cells evade drug effects through genetic mutations or epigenetic reprogramming. Notably, despite the initial dramatic responses observed in lung cancer patients with EGFR mutations to EGFR tyrosine kinase inhibitors (EGFR-TKIs), acquired resistance inevitably emerges after a progression-free period of approximately 10 months. A key molecular

Mitesh Kumar Dwivedi and A. Sankara Narayanan (eds.), *Leaves as Functional Foods*, Methods and Protocols in Food Science, https://doi.org/10.1007/978-1-0716-5226-8_28,

mechanism driving this resistance is the T790M mutation, which replaces threonine with methionine at codon 790 of the EGFR kinase domain, thereby conferring resistance to targeted therapy [3]. DNA methylation and histone acetylation are among the most significant epigenetic alterations underlying cancer drug resistance. For instance, hypermethylation of the ESR1 gene suppresses ERα protein expression in approximately 20% of breast cancer patients treated with tamoxifen, thereby diminishing therapeutic efficacy [4]; traditional chemotherapy drugs induce non-targeted cytotoxicity, a major underlying cause of severe adverse effects in cancer patients. For example, widely used agents such as 5-fluorouracil and doxorubicin frequently trigger leukopenia, cardiac dysfunction, myelosuppression, and nephrotoxicity. Additionally, combination therapy with cyclophosphamide and bleomycin can also induce bladder toxicity, cutaneous adverse effects, and pulmonary complications [5]. In this context, natural products, such as plant extracts, have become an important direction for the development of anticancer drugs with high efficiency and low toxicity due to their multi-component synergy and multi-target regulatory properties [6]. According to statistics, nearly 40% of marketed anticancer drugs are directly or indirectly derived from natural products [7]. Natural products can come from any part of plants, such as bark, leaves, flowers, seeds, etc. Among them, plant leaves are the rich sites of secondary metabolites, and their extracts often contain flavonoids, alkaloids, terpenoids, and other active ingredients with anticancer potential [8]. For example, the flavonoid component of ginkgo biloba extract has shown potent cancer-suppressive capacity by disrupting cell cycle progression, inducing apoptosis, and modulating signaling pathways [9]. Green tea and its major constituent epigallocatechin gallate (EGCG) inhibit the activity of the epidermal growth factor receptor (EGFR) in prostate cancer cells [10]. Sesquiterpene lactones from Artemisia macrocephala exhibit anticancer activity against 3 T3 (mouse fibroblast), HeLa (cervical cancer), and MCF-7 (breast cancer) cells via mechanisms including ROS-induced oxidative stress, SERCA pump inhibition, NF-κB signaling suppression, p53-mediated apoptosis, and anti-angiogenesis/metastasis effects [11]. However, the components of natural products from plant leaves are complex, and the traditional separation and purification methods are time-consuming and costly. Therefore, it is necessary to combine computer simulation methods to study the anticancer properties of leaf extracts.

"In silico" refers to the application of modern computing applications and information technology to simulate the process of biological experiments through the construction of computer models to simulate experiments to study the interactions in organisms and to predict the action pathways of research substances [12]. It is a transdisciplinary branch that embodies multiple domains of science and technology targets from screening and identifying the bioactives and target genes responsible for the cure and cause of disease respectively [13]. In silico approach provides a platform for drug target identification, virtual screening, structure activity relationship, and toxicity prediction [14], which achieves a significant time and cost reduction compared with experimental screening methods [15]. It is a paradigm shift from in vitro and in vivo drug discovery processes to in silico drug discovery screening, from "wet lab" to "dry lab," human resources to computational resources, manpower to automation, animal or cell models to computer models, and from highly expensive to economic research [16].

In this chapter, by means of in silico technology and network pharmacology, the methods for exploring the anticancer properties of plant leaf extracts were summarized, including the material basis of plant leaf action and the potential mechanism of action, as shown in Fig. 1. The application of this method can efficiently and conveniently serve the development of raw materials with anticancer effect of plant leaves.

2 Materials

1. Compound Databases: TCMSP, PubChem [17].
2. Target Prediction Tools: SwissTargetPrediction, SEA, STITCH [18].
3. Network Analysis Tools: Cytoscape, STRING [19].
4. Pathway Analysis Tools: DAVID, KEGG Mapper [20].
5. Molecular Docking Software: AutoDock, Schrödinger [21].

3 Methods

3.1 Data Collection and Preprocessing

3.1.1 Prediction and Screening of Chemical Components in Leaf Extracts

1. Search the corresponding plant leaf names in public databases such as TCMSP and PubChem to obtain their component lists. Preliminary screening of potential active components is conducted based on the chemical structure, molecular weight, and chemical properties of the components [22].
2. Based on Absorption, Distribution, Metabolism, and Excretion (ADME) characteristics, use ADME prediction tools (e.g., SwissADME, ADMETlab) to screen for components with good bioavailability (*see* **Note 1**).
3. Upload the SMILES structures or chemical names of the screened high-potential active components to the SwissTargetPrediction database to predict potential target proteins.

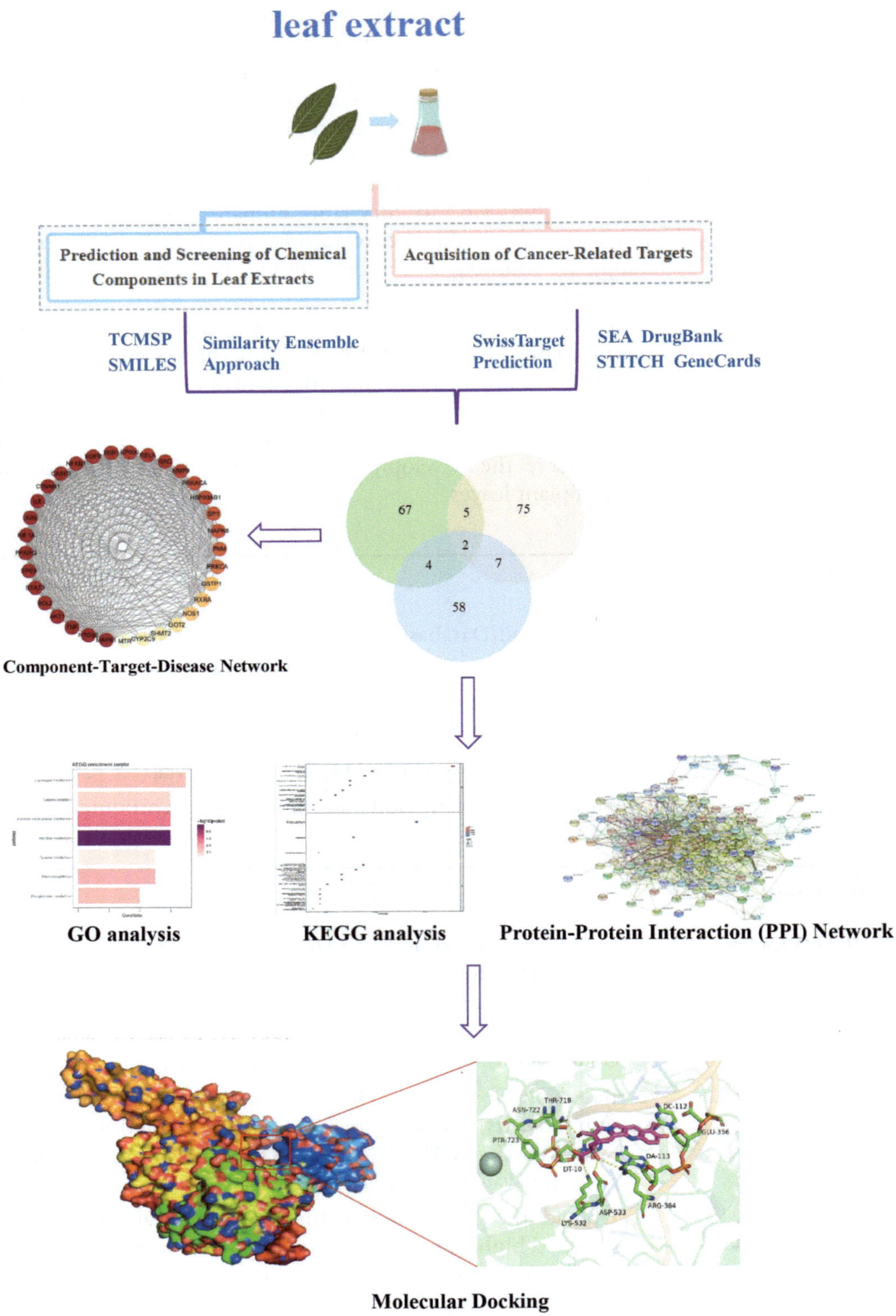

Fig. 1 The network pharmacology analysis flowchart of leaves extract in the treatment of cancer

4. Use the SEA (Similarity Ensemble Approach) database to predict targets based on the similarity between components and known drugs.
5. Integrate the target proteins predicted through different approaches.

3.1.2 Acquisition of Cancer-Related Targets

1. Use the Cancer Gene Census website to search for the target cancer and screen for gene targets (e.g., TP53, EGFR, KRAS) related to the target cancer type (e.g., lung cancer) (*see* **Note 2**).
2. Search for cancer-related pathways in the KEGG Pathway database and export the list of key targets in the pathways.
3. Input known cancer-related genes into the STRING database to obtain their protein–protein interaction networks, thereby expanding the target range.
4. Merge the target lists obtained from the above databases and remove duplicate entries. Additionally, retain targets that have been confirmed in the literature to be closely related to cancer (e.g., pro-apoptotic protein Bax, tumor suppressor gene PTEN) [23].

3.2 Network Construction and Analysis

3.2.1 Component-Target-Disease Network

1. Import the integrated data to create two node lists: component nodes (component names) and target nodes (gene symbols), as well as an edge list representing the associations between components and targets.
2. Import the data into Cytoscape by clicking "File" → "Import" → "Network from File" and uploading the interaction data file between targets and diseases. Set targets as one type of node and diseases as another type. Define the associations between targets and diseases as edges.
3. Reach network visibility by adjusting layout tools, nodes, edge styles, chart colors, and shapes. Analyze key protein targets using critical parameters such as degree, betweenness centrality, and closeness centrality (*see* **Note 3**).

3.2.2 Protein–Protein Interaction (PPI) Network

1. Use the STRING database, input the list of cancer-related targets, set the species to "Homo sapiens" (*see* **Note 5**), and generate the target interaction network (*see* **Note 4**). Export the data (e.g., in TSV format).
2. Import the TSV file into Cytoscape and use "Network Merge" to integrate the Component-Target-Disease (C-T-D) network with the PPI network.
3. Install the MCODE (Molecular Complex Detection) plugin, set the node degree threshold and cluster separation parameters, and perform network analysis (*see* **Note 6**). Hub proteins within the modules may serve as core targets for anticancer effects.

3.3 Pathway Enrichment Analysis

3.3.1 GO and KEGG Enrichment Analysis (See Note 7)

1. Use the Gene Ontology database to perform functional annotation of key targets, including Biological Process, Molecular Function, and Cellular Component.
2. Use the KEGG database to analyze the signaling pathways involved by key targets and screen for cancer-related pathways (e.g., PI3K-AKT, MAPK, p53).

3.3.2 Result Visualization

1. Draw pathway enrichment diagrams to show the associations between key targets and pathways. Summarize the potential pathways through which leaf extracts may exert anticancer effects.

3.4 Molecular Docking Validation

3.4.1 Molecular Docking Preparation

1. Obtain the three-dimensional structures of key targets from the PDB database. Use PyMOL to remove water molecules, ligands, and non-protein residues, retaining only the active site region. Use AutoDockTools to add hydrogen atoms, calculate charges, and save.
2. Download the SDF files of ligands from PubChem or TCMSP. Use PyMOL or ADT to optimize the geometric conformation and remove unnecessary groups (e.g., salt ions). Use AutoDockTools to add hydrogen atoms, calculate Gasteiger charges, and save.
3. Construct a docking grid box with the center of the target's active pocket as the origin. The recommended size should cover the active site. Use AutoDock Vina to generate the .config file.

3.4.2 Molecular Docking

1. Use software such as AutoDock or Schrödinger to perform molecular docking. Set docking parameters (e.g., grid size, energy range) and run the docking calculations.
2. Calculate the binding energy between active components and target proteins. The lower the binding energy, the more stable the binding (*see* **Note 8**).
3. Use software such as PyMOL to visualize the docking results. Analyze the binding modes (e.g., hydrogen bonds, hydrophobic interactions) between the components and target proteins, and generate visualization diagrams.
4. Integrate all data results to predict the high-potential anticancer targets of leaf extracts. Propose possible anticancer mechanisms and signaling pathways, providing data support and guidance for subsequent in vitro and in vivo experimental validation [22, 23] (Fig. 1).

4 Notes

1. Screen and retain components with OB $\geq$ 30% and DL $\geq$ 0.18 as high-potential active components in leaf extracts.
2. Generally, select targets with higher degree values (e.g., Degree $\geq$10) and higher betweenness centrality (e.g., Betweenness Centrality $\geq$0.1).
3. Degree: Calculate the number of edges connected to each node. Components or targets with high Degree are the core of the network. Betweenness Centrality: Measure the importance of nodes in information transfer. Nodes with high values are potential drug targets. Closeness Centrality: The average distance from a node to other nodes. Nodes with high values influence the global network.
4. Highlight key targets in the network diagram (e.g., by changing colors or increasing node size).
5. Generally, after setting the species, set a confidence threshold $\geq$0.7.
6. Typically, select a node degree threshold: $\geq$ 2 (to exclude isolated nodes) and cluster separation: choose high-scoring modules (e.g., MCODE Score $\geq$ 5).
7. GO and KEGG enrichment analysis commonly use tools such as DAVID, Metascape, KEGG Mapper, and ClueGO.
8. Select component-target pairs with binding energy $\leq$ $-$8.0 kcal/mol as potential interaction pairs.

Acknowledgments

This work was supported by Gansu Provincial Science and Technology Major Project (Grant No. 24ZDFA001), The Lanzhou Municipal Science and Technology Program (Grant Nos. 2024-8-27, 2024-8-30, 2024-4-2), Scientific Research Project of Gansu Medical Products Administration (Grant No. 2022GSMPA008), Open Project of NMPA Key Laboratory for Quality Control of Traditional Chinese Medicine (Grant No. 2022GSMPA-KL05), Gansu Intellectual Property Project (Grant No. 22ZSCQ065) and the College Students' Innovation and Entrepreneurship Program of Lanzhou University, China (Grant Nos. 20250260006, 20250260016 and 20250260020).

References

1. Pyarilal S et al (2025) Early detection of carcinoma: correlating quantifiable tumor biomarkers with high-resolution microscopy (HRME) findings. Expert Rev Mol Diagn 25(1–3): 33–45
2. Bray F et al (2024) Global cancer statistics 2022: GLOBOCAN estimates of incidence and mortality worldwide for 36 cancers in 185 countries. CA Cancer J Clin 74(3): 229–263
3. Suda K et al (2009) EGFR T790M mutation: a double role in lung cancer cell survival? J Thorac Oncol 4(1):1–4
4. Karami Fath M et al (2022) The role of epigenetic modifications in drug resistance and treatment of breast cancer. Cell Mol Biol Lett 27(1):52
5. Tanaka T et al (2009) Nanotechnology for breast cancer therapy. Biomed Microdevices 11:49–63
6. Yang Y et al (2021) Natural products with activity against lung cancer: a review focusing on the tumor microenvironment. Int J Mol Sci 22(19):10827
7. Newman DJ, Cragg GM (2020) Natural products as sources of new drugs over the nearly four decades from 01/1981 to 09/2019. J Nat Prod 83(3):770–803
8. Savithramma N, Linga Rao M, Ankanna S (2011) Screening of traditional medicinal plants for secondary metabolites. Int J Res Pharm Sci 2(4):643–647
9. Menezes José CJMDS, Diederich Marc F (2021) Bioactivity of natural biflavonoids in metabolism-related disease and cancer therapies. Pharmacol Res 167:105525
10. Bhatia N, Agarwal R (2001) Detrimental effect of cancer preventive phytochemicals silymarin, genistein and epigallocatechin 3-gallate on epigenetic events in human prostate carcinoma DU145 cells. Prostate 46(2):98–107
11. Mohammad S et al (2017) Sesquiterpene lactone! a promising antioxidant, anticancer and moderate antinociceptive agent from Artemisia macrocephala jacquem. BMC Complement Altern Med 17(1):27
12. Kazmi SR et al (2019) In silico approaches and tools for the prediction of drug metabolism and fate: a review. Comp Biol Med 106:54–64
13. Singla RK, Bhat GV, Gonzalez-Diaz H (2014) Editorial (Thematic issue: from phytochemistry to medicinal chemistry: isolation, semisynthesis, evaluation and computational studies). *Curr Top Med Chem* 14(8):979
14. Sahu R et al (2022) Ayurveda and in silico approach: a challenging proficient confluence for better development of effective traditional medicine spotlighting network pharmacology. Chin J Integr Med 29(5):470
15. Fleming N (2018) How artificial intelligence is changing drug discovery. Nature 557:S55–S57
16. Ekins S, Mestres J, Testa B (2007) In silico pharmacology for drug discovery: methods for virtual ligand screening and profiling. Br J Pharmacol 152(1):9–20
17. Zhang LX et al (2022) TCMSID: a simplified integrated database for drug discovery from traditional chinese medicine. J Cheminform 14(1):89
18. Shaikh F et al (2021) LigTMap: ligand and structure-based target identification and activity prediction for small molecular compounds. J Cheminform 13(1):44
19. Otasek D, Morris JH, Bouças J et al (2019) Cytoscape automation: empowering workflow-based network analysis. Genome Biol 20:185
20. Kadkhoda S, Darbeheshti F, Tavakkoly Bazzaz J (2020) Identification of dysregulated miRNAs-genes network in ovarian cancer: An integrative approach to uncover the molecular interactions and oncomechanisms. Cancer Reports (Hoboken, N.J.) 3(6):e1286
21. Ruiz-Carmona S et al (2014) rDock: a fast, versatile and open source program for docking ligands to proteins and nucleic acids. PLoS Comput Biol 10(4):1003571
22. Lv Q et al (2023) TCMBank: bridges between the largest herbal medicines, chemical ingredients, target proteins, and associated diseases with intelligence text mining. Chem Sci 14(39):10684–10701
23. Ahmad A et al (2021) Marantodes pumilum: systematic computational approach to identify their therapeutic potential and effectiveness. J Ethnopharmacol 283:114751

Part IV

Biosafety of Leaf

Chapter 29

Determination of Toxicological Evaluation of Leaf Extracts: In Vitro Study

Emmanuel Letsyo and Felix Kwashie Madilo

Abstract

Plant leaves are often used for both therapeutic and nutritional purposes as they are rich sources of bioactive secondary metabolites. Despite their benefits, they may be toxic to humans. This chapter provides an overview of in vitro cytotoxicity evaluation of leaf extracts using the MTT [3-(4,5 dimethylthiazol-2-yl)-2,5-diphenyltetrazolium bromide] assay to ascertain their safety and potential applications. The MTT assay is quick, easy, well-established, widely used, and allows many assays to be carried out simultaneously. It involves exposing NIH/3 T3 mouse embryonic fibroblast cell lines to water, methanol, ethanol, or hydroalcoholic leaf extract at various concentrations to measure the reduction of yellow MTT tetrazolium salts by active cells to insoluble purple formazan crystals. The crystals are then solubilized with dimethyl sulphoxide (DMSO) solution, and absorbance is measured by a microplate reading spectrophotometer at a wavelength of 570 nm and compared to the absorbance in a control solution. The concentration of the leaf extract required to achieve 50% growth inhibition (IC_{50}) is used to assess the cytotoxicity of the extract.

Key words Leaf extract, In vitro cytotoxicity, MTT assay, Toxicology, Cell viability assay, NIH/3 T3 cell lines

1 Introduction

Plant leaves are often used in traditional medicines to treat and manage many disease conditions and as food for their nutritional benefits. However, the secondary metabolites in the leaves may exhibit toxicity to humans [1]; hence, their cytotoxic effect must be examined before their application.

The MTT [3-(4,5 dimethylthiazol-2-yl)-2,5-diphenyltetrazolium bromide] assay has been widely used, among the various colorimetric viability assays, to determine in vitro cytotoxicity [2–4]. It is a simple and cost-effective method employed to evaluate cell viability, proliferation, and cytotoxic effects of drugs/chemicals [5]. The assay is based on the principle that only mitochondrial dehydrogenase enzyme in active cells can reduce the yellow-colored

Mitesh Kumar Dwivedi and A. Sankara Narayanan (eds.), *Leaves as Functional Foods*, Methods and Protocols in Food Science, https://doi.org/10.1007/978-1-0716-5226-8_29,

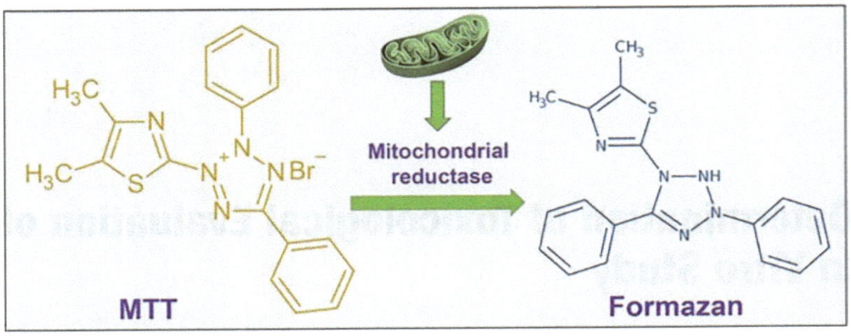

Fig. 1 Reduction of yellow MTT to purple formazan crystals by active cells [7]

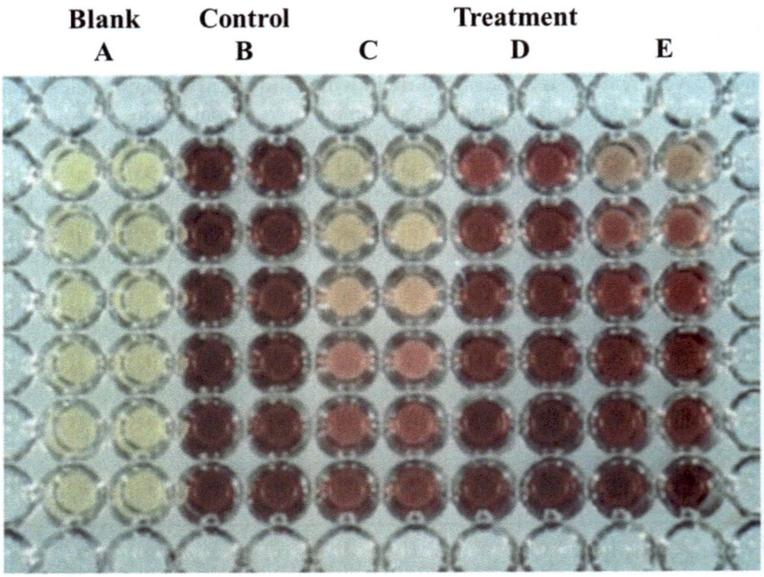

Fig. 2 A 96-well plate after solubilization of formazan crystals (**A**) Blanks control wells, (**B**) untreated cell control wells, and treated wells (**C**), (**D**), and (**E**) at various concentrations [9]

water-soluble MTT reagent into insoluble purple-colored formazan crystals [6] (Fig. 1).

Color formation can therefore be a useful indicator of viable cells. The amount of light absorbed is proportional to the amount of formazan accumulated inside the cell surface. The deeper the purple color, the higher the cell viability and absorbance, while the lighter the purple color, the lower the cell number and cytotoxicity of the given substance [8] (Fig. 2).

A decrease in cell number and absorbance measured at 570 nm with a microplate reader reflects cell growth inhibition [10–12]. This is used to determine the concentration of the extract required to achieve 50% growth inhibition (IC_{50}) compared to the growth of the untreated control.

2 Materials

Prepare all solutions using ultrapure water, and all reagents used should be of analytical grade. Also, all the experiments should be performed in a sterile tissue culture hood.

2.1 Leaf Extract Preparation

1. Wash the collected leaves thoroughly under running tap water to remove sand and other solid contaminants (*see* **Note 1**).
2. Dry in a shed at room temperature (27–30 °C) for 2 to 3 weeks.
3. Pulverize dried leaves into a fine powder using a mechanical grinder.
4. Macerate 500 g of the fine powdered leaves in either 2.5 L of water, or 70% alcohol (i.e. either methanol or ethanol), and shake for 24 h.
5. Filter the mixture through cotton/gauze and then through Whatman filter paper number 1 (pore size: 11 μm) to remove insoluble constituents.
6. Concentrate the entire extracts to dryness using a rotary evaporator at 45 °C under reduced pressure.
7. Prepare leaf extract at varying concentrations using water, alcohol (e.g., methanol or ethanol), or hydroalcohol.

2.2 Prepare MTT Stock Solution

1. Add 1 mL of sterile phosphate-buffered saline (PBS) to 5 mg of MTT (*see* **Note 2**).
2. Vortex the solution in the dark.
3. Filter or centrifuge undissolved particulate material.
4. Store MTT solution at −20 °C until analysis or at 4 °C for immediate use (*see* **Note 3**).

2.3 Equipment

1. Rotary evaporator.
2. 96-well plates.
3. Plateshaker.
4. Pipettes and multichannel pipettes.
5. Microplate reader.
6. CO_2 incubator.
7. NIH/3T3 mouse embryonic fibroblast cell lines.
8. Centrifuge.

3 Methods

3.1 Cell Seeding

1. Seed 100 μL of cells into each well of a 96-well plate.
2. Distribute the cells consistently across each well at a concentration of 5000–10,000 cells/well (*see* **Note 4**).
3. Incubate at 37 °C with 5% CO_2 for 24 h (*see* **Note 5**).

3.2 Leaf Extract (Treatment) Application

1. Treat the cells in the wells with varying concentrations of leaf extract. Include control wells (i.e., without leaf extract treatment) and blank wells (i.e., without cells).
2. Incubate at 37 °C with 5% CO_2 for 24 h.

3.3 MTT Application

1. Add 10 μL of the MTT stock solution to each well.
2. Incubate at 37 °C with 5% CO_2 for 4 h.

3.4 Removal of Medium and MTT Solution

1. After incubation, carefully remove the cell media and unreacted MTT solution from each well.

3.5 Solubilizing Formazan Crystals

1. Add 100 μL of DMSO to each well to dissolve the formazan crystals (*see* **Note 6**).
2. Shake the covered plate on an orbital shaker at room temperature until the formazan crystals are completely solubilized.
3. Incubate in the dark for 30 min at 37 °C with 5% CO_2 (*see* **Note 7**).

3.6 Measurement

1. Measure the absorbance of each well in triplicate using a microplate reader at 570 nm.
2. Calculate the concentration of the leaf extract required for 50% of cell viability (IC_{50}) using the following formula:

$$\%\text{Cell viability}\frac{\text{TG(Mean absorbance)}}{\text{CG(Mean absorbance)}}\times 100$$

Where TG= Treated Group, CG = Control Group

4 Notes

1. Collection of plant leaves, authentication, and the deposited voucher number should be recorded.
2. MTT is light-sensitive, so the experiment should be conducted in dark settings. Cover cells and MTT-formazan crystals from direct light exposure, as this can cause MTT degradation and affect absorbance measurements.

3. Store MTT solution at −20 °C (stable for at least 6 months).
4. Use precise pipetting techniques to evenly distribute reagent volumes to each well, preventing variability and inaccurate results.
5. To prevent variation in cell metabolism, maintain stable cell culture conditions, such as temperature, humidity and CO_2 levels.
6. Be cautious when handling DMSO. It can introduce impurities into your experiment and, at larger doses, can be hazardous to cells.
7. Complete the microplate readings within 1 h of solubilization to avoid precipitation.

References

1. Kundishora A, Sithole S, Mukanganyama S (2020) Determination of the cytotoxic effect of different leaf extracts from *Parinari curatellifolia* (Chrysobalanaceae). J Toxicol 2020. https://doi.org/10.1155/2020/8831545
2. Sowmyashree K, Rajukrishna C, Divakar MS, Ravindra BM, Nityasree BR (2024) Toxicological evaluation of *Pajanelia longifolia* (Willd.) K. Schum methanol leaf extract in Wistar rats: Cytotoxic and Acute Oral Toxic effects. Res J Pharm Technol 17(6). https://doi.org/10.52711/0974-360X.2024.00451
3. Kamiloglu S, Sari G, Ozdal T, Capanoglu E (2020) Guidelines for cell viability assays. Food Front 1:332–349. https://doi.org/10.1002/fft2.44
4. Kasthuri OR, Ramesh B (2018) Toxicity studies on leaf extracts of *Alternanthera brasiliana* (L.) Kuntze and *Alternanthera bettzickiana* (Regel) Voss. J Appl Pharm Sci 8:082–089. https://doi.org/10.7324/JAPS.2018.81011
5. Ghasemi M, Turnbull T, Sebastian S, Kempson I (2021) The MTT assay: utility, limitations, pitfalls, and interpretation in bulk and single-cell analysis. Int J Mol Sci 22:12827. https://doi.org/10.3390/ijms222312827
6. Bahuguna A, Khan I, Bajpai VK, Kang SC (2017) MTT assay to evaluate the cytotoxic potential of a drug. Bangladesh J Pharmacol 12:115–118. https://doi.org/10.3329/bjp.v12i2.30892
7. Albayati TM, Alardhi SM, Khalbas AH, Humdi ZJ, Ali NS, Salih IK, Saady NMC, Zendehboudi S, Abdulrahman MA (2024) Comprehensive review of mesoporous silica nanoparticles: drug loading, release, and applications as hemostatic agents. ChemistrySelect, 9(23). https://doi.org/10.1002/slct.202400450
8. Riss TL, Moravec RA, Niles AL, Duellman S, Benink HA, Worzella TJ, Minor L (2016) In: Markossian S, Grossman A, Arkin M, et al (eds) Cell viability assays. 2013 May 1 [Updated 2016 Jul 1]. Assay Guidance Manual [Internet]. Eli Lilly & Company and the National Center for Advancing Translational Sciences, Bethesda; 2004. https://www.ncbi.nlm.nih.gov/books/
9. van Meerloo J, Kaspers GJL, Cloos J (2011) Cell sensitivity assays: The MTT Assay. In: Cree I (ed) Cancer cell culture: methods and protocols, vol 731. Humana Press. https://doi.org/10.1007/978-1-61779-080-5_20
10. Aydına A, Aktayb G, Erdem Yesilada E (2016) A guidance manual for the toxicity assessment of traditional herbal medicines. Nat Prod Commun 11:1763–1773
11. Riyas CT, Sukumaran ST (2023) In Vitro anticancer and antioxidant potential of *Pittosporum dasycaulon* Leaves extract against the human liver and lymphoma cancer. Indian J Pharm Sci 85:13–144
12. Fonseca AG, Dantas LLSFR, Fernandes JM, Zucolotto SM, Lima AAN, Soares LAL, Rocha HAO, Lemos TMAM (2018) In Vivo and In Vitro toxicity evaluation of hydroethanolic extract of *Kalanchoe brasiliensis* (Crassulaceae) leaves. J Toxicol 2018. https://doi.org/10.1155/2018/6849765

Chapter 30

Determination of Toxicological Evaluation of Leaf Extracts: In Vivo Study

Franciane Pereira Bernardes, Sarah de Jesus Cantarino, Jean Moises Ferreira, Maria do Carmo Pimentel Batitucci, and Jean Carlos Vencioneck Dutra

Abstract

In recent years, given the increase in the consumption of medicinal plants, there has been an increase in studies on the safe use of plant extracts and derivatives. Thus, given the importance of the toxicological evaluation of plant extracts, here we present a step-by-step guide to performing in vivo toxicology tests, focusing on the rodent model. This chapter is aimed to describe the fundamental steps to conduct in vivo rodent studies, from the design of treatment doses to the analysis of toxicological parameters using plant extracts.

Key words Plant extract, Rodent toxicology, Cytotoxicity, Mutagenicity, Micronucleus test, Intragastric gavage

1 Introduction

Folk herbal medicine has been used to treat various diseases since the dawn of human societies, and this long-standing practice has led to the widespread recognition of phytochemicals in modern medicine. Thus, plant bioactive compounds play a crucial role in the discovery and development of natural medicines and synthetic pharmaceuticals, since they serve as valuable sources of therapeutic agents, providing models for drug design, semisynthesis, and synthesis [1].

Plants are rich in numerous bioactive compounds that contribute to human health, such as alkaloids, phenols, saponins, and tannins [2]. Furthermore, it has been a challenge to guarantee the quality and efficacy of plant-based compounds, as their chemical composition can be influenced by plant genetic factors, phenological stages, environmental conditions, and harvesting methods [3]. Hence, plant extract standardization is essential to ensure

Mitesh Kumar Dwivedi and A. Sankara Narayanan (eds.), *Leaves as Functional Foods*, Methods and Protocols in Food Science, https://doi.org/10.1007/978-1-0716-5226-8_30,

consistent bioactive effects, as pharmacokinetic fluctuations may occur due to heterogeneity in the synthesis and storage of bioactive compounds [4].

Drugs derived from natural sources are reported to be free of side effects and are necessary to advance the frontiers of drug discovery [1]. Despite the recent use of emerging biotechnological techniques, fundamental assays for assessing toxicity and cytotoxic effects remain essential, as required by regulatory bodies for rigorous safety and efficacy testing [5].

Both in vitro and in vivo bioassays are essential for assessing a drug's efficacy, safety, and potential mechanisms of action [6]. In vitro studies take place in controlled laboratory environments, typically utilizing cell cultures or isolated biochemical systems to examine the effects of substances at the cellular or molecular level [7]. Conversely, in vivo studies involve testing chemicals in living organisms, offering valuable information about their physiological effects and toxicological properties [1].

Therefore, this chapter aims to present the steps necessary to investigate the toxicity of plant extracts in rodents, considering the establishment of treatment doses to the analysis of toxicity parameters.

2 Materials

Syringe, needle, gavage needle, scissors, scalpel, microscope slides, optical microscope, leaf extract, fetal bovine serum, leishman, digital scale (or similar), wood shavings, and polypropylene cages with metal bars.

3 Methods

To assess in vivo toxicity, it is necessary to prepare and organize all test solutions in advance. It is recommended that only trained technicians or researchers handle the animals, minimizing stress and possible interference in the analyses.

3.1 Rodent Experimental Groups

1. Add wood shavings to clean polypropylene cages and keep them closed with metal bars.
2. Separate the rodents into necessary experimental groups randomly (*see* **Notes 1–4**).
3. Keep the animals at room temperature, day-night cycle, with food and water ad libitum until the beginning of the experiment.
4. It is recommended to acclimatize the rodents before starting the experiment—about 2–7 days.

3.2 Extract Volume per Treatment

1. Weigh the rodents on a digital scale.
2. Calculate the treatment dose (volume of extract administered to each animal). In this step, in order to establish the volume of extract to be administered per animal in each experimental group (treatment group), it is necessary to consider rodent weight and the following formula:

$$\text{Volume of extract (per treatment group)} = \frac{\text{Body weight (g)} \times \text{Treatment dose (mg/kg)}}{1000 \times \text{Extract stock solution (mg/ml)}}$$

where the *body weight (g)* is the rodent body weight in grams; *treatment dose (mg/kg)* is the dose of the extract to be administered per kilogram of weight (in milligrams per kilogram); *extract stock solution (mg/ml)* is the concentration of the extract in *mother solution*, usually in mg per ml (milligrams per milliliter) and prepared on the day of application to rodents (*see* **Note 5**).

3.3 Median Lethal Dose (LD_{50})

1. Weigh the rodents on a digital scale and calculate the treatment dose (volume of extract administered to each animal).
2. Apply the doses of plant extract intraperitoneally using a syringe and needle.
3. Keep the rodents under controlled temperature and day-night cycle, with food and water ad libitum.
4. Count the number of deaths after 48 h of applying the plant extract (*see* **Notes 6–9**).

3.4 Oral Administration (Intragastric Gavage)

1. On a digital scale, weigh the rodents and calculate the treatment dose to each animal.
2. Using a gavage needle attached to a syringe, through the rodent's mouth opening, access the intragastric cavity and apply the doses of plant extract (Fig. 1a).
3. Keep rodents under observation in controlled experimental conditions.

3.5 Intraperitoneal Administration

1. Weigh the rodents on a digital scale and establish the treatment dose for each rodent.
2. Using a needle attached to a syringe, carefully insert the needle into the rodent's abdominal region and apply the doses of the plant extract (Fig. 1b) (*see* **Notes 10** and **11**).
3. Keep the rodents under observation under controlled experimental conditions.

3.6 Micronucleus Test in Mice Peripheral Blood Cells

1. Use a sterile needle to injure the rodent's caudal artery.
2. Collect drops of whole peripheral blood from each animal in the experimental groups and smear them on clean microscope slides.
3. Allow the slides to air dry.
4. Keep the slides immersed in methanol for 15 min to stain the biological material.
5. Stain the cells with Leishman twice in two different concentrations in order to differentiate immature polychromatic erythrocytes (PCE) and mature normochromatic erythrocytes (NCE) [8, 9].
6. Record 2000 NCE per animal, considering the micronucleated normochromatic erythrocytes (MNNCEs), using an optical microscope with an increase of 1000× [9].
7. Establish the cytotoxicity and mutagenicity parameter as described in the literature [10].

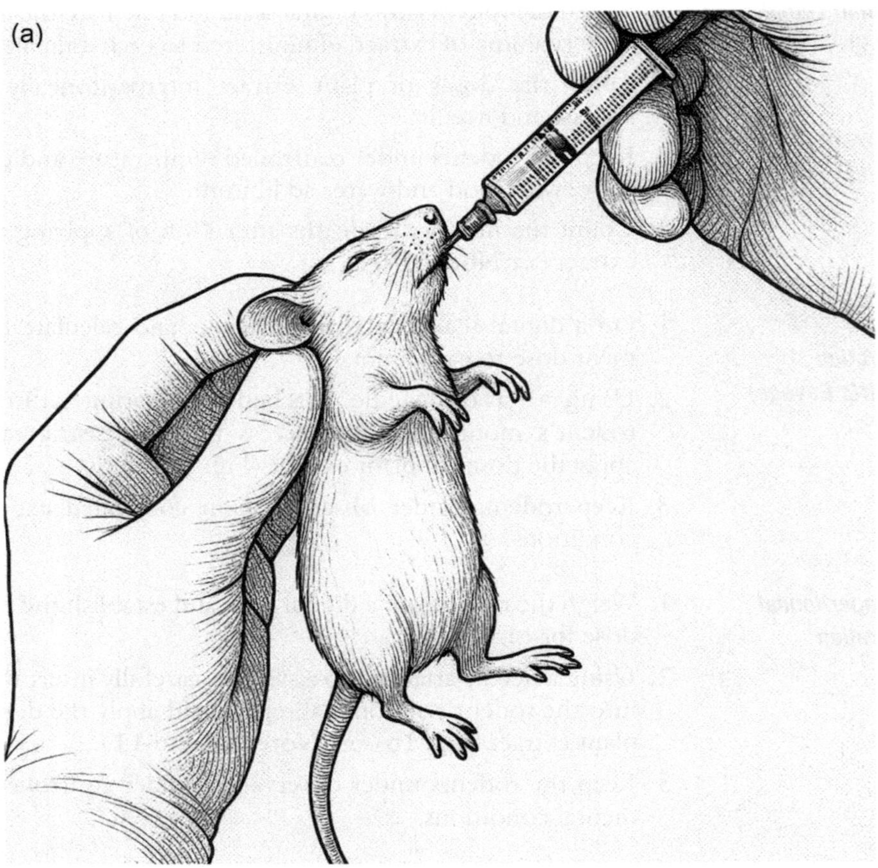

Fig. 1 Rodent management for in vivo experiments. (**a**) intragastric extract administration in rodents; (**b**) intraperitoneal extract administration in rodents; (**c**) dissection of the rodent femur—image from the authors' personal collection edited by artificial intelligence (AI)

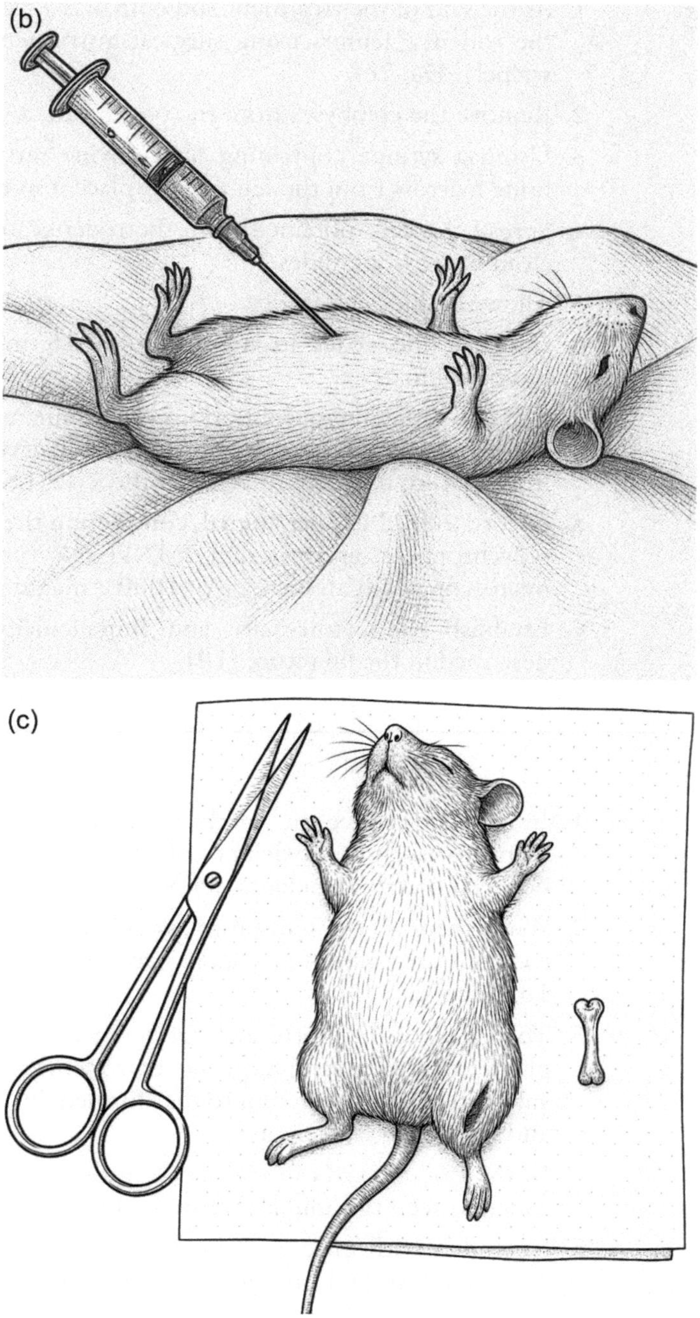

Fig. 1 (continued)

3.7 Micronucleus Test in Mice Bone Marrow Cells

1. At the end of the treatment and euthanasia, proceed to dissect the rodents' femurs using surgical instruments (scissors and scalpel) (Fig. 1c).
2. Remove the epiphyses from the bones with a scalpel.
3. Using a syringe containing fetal bovine serum, remove the bone marrow from the femurs and place it in test tubes.
4. Spread the cells obtained from the rodents' bone marrow on clean microscope slides.
5. Allow the slides to air dry.
6. Immerse the slides in methanol for 15 min to stain the biological material.
7. Use two different concentrations of Leishman to stain the cells and differentiate immature polychromatic erythrocytes (PCE) from mature normochromatic erythrocytes (NCE) [8, 9].
8. Record 2000 PCE per animal, considering the micronucleated polychromatic erythrocytes (MNPCEs), using an optical microscope with an increase of 1000× magnification [9].
9. Establish the cytotoxicity and mutagenicity parameter as described in the literature [10].

4 Notes

1. In order to reduce the number of animals used in the experiment, to assess the toxicity of plant extracts, sample groups of between 6 and 10 rodents can be established.
2. When there is no intention to assess hormonal influence, male rodents are used—a sample group of between 6 and 10 rodents.
3. To assess possible hormonal interference in the metabolism of plant extracts, a sample group of rodents ($n = 10$) can be subdivided into two subgroups, formed by female ($n = 5$) and male ($n = 5$) rodents.
4. In the case of studies with sample groups of male and female rodents, keep the animals separated by gender.
5. In studies with mice, a volume of less than 1 mL should be considered for oral (intragastric) administration.
6. In order to assess in vivo toxicity, it is necessary to conduct a broad bibliographic survey. This investigation allows us to verify the lethal dose for the plant studied has been established in the literature. If no records are found, the in vivo toxicity investigation should begin by establishing the lethal dose for fifty percent of a sample group of the species studied (LD_{50}).

7. To establish the acute, subchronic or chronic toxicity of the plant extract, the extract may be administered in a single dose (acute toxicity) or in repeated doses (subchronic toxicity >30 days of treatment < chronic toxicity)—the number of days to consider a treatment as chronic may depend on the model studied or protocol used in each study.
8. Acute toxicity tests are considered to be single dose and LD_{50} tests; on the other hand, in subchronic or chronic toxicity tests there is repeated administration of a substance—in both cases the clinical signs, physiological and behavioral changes of the animal model studied must be observed.
9. To determine the LD_{50}, it is recommended to use at least 10 rodents.
10. In intraperitoneal administration using the syringe, it is necessary to perform suction beforehand and check whether any of the rodent's blood vessels have been reached (rupture of vessels should be avoided during this procedure).
11. For intraperitoneal applications, application in the lower quadrants of the rodent's abdomen is preferable.

Acknowledgement

Authors are grateful to Universidade Federal do Espírito Santo (UFES) and to Instituto Federal do Espírito Santo (IFES).

References

1. Chaachouay N, Zidane L (2024) Plant-derived natural products: a source for drug discovery and development. Drugs and Drug Candidates 3(1):184–207
2. Rana P, Kumar A, Choudhary A, Kaur H, Singh R (2021) The wisdom of prevention: holistic, preventive herb approach for healing of the globe. J Pharm Innov 10:29–46
3. Yang L, Wen KS, Ruan X, Zhao YX, Wei F, Wang Q (2018) Response of plant secondary metabolites to environmental factors. Molecules 23(4):762
4. Lan K, Jia W (2010) An integrated metabolomics and pharmacokinetics strategy for multicomponent drugs evaluation. Curr Drug Metab 11(1):105–114
5. Fan TP, Deal G, Koo HL, Rees D, Sun H, Chen S et al (2012) Future development of global regulations of Chinese herbal products. J Ethnopharmacol 140(3):568–586
6. Araújo ASF, Monteiro RTR (2005) Plant bioassays to assess toxicity of textile sludge compost. Sci Agric 62:286–290
7. Grant WF (1994) The present status of higher plant bioassays for the detection of environmental mutagens. Mutat Res 310(2):175–185
8. Vencioneck Dutra JC, Moisés Ferreira J, Costalonga Pereira PR, Ben-Hur de Oliveira J, Vitorino Gervásio S, Bernardes Xavier M et al (2018) Cereus jamacaru DC hydroalcoholic extract promotes anti-cytotoxic and antitumor activity. Pharmaceuticals 11(4):130
9. Krishna G, Hayashi M (2000) In vivo rodent micronucleus assay: protocol, conduct and data interpretation. Mutat Res 455(1–2):155–166
10. Holden HE, Majeska JB, Studwell D (1997) A direct comparison of mouse and rat bone marrow and blood as target tissues in the micronucleus assay. Mutat Res 391(1–2):87–89

Index

A

B

C

D

E

F

Mitesh Kumar Dwivedi and A. Sankara Narayanan (eds.), *Leaves as Functional Foods*, Methods and Protocols in Food Science, https://doi.org/10.1007/978-1-0716-5226-8,

Index

Q

R

S

T

U

V

W

MIX
Papier aus verantwortungsvollen Quellen
Paper from responsible sources
FSC® C105338

If you have any concerns about our products,
you can contact us on
ProductSafety@springernature.com

In case Publisher is established outside the EU,
the EU authorized representative is:
Springer Nature Customer Service Center GmbH
Europaplatz 3, 69115 Heidelberg, Germany

Printed by Libri Plureos GmbH
in Hamburg, Germany